INTRODUCTION TO MESOSCOPIC PHYSICS

Mesoscopic Physics and Nanotechnology

1. Y. Imry. *Introduction to Mesoscopic Physics*

INTRODUCTION TO MESOSCOPIC PHYSICS

Yoseph Imry

New York *Oxford*
Oxford University Press
1997

Oxford University Press

Oxford New York
Athens Auckland Bangkok Bogota Bombay Buenos Aires
Calcutta Cape Town Dar es Salaam Delhi Florence Hong Kong
Istanbul Karachi Kuala Lumpur Madras Madrid Melbourne
Mexico City Nairobi Paris Singapore Taipei Tokyo Toronto

and associated companies in
Berlin Ibadan

Published by Oxford University Press, Inc.
198 Madison Avenue, New York, New York 10016

Library of Congress Cataloging-in-Publication Data
Imry, Yoseph.
Introduction to mesoscopic physics / Yoseph Imry.
p. cm.—(Mesoscopic physics and nanotechnology)
Includes bibliographical references and index.
ISBN 0-19-510167-7
1. Mesoscopic phenomena (Physics) I. Title. II. Series.
QC176.8.M46I47 1997
537.6—dc20 95-47793

9 8 7 6 5 4 3 2

Printed in the United States of America
on acid-free paper

To CYLA

Preface

Mesoscopic physics is a rather young branch of science. It started about 15 years ago and has already had several exciting and instructive achievements. It enjoys the unique combination of being able to deal with and provide answers on fundamental questions of physics while being relevant for applications in the not-too-distant future. In fact, some of the experimental possibilities in this field have been developed with an eye to reducing the sizes of electronic components. It can be hoped that cross-fertilization between physics and technology will continue and go both ways. We now already understand much more about the realm intermediate between the microscopic and macroscopic. Basic questions about how the quantum rules operate and go over into the classical macroscopic regime have been and are being answered. It is hoped that the whole regime between man-made structures and naturally occurring molecules, with their modifications, will be approached and understood soon. Impressive nanoscale techniques for that future stage are being developed.

This book is written in an attempt to make these interesting issues clear to physicists, chemists, and electronic and optical engineers and technologists. The reader should have a solid background in physics, but not necessarily be conversant with advanced formal theoretical methods. The understanding of the underlying physical ideas and the ability to make quite accurate estimates should be of help to both experimental researchers and technologists. At the same time, the study of this material should be helpful to graduate physics and chemistry students for integrating and solidifying their studies of quantum mechanics, statistical mechanics, electromagnetism, and condensed-matter physics.

The author is indebted to many colleagues for collaborations related to these subjects over the years, from which much was learned and the results obtained from which constitute much of the material covered. These colleagues include: Y. Aharonov, A. Aharony, B. L. Altshuler, N. Argaman, the late A. G. Aronov, M. Ya Azbel, D. J. Bergman, M. Büttiker, G. Deutscher, O. Entin-Wohlman, B. Gavish, Y. Gefen, L. Gunther, C. Hartzstein, I. Kander, R. Landauer, N. Lang, I. Lerner, Y. Levinson, S. Mohlecke, G. Montambaux, M. Murat, Z. Ovadyahu, J. L. Pichard, S. Pinhas,

E. Pytte, A. Shalgi, D. J. Scalapino, A. Schwimmer, N. S. Shiren, N. Shmueli, U. Sivan, U. Smilansky, A. Stern, A. D. Stone, M. Strongin, D. J. Thouless, A. Yacoby, and N. Zanon.

Many other colleagues contributed by instructive discussions for which the author is extremely grateful. They include: E. Abrahams, E. Akkermans, S. Alexander, E. L. Altshuler, A. Altland, V. Ambegaokar, T. Ando, Y. Avishai, Y. Bar-Joseph, A. Baratoff, C. Beenakker, E. Ben-Jacob, A. Benoit, M. Berry, the late F. Bloch, H. Bouchiat, E. Brezin, M. Brodsky, C. Bruder, J. Chalker, P. Chaudhari, C.-s. Chi, M. Cyrot, D. Divincenzo, V. Eckern, K. B. Efetov, A. L. Efros, W. A. B. Evans, A. Finkel'stein, A. Fowler, E. Fradkin, H. Fukuyama, N. Garcia, L. Glazman, G. Grinstein, D. Gubser, B. I. Halperin, M. Heiblum, S. Hikami, A. Houghton, A. Kameneev, M. A. Kastner, D. E. Khmel'nitskii, S. Kirkpatrick, S. Kivelson, S. Kobayashi, W. Kohn, B. Kramer, A. Krichevsky, the late R. Kubo, J. Langer, A. I. Larkin, D.-H. Lee, P. A. Lee, A. J. Leggett, S. Levit, L. P. Levy, H. J. Lipkin, D. Loss, the late S.-k Ma, A. MacDonald, A. MacKinnon, D. Mailly, R. S. Markiewicz, Y. Meir, P. A. Mello, U. Meirav, M. Milgrom, J. E. Mooij, B. Mühlschlegel, D. Mukamel, D. Newns, Y. Ono, D. Orgad, the late I. Pelah, J. P. Pendry, M. Pepper, D. Prober, N. Read, H. Rohrer, T. M. Rice, M. Sarachik, M. Schechter, A. Schmid, G. Schön, T. D. Schultz, Z. Schuss, M. Schwartz, S. Shapiro, B. I. Shklovskii, N. Sivan, C. M. Soukoulis, B. Z. Spivak, F. Stern, C.-c. Tsuei, D. C. Tsui, B. van Wees, D. Vollhardt, K. von Klitzing, S. von Molnar, R. Voss, S. Washburn, R. Webb, F. Wegner, H. Weidenmüller, R. Wheeler, P. Wiegman, J. Wilkins, N. Wingreen, S. Wolf, and P. Wölfle.

Special thanks are due to the author's most recent four Ph.D. students (in chronological order): Yuval Gefen, Uri Sivan, Ady Stern, and Nathan Argaman, and to Amir Yacoby. All of them quickly became colleagues and friends and contributed immensely to the work and to the physical understanding of the subject. The collaborations with the late A. G. Aronov, S.-k. Ma, and I. Pelah are especially remembered. The person whose ideas and insights have contributed the most to the author's understanding of the related physics is Rolf Landauer, who deserves special thanks and whose contribution is deeply appreciated. The responsibility for errors, omissions, and misunderstandings rests solely on the author. R. Landauer, C. Bruder, M. Heiblum, D. Orgad, U. Sivan, U. Smilansky and A. Stern are also thanked for pertinent comments on the manuscript.

Various phases of this research were done in a number of laboratories and institutes for which the author is grateful for support. These include Soreq Nuclear Research Center, Cornell University, Tufts University, Tel-Aviv University, the University of California at Santa Barbara and San Diego, Brookhaven National Laboratory, the IBM Yorktown Research Center, CEN Saclay, the University of Karlsruhe and the Humboldt Foundation, the Wissenschaftskolleg of Berlin, Ecole Normale Supérieure and, last but not least, the Weizmann Institute of Science. The following agencies are thanked for recently supporting parts of this research: BSF (U.S.–Israel Binational Science Foundation), GIF (German–Israeli Binational Science Foundation), the Israel Academy of Sciences, and the Minerva Foundation. Mrs. Naomi Cohen is thanked for expert typing of the manuscript.

Contents

1 Introduction and a Brief Review of Experimental Systems **3**

1 Generalities 3

2 A Brief Description of Systems and Fabrication Methods 6

2 Quantum Transport, Anderson Localization **12**

1 Basic Concepts 12

Localization ideas 14

2 Thermally Activated Conduction in the Localized Regime 18

3 The Thouless Picture, Localization in Thin Wires and Finite Temperature Effects 21

4 The Scaling Theory of Localization and its Consequences 26

General 26

The case $d \leq 2$ 27

The case $d > 2$, the metal–insulator (M–I) transition 30

5 The Weakly Localized Regime 34

3 Dephasing by Coupling with the Environment, Application to Coulomb Electron–Electron Interactions in Metals **38**

1 Introduction and Review of the Principles of Dephasing 38

2 Dephasing by the Electron–Electron Interaction 46

3 Review of Results in Various Dimensions 50

4 Dephasing Time vs. Electron–Electron Scattering Time 56

4 Mesoscopic Effects in Equilibrium and Static Properties **60**

1 Introductory Remarks, Thermodynamic Fluctuation Effects 60

2 Quantum Interference in Equilibrium Properties, Persistent Currents 65

Generalities, simple situations 65

Independent electrons in disordered systems 71

The semiclassical picture 75

General results on ensemble-averaged persistent currents for constant N 80
Semiclassical theory of spectral correlations, applications to rings 82
Interaction effects on the persistent currents 84

5 Quantum Interference Effects in Transport Properties, the Landauer Formulation and Applications **89**
1 Generalities, Remarks on the Kubo Conductivity for Finite Systems 89
2 The Landauer-type Formulation for Conductance in a Mesoscopic System and Some of its Generalizations 93
Introduction: the "single-channel" case 93
The multichannel Landauer formulation 96
The Onsager-type relationship in a magnetic field: generalized multiterminal conductance formulas 103
3 Applications of the Landauer Formulation 107
Series addition of quantum resistors, 1D localization 107
Parallel addition of quantum resistors, A-B oscillations of the conductance 109
On the universality of the conductance fluctuations 120

6 The Quantum Hall Effect (QHE) **124**
1 Introduction 124
2 General Arguments 129
3 Localization in Strong Magnetic Fields and the QHE 133
4 Brief Remarks on the Fractional Quantum Hall Effect (FQHE) 139

7 Mesoscopics with Superconductivity **147**
1 Introduction 147
2 Superconducting Rings and Thin Wires 151
3 Weakly Coupled Superconductors, the Josephson Effect and *SNS* Junctions 160
The Bloch picture 160
The Josephson junction and other weak links 163
4 Brief Remarks on Vortices 165
5 The Andreev Reflection, More on *SN* and *SNS* Junctions 167

8 Noise in Mesoscopic Systems **176**
1 Introduction 176
2 Shot Noise for "Radiation" from a Reservoir 178

3 The Effect of Fluctuations in the Sink, the Equilibrium Limit 181
4 Low-Frequency (1-*f*) Noise 185

9 Concluding Remarks **191**

Appendices

A The Kubo, Linear Response, Formulation 195
B The Kubo–Greenwood Conductivity and the Edwards–Thouless Relationships 198
C The Aharonov–Bohm Effect and the Byers–Yang and Bloch Theorem 200
D Derivation of Matrix Elements in the Diffusion Regime 201
E Careful Treatment of Dephasing in 2D Conductors at Low Temperatures 201
F Anomalies in the Density of States (DOS) 202
G Quasiclassical Theory of Spectral Correlations 204
H Details of the Four-Terminal Formulation 207
I Universality of the Conductance Fluctuations in Terms of the Universal Correlation of Transmission Eigenvalues 208
J The Conductance of Ballistic "Point Contacts" 209

References **211**

Index **231**

INTRODUCTION TO MESOSCOPIC PHYSICS

1

Introduction and a Brief Review of Experimental Systems

1. GENERALITIES

Much of solid state theory and statistical physics is concerned with the properties of macroscopic systems. These are often considered while using the "thermodynamic limit" (system's volume, Ω, and particle number, N, tending to infinity with $n = N/\Omega$ kept constant), which is a convenient mathematical device for obtaining bulk properties. Usually, the system approaches the macroscopic limit once its size is much larger than some correlation length ξ (or, more generally, than all such relevant lengths). In most cases, ξ is on the order of a microscopic length (e.g., $\sim n^{-1/3}$), but in some special cases, such as in the vicinity of a second-order transition, ξ can become very large and one may observe behavior which is different from the macroscopic limit for a large range of sample sizes (Imry and Bergman 1971). Another case where the effective length scale dividing microscopic from macroscopic behavior is very large is that of small conducting (or semiconducting) systems at low temperatures. Here, two important new elements occur: First, the spectrum of electronic states is discrete (although the interaction with the outside world may broaden the levels enough to make that less relevant, see below). Second, the motion is coherent in the sense that once an electron can propagate across the whole system without inelastic scattering, its wavefunction will maintain a definite phase. The electron will thus be able to exhibit a variety of novel and interesting interference phenomena. In what follows, we shall concentrate on the study of the latter type of systems.

The interest in studying systems in the intermediate size range between microscopic and macroscopic (sometimes referred to as "mesoscopic"—a word coined by Van Kampen 1981) is not only in order to understand the macroscopic limit and how it is achieved by, say, building up larger and larger clusters to go from the "molecule" to the "bulk." Many novel phenomena exist that are intrinsic to mesoscopic systems. A mesoscopic system is really like a large molecule, but it is always, at least weakly, coupled to a much larger, essentially infinite, system—via phonons, many-body excitations, and so on. Sometimes such a coupling can be controlled. Ideally, one would like to interpolate between open and closed systems, as far as energy, particle number, and so on are concerned, by varying some coupling strengths. The special phenomena that exist in this range are of great interest by themselves. We shall see how fundamental principles of quantum mechanics (related to the concept of the phase of the wavefunction) and statistical physics (brought about by the small specimen size and by the slowness of inelastic scattering and thermalization) appear and are amenable to theoretical clarification and experimental examination in these systems.

An important concept is that of the "impurity ensemble"—the collection of systems having the same *macroscopic* parameters (e.g., average concentrations of various defects) but differing in the detailed arrangement of the resulting disorder. In the macroscopic limit an average over this ensemble is usually performed, which restores various symmetries on average. However, a principal interesting aspect of mesoscopics is the distinction (Landauer 1970, Azbel 1973, Imry 1977, Anderson et al. 1980, Azbel and Soven 1983, Gefen 1984 personal communication) between ensemble-averaged properties, and those specific to a particular given small system taken from the ensemble. The specific "fingerprint" of such a small system is of interest and may sometimes be used to obtain some statistical information on the particular arrangement of the constituents in the system (Azbel 1973). Alternatively, changes with time (usually on long scales) of the disorder configuration may lead to low-frequency noise (Feng et al. 1986).

Many of the usual rules that one is used to in macroscopic physics may not hold in "mesoscopic" systems. For example, the rules for addition of resistances, both in series (Landauer 1970, Anderson et al. 1980) and in parallel (Gefen et al. 1984a,b) are different and more complicated. The electronic motion is wavelike and it is not dissimilar to that of electromagnetic radiation in waveguide structures, except for very interesting complications due to disorder. These effects may set fundamental limits on how small various electronic devices can eventually be made. On the other hand, ideas for new devices such as those operating in analogy with various optical and waveguide ones, as well as with SQUIDS (superconducting quantum interference devices) and other Josephson-effect systems (see, e.g., Hahlbohm and Lübbig 1985) may emerge for small normal conductors.

The technology (see, e.g., Howard and Prober 1982, Prober 1983, Laibowitz 1983, Broers 1989) for the fabrication of supersmall structures is progressing very quickly and has reached the stage where many theoretical predictions can now be confronted by experiments. One uses controlled growth, such as in the MBE method (see, e.g., Herman and Sitter 1989), and advanced optical, x-ray, or electron-beam lithographic techniques. In semiconducting systems based on quantum-well concepts, an excellent restriction in one direction exists (Ando et al. 1982), so that creating small structures parallel to the 2D (two-dimensional) layer may achieve systems with a rather small number of active electrons or quantum states. (See the next section for a brief summary and references on available systems and fabrication methods.) On the other hand, we have the recent STM breakthrough (Binning et al. 1982), which provides a novel tool for atomic-scale fabrication, analysis, and measurement (for a recent example see Crommie et al. 1993). One may soon reach the stage of having large conducting artificial molecules on which macroscopic-type experiments can be performed, in the same size range as ordinary macromolecules, or smaller. The latter may be addressed and are of course also of great interest.

It should be noted that photons may also be well "guided" in such systems and similar phenomena may thus occur for these electromagnetic waves, not to mention ideas for electron–photon coupling in various combinations. Both subjects of mesoscopics and light propagation in disordered media (John et al. 1983, Feng and Lee 1991, Sheng 1995) have undergone real advances due to analogies and mutual fertilization; see van Haeringen and Lenstra (1991) for pertinent papers.

This book will deal mostly with theory. However, an attempt will be made to present the most economical explanation/semiquantitative calculation for physical effects, rather than to "demonstrate the power" of some formal method. It is hoped that the qualitative understanding of the physical principles involved that is gained might be useful in conceiving new experiments.

One of the powerful and useful concepts which appear is that of "universality," namely, that various measurable quantities do not depend on most microscopic details of the system. This was originally introduced by Kadanoff in the context of critical phenomena, where large-scale physical properties do not depend on most microscopic details. Older examples are: (a) the Hall coefficient, which does not depend in the simplest picture on the effective mass and on the scattering time; and (b) the Debye T^3 specific heat. The latter is a good example of universality due to general properties of the spectrum (here density of states (DOS)) of certain operators (the phonon Hamiltonian, for the Debye law). Such universalities are very relevant to our subject. In a sense, it is even more remarkable that dirty systems display such universalities, compared to the ultra-clean and perfect systems needed for studies of critical phenomena.

Experimental techniques will be referred to very briefly, mainly in order to understand what can be done and what the limitations are. The theoretical task is becoming more difficult (and interesting) as the understanding of single-electron properties has advanced and one now has to delve into many-body physics, where interactions are important and the theory is much more sophisticated (for a recent account, see Imry and Sivan 1994). Our emphasis will be on equilibrium as well as on various electronic transport phenomena. For a good review on optical effects, we refer to Schmitt–Rink et al. (1989). Bastard et al. (1991) contains a good review of related aspects of electronic properties of semiconducting heterostructures and some of their optical properties. Analogies between optics and electronic phenomena are discussed in van Haeringen and Lenstra (1991). We shall treat only some aspects of the interesting case of ballistic transport (Heiblum et al. 1985, Beenakker and van Houten 1991d) and just mention briefly the topic of "Coulomb blockade" (see problem 5 of chapter 5 and references in Grabert and Devoret 1992, Hekking et al. 1994). A recent hydrodynamic analogy (de Jong and Molenkamp 1995) of electronic transport is also worth noting.

In the next section, systems and fabrication techniques will be briefly described. In Chapter 2 the limitations of the ordinary quasiclassical transport will be discussed and the opposite limit, of Anderson localization, introduced. Since phase coherence is so important in mesoscopics, we shall devote Chapter 3 to elucidating what it takes to destroy phase coherence, with some examples. Chapter 4 will consider equilibrium properties and Chapter 5 will discuss transport phenomena in mesoscopic systems. Chapter 6 will be devoted to high magnetic fields and the quantum Hall effect, Chapter 7 to mesoscopics with superconducting components and Chapter 8 to various noise phenomena. Concluding remarks will be given in Chapter 9. Various details are discussed in the appendices.

2. A BRIEF DESCRIPTION OF SYSTEMS AND FABRICATION METHODS

For experiments in conducting mesoscopic systems, one may in principle use members of the three principal classes of conductors:

1. *Metals*, having high charge-carrier densities in the range of $10^{22}/\text{cm}^3$ and a wide range of purities and mean free paths. Many metals become superconductors at low temperatures—which provides another interesting degree of freedom.
2. *Semiconductors*, where the carrier densities can range practically between 10^{15} and $10^{19}/\text{cm}^3$ and may be controlled, including the type of carrier, by doping, optical excitation, or electrostatic "gates." Special methods may be used to produce high mobilities and heterojunctions—interfaces between different semiconductors with interesting properties (see below).
3. In special cases, *semimetals* having intermediate carrier densities of 10^{19} to $10^{20}/\text{cm}^3$ are useful. They contain electrons and holes concurrently. In some cases, notably bismuth on which many of the quantum oscillation effects have very early been demonstrated, these materials can have

very long, essentially macroscopically large, mean free paths at low temperatures.

To limit the sizes (for general references see the special issue of IBM J. Research and Development, **32**, 4 (1988)) of the conducting systems and make them low-dimensional, one should distinguish between the "thickness" vs. the lateral (parallel to the thin direction) dimensions. Films, including very thin ones, may be prepared by standard deposition techniques such as evaporation or sputtering. This applies to both conducting and insulating layers. Extra-high-quality semiconductor systems including the two-dimensional (2D) case are presently achieved by growing individual lattice layers with a very impressive control of parameters by the "molecular beam epitaxy" (MBE) method briefly discussed below (see, e.g., Esaki 1984, 1986, Gossard 1986, Herman and Sitter 1989). This can be used to grow two different semiconductors on top of each other, especially[1] if their lattice parameters are matched. The by now classic example is GaAs and AlAs and their mixtures, but other combinations are possible. Recently, epitaxial growth of SiGe on Ge was achieved (Meyerson et al. 1990, Ismail et al. 1991).

The different band structures (mainly the energy gap, E_g) and work functions (the energy difference between the Fermi levels in the bulk and the vacuum level outside) usually cause some charge to be transferred between the two adjacent materials in order to equalize the electrochemical potentials. Electrons are attracted to the remaining holes and the well-known dipole layer is formed at the interface (see, e.g., Ashcroft and Mermin 1976), which leads to a space-dependent potential energy or "band-bending" near the interface. Thus, potential wells, barriers, and so on can be formed. One can get accumulation layers or inversion layers, effectively including strictly 2D situations By using sandwiches of one type of semiconductor between two layers of another type, further types of wells and barriers can be made. Such structures can be repeated to form, for example, a periodic superlattice in the growth direction (Esaki 1984, 1986). Together with charge control by electrostatic gates, this leads to an impressive control over the design of man-made materials and their combinations. For reviews see Ando et al. (1982), Sze (1986), Esaki (1986) and Gossard (1986).

With metallic systems, purities and control of thin films grown in high-vacuum systems approaching the range of those of MBE have already been achieved (Haviland et al. 1989). There, of course, one does not have the variability due to parameters such as doping, several related material systems, and electrostatic gates (although the latter is possible in principle). Another useful property of many semiconductors is the small effective mass for small band gaps, which does not occur in usual metals. We remark that other, less expensive, growth techniques which currently produce lower quality than MBE do exist and may often be sufficient (e.g., Razeghi 1989).

[1] But not exclusively, see van der Merwe 1963. This leads to the interesting possibility of "strained layers."

The idea of MBE growth starts with an extremely high-vacuum system (10^{-11} torr is common) and with a good-quality single-crystal substrate to grow the sample on. The substrate is held at rather high temperatures to cause a high mobility (parallel to the substrate) of atoms impinging on it from sources that are well-controlled with shutters (often it is enough to control one kind of atom (e.g., Ga) and the other type (e.g., As) will stick only upon finding a counterpart—thus giving "automatic" stoichiometry). The structure is monitored at real times with high-energy electron diffraction, enabling a controlled layer by layer growth of atomically smooth interfaces and excellent purities. This can lead to transport mean free paths of up to tens of micrometers. Thus effectively "ballistic" electron motion both parallel and perpendicular to the layer is achievable over significant size ranges.

The structuring of the system in the lateral direction, parallel to the 2D layer, is achieved by a variety of lithographic techniques (see, e.g., Prober 1983, Howard and Prober 1982, Broers 1989). We shall explain later, rather schematically, the method and its possibilities and limitations.

The underlying principle is to coat the planar surface, using a spinning and "baking" process, with a layer of polymer called a "photo-resist" in the case of optical lithography (a typical substance being PMMA—poly(methyl methacrylate)—for electron beam techniques) which is very sensitive to being irradiated with appropriate radiation such as light, x-rays, electrons, or ions. Thus the method uses the radiation damage, or "photochemistry" in the resist. The irradiated resist undergoes some physicochemical changes. For example, bonds are broken and a shortening of the chains and some cross-linking takes place. The affected material may dissolve in a solvent called the "developer" which dissolves the virgin resist at a significantly slower rate. Thus, if a pattern (e.g., a thin wire, a Hall bar, a ring) is projected onto the resist, say optically, it is then possible to wash away the irradiated part while the virgin part remains.

Now, there are many ways to proceed. Let us consider a specific one. If the resist has been deposited on a metal film, as an example, it is then possible to dissolve or etch the exposed metal with, for example, an appropriate acid, while the part under the unprocessed resist remains intact. Alternatively, one may coat the whole processed resist with a metal film and an appropriate solvent, which may be called a "second developer," will wash away ("lift-off") the metal that is above the existing resist, while the other part of the metal film, which was deposited where there was no resist, will stay.

Other processing possibilities using similar lithography techniques are also feasible. Examples include: (a) reactive ion etching (RIE) for producing very high-aspect-ratio etching, damaging specified regions, and removal and deposition of oxides; and (b) ion implantation, which can be used to dope, alloy, or damage the sample with some control over the depth. It is clear that by using well-known optical techniques, lateral structures with a size resolution determined by the wavelength of light (diffraction limit) can be made. If a better resolution is desired, shorter-wavelength radiation is necessary. UV light can be used to yield a modest increase of resolution. For the best resolution, x-rays, electrons, and ions can be invoked with the appropriate resists.

It turns out that such techniques can go down to a lateral resolution approaching 100 Å, the limit being determimed not by the wavelength of radiation used, which may be on the atomic scale, but by the resist itself. The resolution limit may be determined by the range of "secondary" electrons produced by the high-energy radiation. These electrons damage the resist not dissimilarly to the original radiation and they produce the same effect as a

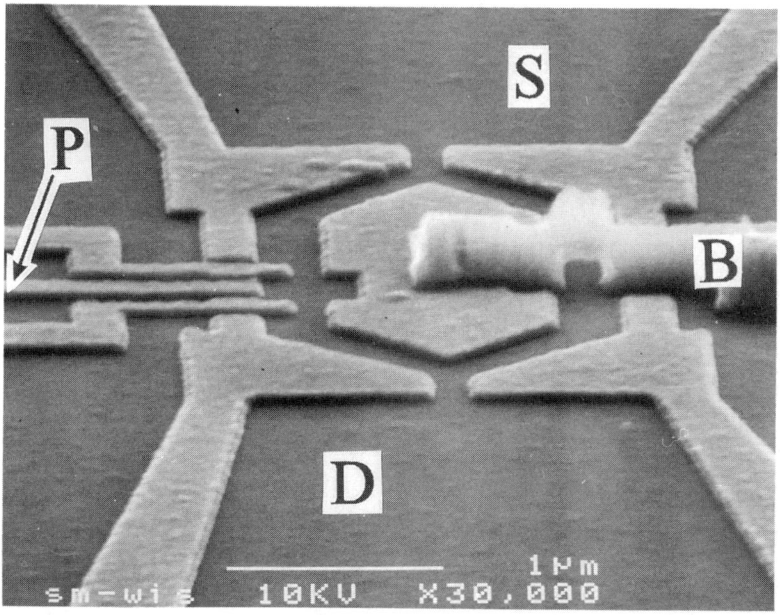

Figure 1.1 An example of an electron beam-made state-of-the-art nanostructure formed to demonstrate a double-slit type interference in a ring geometry, by Yacoby et al. (1995). Electrons pass from a "source" (S) to a "drain" (D) in the 2D gas (black area) through the two arms of the ring defined electrostatically by the gates (brighter shade). On the left-hand arm of the ring a "quantum dot" is formed by the two unmarked thin gates that isolate the dot from the arm. In addition, another "plunger" gate, marked P, modulates the electrostatic potential of the dot. An additional gate (B) modulates the central section ("hole of the conducting ring") and thus controls the width of the arms. This gate is connected to the outside by an "air bridge" (B) avoiding contact with the gate below it by going through a higher level. To make this structure, a very nontrivial good realignment of the higher to the lower level was achieved. The interference oscillations (see page 110 for a related theoretical discussion) are observed as function of the plunger voltage and/or a magnetic field that contributes an Aharonov–Bohm flux through the opening of the ring. This experiment demonstrated the coherence of electron waves passing by resonant tunneling through the dot, over dwell times of about 3 ns.

wider beam of radiation. Keeping in mind that 100 Å is about 30 Fermi wavelengths in a metal and less than a Fermi wavelength in GaAs, this is quite impressive. Especially with semiconducting systems, it is often desirable to perform such processes, but with differing patterns on different layers of the heterostructure. Techniques exist to bring back the electron beam, for example, in a reproducible way to the same spot, within a resolution similar to the above, after the sample has been removed, treated, and returned to the microscope chamber. Thus, a great variety of mesoscopic structures with the above resolution can be produced with present state-of-the-art technology. Modifications of regular scanning electron microscopes have been quite successful, with resolutions better than 500 Å, for single-layer structures. For a critical review, see Broers (1989). A recent example of an electron beam-formed structure is shown in Fig. 1.1.

While this technology is extremely impressive, it should be recognized that the structures produced are far from being perfect. For example, a wire a few hundred angstroms wide produced by etching will typically have a very irregular edge, which will produce strong diffuse scattering, substantially

Figure 1.2 A "quantum corral" demonstrating the power of STM-type techiques (from Crommie et al. 1993). A ring of 48 Fe atoms, with a radius of about 140 Å, was constructed on the 111 surface of copper. The radial height variations were measured with the same technique and demonstrate the almost ideal nature of the surface on the inside of the ring. They result from the eigenstate of the corral.

reducing the mobility, especially in semiconducting systems. One way to counteract this is by forming the semiconducting structures under electrostatic gates. These can, for example, deplete the 2D electron gas beneath them and also including a fringing-field adjacent area, with a strong enough negative bias potential. Thus, a narrow channel whose width can be somewhat varied can be formed between these "split gates." It turns out that electrostatic effects tend to smear the microscopic irregularities of the edge of the gate, thus producing narrow channels of good mobility (Thornton et al. 1986). This can be used to produce good-quality channels down to widths of a few wavelengths. Getting to even narrower "pure 1D" channels necessitates other techniques. A recent novel 1D structure has been produced by growing it mainly by the MOCVD (Razeghi 1989) technique inside a suitable groove in the substrate (Kapon et al. 1989). Another useful technique is that of cleaved edge overgrowth (Pfeiffer et al. 1993, Yacoby et al. 1996).

The utmost atomic resolution of structures can be achieved by scanning tunneling microscope-related (STM; Binning et al. 1982) techniques. These have reached the stage where individual atoms can be manipulated on the substrate to form the desired structure (see Fig. 1.2). The same technique will be used to contact the small structures and make a variety of measurements on them, which may well lead to the microscopic limit of mesoscopics (see, e.g., Crommie et al. 1993, Avouris and Lyo 1994). Since the method is slow and treats structures on an individual basis, attempts to automate it and increase its speed are already being made.

2

Quantum Transport, Anderson Localization

1. BASIC CONCEPTS

The Bloch–Boltzmann quasiclassical theory of electronic transport in lightly disordered conductors has been quite successful in describing the impurity and temperature dependences of the conductivity in ordinary relatively pure conductors. Further transport properties such as magnetoconductivity, Hall effect, thermal conductivity, and thermopower can also be handled, often with success. However, when the amount of disorder (or impurity concentration) becomes very large, novel phenomena occur which are unexplainable within the weak-scattering theory. In particular, the temperature dependence of the resistivity, ρ, becomes much weaker and eventually changes direction for high enough disorder (Weismann et al. 1979). The Mathiessen rule, according to which the contributions to ρ due to disorder and temperature are additive, thus breaks down. An extremely interesting correlation was found by Mooij (1973), namely, that $d\rho/dT$ becomes negative when ρ becomes larger than a value ranging around 80–180 μΩ cm in something like a hundred disordered dirty systems. This almost "universal" trend must have an explanation which is only weakly dependent on material properties, and we remember that $d\rho/dT > 0$ *always* in weak-scattering theory. To understand the limitation of the quasiclassical transport picture we remind ourselves that the ordinary Drude expression for the weak-scattering conductivity, σ_0,

$$\sigma_0 = \frac{ne^2\tau}{m}, \tag{2.1}$$

where n is the electron density, τ the transport time, and m the (effective) mass, is valid only when the electron wavelength $\lambda_F = 2\pi/k_F$ is much smaller than the mean free path $l = v_F\tau$, that is,

$$k_F l \gg 1 \quad ; \quad E_F\tau \gg 1. \tag{2.2}$$

We rewrite (2.1), using $n = k_F^3/3\pi^2$:

$$\sigma_0 = \frac{e^2}{3\pi^2\hbar}k_F^2 l. \tag{2.1'}$$

Thus, eq. 2.2 becomes, remembering[1] that $\hbar/e^2 \cong 4.1\ \mathrm{k\Omega}$ (so that $(e^2/\hbar)k_F$ is a conductivity unit appropriate to the microscopic length λ_F)

$$\sigma_0\lambda_F \cong 5\times 10^{-5}(k_F l)/\Omega \gg \frac{5\times 10^{-5}}{\Omega}, \tag{2.3}$$

or, for the resistivity $\rho_0 = 1/\sigma_0$:

$$\rho_0 \ll 200\ \mu\Omega\ \mathrm{cm}\cdot\lambda_F\ (\text{in Å}). \tag{2.3'}$$

For $\lambda_F \sim 5$ Å, and λ_F does not vary much in most metals, this yields $\rho_0 \ll 10^{-3}\ \Omega$ cm. We see that the quasiclassical theory becomes problematic for ρ approaching 100 μΩ cm. Much larger values apply to semiconductors and semimetals; and granular metals require a small modification (Imry 1981a, see discussion around eq. 2.41 below) of the above. But it is clear that $k_F l \lesssim 1$ or $\rho_0 \gtrsim 10^{-3}\ \Omega$ cm for metals, brings us into a totally different regime. The above condition is called the Yoffe–Regel criterion (Yoffe and Regel 1960; see Mott and Davis 1979). We shall see that a better starting point for this new regime is that of Anderson-localized states. More recently (and, in fact, following ideas from localization theory: for general references on localization see, e.g., Lee (1980, 1984) and articles in the books edited by Friedman and Tunstall (1978), Balian et al. (1979), Stern (1982), Castellani et al. (1981), Nagaoka and Fukuyama (1982), as well as the book by Mott and Davis (1979) and the reviews by Lee and Ramakrishnan (1985) and by Aronov and Sharvin (1987)) it has been found that "restricted geometry" systems, that is, thin films and wires, appear to always have $d\rho/dT < 0$ at low enough temperatures. This behavior of effectively one-dimensional (1D)

[1] To understand this, note that $(e^2/\hbar)$ is a *conductance*, which happens to be equal to $\sim(4\ \mathrm{k\Omega})^{-1}$ in MKS (this should not be surprising, since the fine structure constant is $\alpha = e^2/\hbar c \cong 1/137$, and $c^{-1} \sim 30\ \Omega$, related to the "free space impedance").

and 2D systems is also quite universal. A growing amount of evidence has indicated (Dolan and Osheroff 1979; Giordano et al. 1979, Bishop et al. 1980, Pepper and Uren 1982) that these latter systems are always (Thouless 1977, Abrahams et al. 1979) *insulators*, that is, $\rho \to \infty$ as $T \to 0$. However, there is still no really sound theory in 2D, and several caveats exist (e.g., the case of a strong spin–orbit interaction, where an "ideal" conductivity seems to follow theoretically as $T \to 0$).

As mentioned earlier, ordinary bulk "3D" systems will also become insulators in the above sense when the disorder is strong enough (Anderson 1958). Thus, the disorder is another important ingredient for the *transition from the metallic to the insulating state*. Other mechanisms for this transition being (Mott 1974) electronic (band) structure effects, electron–electron interactions (Mott 1974, Hubbard 1964) or excitonic mechanisms (Knox 1963, des Cloizeaux 1965, Keldysh and Kopaev 1965, Kohn 1965) and, possibly, self-trapping of the electron by the phonons (Holstein 1959, Toyozawa 1961). These mechanisms may strongly couple and influence each other, as we shall see, but we would like to start with the problem of noninteracting electrons in a given, static, aperiodic potential.

The plan of this chapter is as follows: In the next subsection the general localization phenomenology will be discussed. Transport in the localized phase will be treated in section 2. The Thouless picture, starting with the thin wire case, will be presented in section 3. This is the basis of the scaling theory of the dependence of the conductance on the length, which will be described in section 4, along with its many consequences. The case of weak localization will be briefly reviewed in section 5. In this chapter we shall not use the Landauer scattering approach, although it is extremely useful for the study of some aspects of localization. It will be developed in detail in chapter 5 and applied to both localization and (mainly) mesoscopic phenomena.

Localization Ideas

The solution for the electronic transport for the problem with strong disorder is called "Anderson localization theory." It appears to be a good candidate for the new discipline needed for discussing the above-mentioned problems (Jonson and Girvin 1979, Imry 1980a, 1981a,b). It explains much of the temperature and magnetic field (B) dependence of ρ for not too dirty systems (Hikami et al. 1981, Kawaguchi and Kawaji 1982; for reviews see Fukuyama 1981b, Altshuler et al. 1982a, Bergmann 1984, Altshuler and Aronov 1985, Lee and Ramakrishnan 1985). It predicts that "1D and 2D systems are not true metals" and gives useful indications on the metal–insulator transition in 3D. Interesting insights are obtained on disordered magnetic metals, superconductivity in disordered metals, and so on. The electron–electron (and electron–phonon) interaction is also important and should be considered (Schmid 1974, Abrahams et al. 1981, Altshuler and Aronov 1979) once the pure "localization" part is understood.

The model usually considered for the disorder has a random potential $V(r)$ such that

$$\langle V(R)\rangle = 0,\ \langle V(r)V(r')\rangle = C(|r-r'|), \tag{2.4}$$

where the range of the function C, called a, is the microscopic length in the problem and the size of C ($C(0) > 0$) is related to the strength of the potential fluctuations. The case $C(x) \propto \delta(x)$ is referred to as the white-noise potential (since the Fourier transform, $\tilde{C}_k$ is constant). Another useful model potential is a periodic one with a random modulation, which is called the Anderson model (Anderson 1958), when taken as a nearest-neighbor tightbinding model, that is,

$$\mathcal{H} = \sum_i \epsilon_i c_i^\dagger c_i + \sum_{\langle ij\rangle} t_{ij} c_i^\dagger c_j + \text{h.c.} \tag{2.5}$$

Here $\langle ij\rangle$ means that i and j are nearest neighbors and $c_i^\dagger$ creates an electron on the "atomic" state of the ith site of a simple lattice. The site energies ϵ_i (diagonal disorder) or the t_{ij} (nondiagonal disorder), or both, can be taken to be random In the former case, if $t_{ij} = V$ and the width of the ϵ distribution is $2W$, the ratio W/V is a convenient measure for the disorder. Solving for the eigenvalues of eq. 2.5 amounts to diagonalizing a random matrix. Obviously, this will solve a number of other physical problems, such as phonons in a disordered crystal.

In his pioneering paper of 1958 (preceded in some respects only by Landauer and Helland (1954) and Landauer (1957)) Anderson considered the Hamiltonian of eq. 2.5. A strong enough disorder can localize a state, as obviously happens when ϵ_i is very large or very small, with respect to typical values of ϵ_j, analogously to the formation of a bound state (or a local vibration in the phonon case, see Economou (1990)). This means that the envelope of the corresponding wavefunction ψ decays strongly (exponentially in typical cases) at large distances from the localization center (e.g., site i). More general types of localization for the random potential, where ψ extends over a characteristic length ξ and sometimes having bulges and oscillations after being small for a while, also exist. The formation of localized states, taking orthogonalization with different states into account, has presumably some, as yet not fully understood, similarities to bound state formation (Economou 1990; for example, both happen very easily in 1D and 2D). In Anderson's paper, a perturbation theory (locator expansion) for the "self-energy" was constructed, taking the uncoupled sites as the zero-order problem and the t_{ij} as the perturbation. Anderson proved that for large enough W/V, that is, $W/V > W/V|_c$, this perturbation theory converges "in probability." An upper limit for $W/V|_c$ is given by

$$\frac{2eVK}{K}\ln\frac{W}{2V} = 1, \tag{2.6}$$

K being the connectivity of the lattice (for a recent review, see Shalgi and Imry 1995). On a heuristic level, this means that starting from a site, 0, the contributions of very distant sites fall off sufficiently strongly. More formally, it follows from the above convergence (see also Thouless 1970) that the singularities of $G_{00}(E)$, the local Green's function at site 0, are just a dense set of poles, where most of them (due to states localized far away) have exceedingly small residues at 0. These poles are the zeros of the self-energy, at which it is analytic and therefore has a vanishing imaginary part. This is to be contrasted with what happens with delocalized states, where the self-energy and G_{00} *must* (and do) have a branch cut on the real energy axis, in order for the former to have a finite imaginary part. Such an imaginary part is, of course, necessary in order for a wave packet started at $t = 0$ around site 0 to truly decay into the whole crystal.

Following further analytical arguments and much numerical work (Liciardelo and Thouless 1978, Stein and Krey 1979, 1980, Domany and Sarker 1979, Kramer et al. 1990, Kramer and MacKinnon 1993) it is generally agreed that for a large enough disorder all (or almost all) states are localized. For an intermediate disorder the situation (at dimensionality $d > 2$) is thought to be as follows (Mott 1966). Due to the disorder, states are created in the gap. The states in the middle of the band are not localized (i.e., they are "extended"), while the states near the band extremities may be localized. The extended and localized states are separated by the mobility edge E_{m1}, E_{m2} (Mott 1966) (see Fig. 2.1). The existence of these follows from the physically reasonable strengthening of the tendency for localization as E gets further from the band center and from an intuitive argument by Mott, according to which extended and localized states cannot coexist at the same energy since they will be mixed

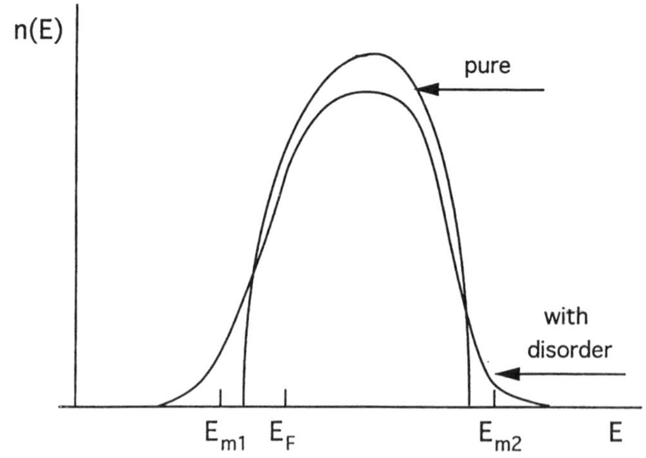

Figure 2.1 The density of states with and without disorder and the mobility edges in the former case (schematic). Note the smooth behavior near E_{mi}. See, however, Appendix F for the effect of interactions near E_F.

by any interaction, however small.[2] When W/V is increased, the mobility edges approach each other and coalesce at some limiting value $(W/V)_c$, where all states become localized, and the Anderson transition has occurred. We emphasize that the density of states (DOS) at the Fermi energy is not expected to develop any singular structure due to the localization process alone (S. Kirkpatrick private communication, Wegner 1976, 1979).

Some aspects of the physical relevance of the localization follow from the observation that if all the physically relevant states—such as the states near the Fermi energy E_F—are localized, the system will be insulating at $T = 0$. This physically obvious assertion can be proved by showing that the diffusion constant D vanishes. From the Einstein relation (Kubo 1957) at $k_B T \ll E_F$

$$\sigma = e^2(dn/dE)D, \tag{2.7}$$

it will then follow that σ vanishes. To demonstrate the vanishing of D, form a narrow minimal wavepacket at $t = 0$, $r = 0$; $\psi(t = 0) = \sum a_j\psi_j$, from the available eigenstates $\psi_j = |j\rangle$. a_i will be exponentially small for states $|i\rangle$ localized many localization lengths ξ from $r = 0$. At finite times

$$\psi(t) = \sum a_j e^{-i(E_j/\hbar)t}\psi_j, \tag{2.8}$$

where $a_j = \langle 0 \mid j\rangle$, will also decay exponentially with r at any time. Therefore $\langle r^2\rangle$ which should be given at $t > 0$ by $2Dt$ for diffusion, is never (even when $t \to \infty$) larger than $O(\xi^2)$, so that $D = 0$. It is easy to see from (2.8) that the "staying probability" at the initial site (the long-time average of $|\langle 0 \mid \psi\rangle|^2$) is given by $\sum_j |\langle 0 \mid j\rangle|^4$. This is often referred to as the "inverse participation ratio" (IPR). Remembering that $\sum_j |\langle 0 \mid j\rangle|^2 = 1$, we can see that the IPR is proportional to the inverse localization volume and it thus vanishes for delocalized states. It is therefore a useful diagnostic for localization.

The above observation, that $\sigma = 0$ in the localized regime, gives us a very simple mechanism (see Mott 1974) for the metal–insulator (henceforth abbreviated as M–I) transition. The energies E_F and E_{mi} can be shifted by modifying the electron density or disorder, respectively.[3] Whenever E_F passses from the extended to the localized range, the system will go from the metallic to the insulating phase. Since in the insulating phase $\rho(T = 0) = \infty$ and ρ *decreases* with increasing temperature, it is not surprising that $d\rho/dT$ can be negative near the transition, certainly in the "poor" insulator—and, by

[2] This argument is not rigorous, since for a system with linear size L, the characteristic difference in energies of the states that are close in energy is $O(L^{-d})$, while if the extended and localized states reside in different parts of the system, the interaction may be $O(e^{-O(L)})$. This can probably happen in very inhomogeneous, e.g., percolating, systems, but it is assumed not to happen if the system is homogeneous enough.

[3] In more realistic cases the screening in the metallic phase becomes stronger for a larger carrier density, which may decrease the effective disorder as well, in a self-consistent picture.

continuity, in the "poor" metal. One thus already sees that localization theory may be useful in explaining the anomalous properties of disordered conductors as well as the disorder-induced metal–insulator transition.

When the disorder, for example, W/V, becomes very small, the usual weak-scattering theory should apply for dimensions above 2 (and, as we shall see, also at 1D and 2D for small sizes or high temperatures). A convenient dimensionless parameter to express this is $k_F l$ or $E_F \tau$, which are of the same order of magnitude. τ is the mean (elastic) free time and l the mean free path, $l = v_F \tau$ (we do not consider here the case of small-angle scattering, where the transport time is much longer than the scattering time). The small parameter as discussed above in the weak-scattering theory is $1/(k_F l)$. Note that the order of magnitude of the diffusion constant is $D \sim v_F^2 \tau = v_F l$ and the usual weak-scattering conductivity is given by eq. 2.1. The situation in 2D is more interesting, since σ (or the conductance of a thin square film divided by its thickness) $\propto (e^2/\hbar)(k_F l)$—which is just a *universal* constant times $k_F l$. Of course, as discussed in section 1, the weak-scattering theory will break down when the parameter $k_F l$ is no longer $\gg 1$. This defines the concept of the minimum metallic conductivity at $d > 2$ (Yoffe and Regel 1960, Mott 1966), obtained when $k_F l \sim l$. In 3D (see eq. 2.2)

$$\sigma_{min} = C \frac{e^2}{\hbar} k_F, \tag{2.9}$$

where C is a constant estimated by Mott to be on the order of 0.01–0.05. For metallic systems where k_F is a few inverse angstroms, this yields, as discussed before, resistivities of 10^{-3} Ω cm, somewhat larger but of the same order of magnitude as the above Mooij value of $1\text{–}2 \times 10^{-4}$ Ω cm. (Thus $d\rho/dT$ may become negative when, roughly, $k_F l \sim 5\text{–}10$.) $\sigma_{\min}$ is clearly the range where localization is very relevant. Mott has argued that no metals can exist with $\sigma < \sigma_{\min}$. This, and necessary modifications, will be discussed later in some detail. In 2D, one has a "maximum universal metallic resistance" of $h/e^2 \sim 30$ k Ω, and a similar but somewhat smaller value in 1D.

2. THERMALLY ACTIVATED CONDUCTION IN THE LOCALIZED REGIME

If the states at the Fermi energy, E_F, are localized, $E_F < E_m$ (we assume, for definiteness, that E_F is in the lower half of the band, $E_m \equiv E_{m_1}$), then the quantum conduction at $T = 0$ is zero. At low but finite temperatures the electron can gain thermal energy (typically from other excitations, e.g., phonons) to perform a number of possible processes (Mott and Davis 1979).

1. *Activation to* (and above) *the mobility edge*, which will yield

$$\sigma_1 \propto e^{-(E_m - E_F)/k_B T},$$

where the coefficient is proportional to the square of the electron–phonon coupling strength.

2. *Activation to a neighboring localized state.* If the localization length is ξ and the density of states (DOS) at the Fermi level is $n(0)$ then the number of states in a volume of linear dimension ξ in d dimensions is $n(0)\xi^d$; hence the typical energy separation between such states is

$$\Delta_\xi \sim [n(0)\xi^d]^{-1}, \tag{2.10}$$

which would yield a "nearest neighbor" activated conductivity of the form (see, however, the discussion following eq. 2.27):

$$\sigma_2 \propto e^{-\Delta_\xi/k_B T}. \tag{2.11}$$

However, as suggested by Mott (1966, 1970), it pays sometimes for the electron to hop a larger distance, thereby reducing the necessary inelastic energy transfer. This introduces the next type of activated conductivity:

3. *Variable range hopping* (VRH). We assume that the contribution to the hopping conductivity to a state localized a length $L \gg \xi$ away is proportional to the overlap matrix element squared, which goes like $I^2 e^{2L/\xi}$, where I is a characteristic energy of order Δ_ξ. On the other hand, the energy needed now, Δ_L, is obtained by generalizing the argument leading to eq. 2.11 and noting that Δ_L decreases with L:

$$\Delta_L \sim (n(0)L^d)^{-1} \sim \Delta_\xi \left(\frac{\xi}{L}\right)^d \qquad (L \gg \xi). \tag{2.12}$$

The hopping over a length L is controlled by $e^{-2L/\xi - \Delta_L/k_B t}$. At low temperatures (see below) it pays to make hops with $L \gg \xi$. The optimal L, L_M, for such jumps is given by minimizing the exponent:

$$L_M \sim \left(\frac{\xi}{n(0)k_B T}\right)^{1/(d+1)} \tag{2.13}$$

and this mechanism is relevant as long as $L_M \gtrsim \xi$, that is, when the temperature is low enough so that $T \ll T_0$, where

$$k_B T_0 \sim \Delta_\xi. \tag{2.14}$$

At such low temperatures, the VRH conductivity, σ_3, is given by

$$\sigma_3 \propto e^{-C(T_0/T)^{1/(d+1)}}, \tag{2.15}$$

where C is a dimensionless constant. For $T > T_0$, the nearest-neighbor hopping, σ_2 of eq. 2.10 (or its modification, see eq. 2.27), is obtained. One has then to consider the competition between σ_2 and σ_1. It apears that σ_2 usually wins, especially near the transition, when Δ_ξ vanishes faster (see, e.g., the discussion in section 3.3 of Shalgi and Imry 1995) than $(E_m - E_F)$. For $(E_m - E_F) < \Delta_\xi$, σ_1 will be dominant in the simple activated region and the crossover to $\exp[-(-C'/T^{1/(d+1)})]$ will occur at a temperature below T_0.

We also remark that in the variable range hopping regime, L_M is an important length scale which determines, for example, the effective dimensionality of a thin film or wire (Fowler et al. 1982).

We do not discuss here in any detail the effect of Coulomb interactions on the hopping conductivity. Even in a Hartree-type approximation, this was argued (Pollak 1970, Shklovskii and Efros 1971, 1984) to produce a "Coulomb gap" near E_F for the available energies to hopping and to change the exponent in a relation such as eq. 2.15, from $1/(d+1)$ to $\frac{1}{2}$. Such a behavior is indeed observed in many different cases. The theory is still being debated.

A different and extremely instructive and useful way to consider the VRH process was suggested by Ambegaokar et al. (1971), Shklovskii and Efros (1971) and Pollak (1972). Each bond along which hopping occurs can be viewed, following ideas of Miller and Abrahams (1960), as a resistor. The whole lattice is therefore equivalent to a "random resistor network," and the resistances vary greatly within the network due to their exponential dependence on parameters. Conduction will occur mostly due to the smaller resistances. If the largest resistances are eliminated, the resistance of the whole network is almost unaffected. This process can be continued, going to smaller and smaller resistances, until the percolation limit of the network having only resistances smaller than some critical R_0, is reached. Beyond that, cutting off more resistances will produce a disconnected (nonpercolating) structure. After stopping at R_0, one may also say that all the resistances that are much smaller than R_0 behave essentially like shorts. Thus, the whole network resistance is to a good approximation determined by the "just percolating resistance" R_0. From this idea the VRH theory follows with specific values of the numerical coefficients.

The above percolation idea is also useful to obtain the *small field magnetoresistance* (MR) in the VRH regime. The reason that the resistance of each bond depends on B is that the overlap matrix element between the localized states is not only due to the direct path, but to the whole sum over all indirect hopping paths, so it involves a quantum interference phenomenon. An interesting model for the latter has been suggested by Nguyen et al. (1985a,b); its treatment (Entin-Wohlman et al. 1989) using the percolation model gave acceptable results for the MR.

At larger fields, the important element is the change of the localization length with B. Understanding the latter (see, for example, Lerner and Imry 1995), and the Hall effect (Holstein 1959, 1961) in the localized phase are two outstanding problems in that regime.

Mott (1970) also gave an argument showing that the $T = 0$ *frequency-dependent conductivity*, $\sigma(\omega)$ in the localized phase would behave like

$$\sigma(\omega) \propto \omega^2 \ln^{d+1}(I/\omega). \tag{2.16}$$

The physical idea for understanding the logarithmic factor in eq. 2.16 is the following (for a detailed explanation see Sivan and Imry 1987). Given $\hbar\Omega \ll \Delta_\xi$, we look for two initial states localized a distance $R \gg \xi$ apart so that for them, using eq. 2.11, and $I \sim \Delta_\xi$

$$Ie^{-R/\xi} \sim \hbar\omega, \qquad \text{i.e.} \qquad \frac{R}{\xi} = \ln\frac{\Delta_\xi}{\omega}. \tag{2.17}$$

For this distance R one now looks for the pairs of states for which their energy separation (prior to switching on the tunneling of eq. 2.17) is smaller than or on the order of $Ie^{-R/\xi}$; such "resonating" states will be mixed very well by the tunneling and become "double hump" states. The dipole matrix elements between them will be $\sim R \sim \xi \ln(\Delta_\xi/\omega)$. The number of such pairs is proportional to $R^{d-1}\xi$—the volume of a shell of radius R and thickness ξ. The combination of these two factors is the reason for the $|\ln\omega|^{d+1}$ in eq. 2.16. The ω^2 comes from the usual counting of states and the Kubo expression (see eq. A.7 of appendix A). An issue which is possibly relevant for such considerations is that of "level repulsion," that is, that levels usually do not come too close to each other. It turns out that $Ie^{-R/\xi}$ *is* in fact the level repulsion in this case, so it is taken into account here (Sivan and Imry 1987). Since larger distances are involved with low frequencies, interesting frequency–size crossovers can be obtained for $\sigma(\omega)$ in the mesoscopic range.

3. THE THOULESS PICTURE, LOCALIZATION IN THIN WIRES AND FINITE TEMPERATURE EFFECTS

We start this section by briefly reviewing the tunnel-junction picture of conduction which also paves the way to the material of Chapter 5. Consider two pieces (later referred to as "blocks") of a conducting material, connected through a layer of insulator (usually an oxide) which is thin enough to allow for electron tunneling. The interfaces are assumed rough, so there is no conservation of the transverse momentum: each state on the left interacts with each state on the right with a roughly uniform matrix element t. The lifetime τ_L for an electron on one block for a transition to the other block is given by the Fermi golden rule (when tunneling is a weak perturbation, i.e., a weak interblock transmission):

$$\tau_L^{-1} = \frac{2\pi}{\hbar}\overline{t^2} N_r(E_F), \tag{2.18}$$

where $\overline{t^2}$ is the average of the tunnelling matrix element squared and $N_r(E_F)$ is the density of states on the final (right-hand) side. Taking the DOS in the initial side to be $N_l(E_F)$, we find that when a voltage V is applied, $eVN_l(E_F)$ states are available, each decaying to the right with a time constant τ_L, so that the currrent is $I = eN_l(E_F)\tau_L^{-1}V$ and the conductance is

$$G = e^2 N_l(E_F)/\tau_L = \frac{2\pi e^2}{\hbar}\overline{t^2} N_l(E_F), \tag{2.19}$$

which is an extremely useful result. The second equality is well known in the tunnel junction theory (Bardeen 1961, see also Harrison 1970). Note that eqs. 2.18, 2.19 are valid in any number of dimensions. An important remark is that eq. 2.18 strictly needs a continuum of final states, while the final (r.h.s.) bock is finite and has a discrete spectrum. One may make the assumption that the interaction of that system with the outside world leads to a level broadening larger than, or on the same order as, the level spacing,. This is the case in most mesoscopic systems. Otherwise, when levels really become discrete, one gets into the *truly microscopic* (*molecular*) *level.*

The first equality in eq. 2.19 is very general. Divide a large sample into (hyper) cubes or "blocks" of side L. We consider the case $L \gg l$, a, where l is the elastic mean free path and a the microscopic length. The typical level separation for a block at the relevant energy (say, the Fermi level), Δ_L, is given by the inverse of the density of states (per unit energy) for size L, $N_L(E_F)$. Defining an energy associated with the transfer of electrons between two such adjacent systems by $V_L \equiv \pi\hbar/\tau_L$ (τ_L is the lifetime of an electron on one side against transition to the other side), the dimensionless interblock conductance $g_L \equiv G_L/(e^2/\pi\hbar)$ is

$$g_L = V_L/\Delta_L, \tag{2.20}$$

i.e., g_L is the (dimensionless) ratio of the only two relevant energies in the problem. The way Thouless argued for this relation is by noting that the electron's diffusion on the scale L is a random walk with a step L and characteristic time τ_L, thus

$$D_L \sim L^2/\tau_L. \tag{2.21}$$

Note that as long as the classical diffusion picture holds, D_L is independent of L and $\tau_L = L^2/D$, which is the diffusion time across the block. It will turn out that the localization or quantum effects when applicable, cause D_L to decrease with L. For metals the conductivity, σ, on the scale of the block size L, is given by the Einstein relation (eq. 2.7), and the conductance in d dimensions is given by $G_L \sim \sigma_L L^{2-d}$. Putting these relations together and remembering that $N_L(E_F) \sim L^d\, dn/d\mu$, yields eq. 2.20. To get some physical feeling for the energy h/τ_L we note again that, at least for the weak coupling case, the Fermi golden rule yields eq. 2.18 or

$$V_L = 2\pi\overline{t^2}/\Delta_L. \tag{2.22}$$

Thus, V_L is defined in terms of the interblock matrix elements. Clearly, eq. 2.22 is also related to the order of magnitude of the perturbation theory shift of the levels in one block by the interaction with the other. For a given block this is similar to a surface effect—the shift in the block levels due to changes in the boundary conditions on the surface of the block. Indeed, Thouless has given appealing physical arguments for the equivalence of V_L with the sensitivity of the block levels to boundary conditions. This should be valid for L much larger than l and all other microscopic lengths, except perhaps for $g_L \ll 1$, where the sensitivity to boundary conditions might at least be an upper bound for $g_L\Delta_L$.

Since in this scaling picture the separations among the blocks are fictitious for a homogeneous system, it is clear that the interblock conductance is just the conductance of a piece whose size is of the order of L; that is, this is the same order of magnitude as the conductance of the block itself.

The latter can also be calculated using the Kubo linear response expression briefly reviewed in appendix A. It has to be emphasized that the Kubo formulation applies strictly only for an infinite system whose spectrum is continuous. For a finite system, it will be argued later that a very small coupling of the electronic system to some large bath (e.g.,the phonons, or to a large piece of conducting material) is needed to broaden the discrete levels into an effective continuum. Edwards and Thouless (1972), using the Kubo–Greenwood formulation, made the relationship of V_L with the sensitivity to boundary conditions very precise. This is discussed in appendix B.

The above picture can be used also for numerical calculations of $g(L)$, which is a most relevant physical parameter of the problem, for noninteracting electrons, as we shall see. Alternatively, eq. 2.22 as well as generalizations thereof can and have been used for numerical computations, and other powerful numerical methods exist too (Fisher and Lee 1981). It is important to emphasize that $g_L \gg 1$ means that states in neighboring blocks are tightly coupled, while $g_L \ll 1$ means that the states are essentially single-block ones. g is therefore a good general dimensionless measure of the strength of the coupling between two quantum systems. Thus, if $g_L \to 0$ for $L \to \infty$, then the range of scales L where $g_L \sim 1$ gives the order of magnitude of the localization length, ξ.

Although the above analysis was done specifically for noninteracting electrons, it is obviously of much greater generality. The ratio V_L/Δ_L is a general dimensionless measure for the coupling of two quantum systems. g_L will play the role of a conductance also when a more general entity (e.g., an electron pair) (Imry 1995) is transferred between the two blocks.

The analysis by Thouless (1977) of the consequences of eq. 2.20 for a long thin wire has led to extremely important results. First, it showed that 1D localization should manifest itself not only in "mathematically 1D" systems but also in the conduction in realistic, finite cross-section, thin wires, demonstrating also the usefulness of the block-scaling point of view. Second, the understanding of the effects of finite temperatures (as well as other experimental parameters) on the relevant scale of the conduction, clarifies the relationships between $g(L)$ and experiment in any dimension. Thouless gave a simple analysis of what happens in a wire of a given cross-section A as a function of L. This has far-reaching consequences. For a real wire, $\sqrt{A}$ is many atomic distances a, but still much less than "macroscopic" sizes (the precise requirements will become clear later). The usual Ohm's law $G_L \propto L^{-1}$ can, *at best*, hold only for a limited range of Ls. Indeed, suppose $G_L \propto L^{-1}$ for some range of Ls. Once $l \gtrsim L_c$, where L_c is defined by $G_{L_c} \cong e^2/2\hbar$, one would obtain $g \ll 1$ for $L \gg L_c$. This means that localization occurs at the scale L_c, which is therefore the localization length ξ in this case. Stated simply, Ohm's law (which does appear to hold for ordinary thin wires) can hold *only* as long (for $T \to 0$) as the $T \to 0$ resistance of the wire is less than about 10 kΩ. For lengths larger than this length, ξ, the $T \to 0$ resistance should increase exponentially with L (see the discussion following eq. 2.8). ξ is easily estimated assuming $G_L \propto L^{-1}$ for $a, l < L < \xi$ (a being the relevant microscopic length and l the elastic mean free path, as before). For a given wire resistivity ρ, the condition for ξ is $2\hbar/e^2 \cong \rho\xi/A$ or

$$\xi \cong \frac{2\hbar}{e^2} A\sigma \cong \frac{2}{3\pi^2}(Ak_F^2)l, \qquad (2.23)$$

where eq. 2.1′ was used to obtain the second approximate equality. Thus, the order of magnitude of the length ξ is given by the elastic mean free path l times the number of electrons in the wire's cross-section. For a cross-section of atomic dimension this yields just l, in agreement with the "purely 1D" case. The assumption of $G \sim L^{-1}$ for $L \ll \xi$ at least agrees with our intuition on wires. Theoretically it means that $\overline{t^2} \sim 1/L^3$, since $\Delta \sim 1/L$ (which is not unreasonable for a surface effect), as long as $L \ll \xi$.

Obviously, we know that the resistance of thin wires does *not* ordinarily increase exponentially with their length. However, one never measures the *zero-temperature* resistance. For the strong localization result to hold, T must be low enough so that the electron should not feel the effects of temperature during its motion on scales that would be even larger than ξ. Thouless' work indicated the following very plausible physical consideration that was later made very precise: When T approaches zero the characteristic time, τ_ϕ, between inelastic (or, more generally, any phase-breaking, see Chapter 3) events, becomes very large. For example, in many cases one can write

$$\tau_\phi \propto T^{-p}, \qquad \text{where } p \text{ is a positive exponent}, \qquad (2.24)$$

where much will have to be said later on both the exponent p and the prefactor. Consider a diffusing electron. After a time t, it covers a length $\sqrt{Dt}$. To feel strongly the localization effects, one needs that the nominal phase coherence length, defined using the diffusion coefficient, D_1, in the nonlocalized, or the $L \lesssim \xi$, regime,

$$L_\phi \equiv \sqrt{D_1 \tau_\phi}, \qquad (2.25)$$

be much larger than ξ. In the opposite case, of course, the electron will perform many inelastic collisions before it will feel the localization, and one expects that the effects of the latter will not be very strong. Thus, we are led to viewing L_ϕ as an important length scale for the electron's motion. For the wire to be effectively 1D, one is likewise led to the analogous necessary condition,

$$L_\phi \gg \sqrt{A}, \tag{2.26}$$

$\sim\sqrt{A}$ being the geometrical mean of the thickness and width of the wire. For a very flat wire, L_ϕ has to be larger than both of the above.

Thouless has also made a rather complete analysis of the low-temperature transport in the thin wire. Once T is so small that $\tau_\phi > \xi^2/D$ (we shall denote the crossover temperature, where $\tau_\phi = \xi^2/D$, by T_ξ), the motion of the electron starts to be a diffusion process controlled by localization, i.e., a random walk with a step ξ and characteristic time τ_ϕ, thus

$$D \sim \xi^2/\tau_\phi, \qquad \sigma \propto \tau_\phi^{-1} \propto T^p, \tag{2.27}$$

that is, a vanishing of the conductivity for small T like a power of T! This very important result has still not been noticed or appreciated by all localization practitioners (see, however, Shapiro 1983a,b). It appears to have been confirmed by Imry and Ovadyahu (1982b) and Ovadyahu and Imry (1985). Equation 2.27 assumes that an inelastic event gives the electron enough energy to move from one block of size ξ to the next. The condition for this is that $\Delta_\xi < T$. This defines a further crossover temperature, T_0, to exponential temperature dependence (see section 3) for $T \ll T_0$. T_0 is given by (as in eq. 2.14)

$$k_B T_0 = [n(0)A\xi]^{-1}.$$

Thus, the condition for observing the T^p behavior is $T_\xi \gg T_0$. This analysis is valid for localized states in any dimension. It appears to imply that in nearest-neighbor hopping the behavior with temperature should be a power-law one.

Let us concentrate now on the range $T > T_\xi$, where the effects of localization are weak. Here, the length L_ϕ is physically meaningful—it is the scale up to which the electron diffuses quantum mechanically. From then on the motion is classical and controlled by the inelastic scattering. Thus, even if we know the quantum-mechanical $g(L)$ (at $T = 0$), it is only relevant for $L \lesssim L_\phi$. Since one believes in the classical intuition (which simply asserts that for a given current the voltages along consecutive segments add in 1D) in the appropriate $L \gtrsim L_\phi$, range, it is suggested that the macroscopic conductivity of the sample will be determined by Ohm's law using the $T = 0$ conductance on scale L_ϕ (see eq. 2.28 below). This important observation also follows from the Landauer picture, that will be described in Chapter 5—where the electron becomes incoherent in the connected reservoirs. The length L_ϕ over which this happens in the long system defines the maximum length over which the $T = 0$ theory is valid. At larger scales, classical conduction sets in. This also forms the basis for obtaining the so-called weak localization effects which occur when $L_\phi \lesssim \xi$.

4. THE SCALING THEORY OF LOCALIZATION AND ITS CONSEQUENCES

General

The conductance of a (hyper) cube of size L, at $T = 0$, can be calculated, in principle, numerically using the Thouless relation (2.20) in any number of dimensions, employing a variety of methods (e.g., MacKinnon and Kramer 1981) to determine V_L. Using generalizations of the Landauer formula (chapter 5) provides another method which appears to be more effective (Fisher and Lee 1981). Knowing how g scales with L, for the appropriate range of L values, and understanding that L_ϕ (or ξ and τ_ϕ in the localized regime, for $L_\phi \gg \xi$) determines the relevant scale for the temperature dependence of g enables us to obtain the temperature dependence of the macroscopic conductivity, σ:

$$\sigma(T) \cong \frac{e^2}{\pi\hbar} g(L) L^{2-d}|_{L=L_\phi}. \tag{2.28}$$

That is, σ is evaluated from G using the geometry and the scale $L = L_\phi$. The generalization of this to thin films and wires is straightforward. This is valid for $L_\phi \lesssim \xi$ and also in the whole metallic range. In the localized regime, for $L_\phi > \xi$, eq. 2.27 and its appropriate counterparts at low temperatures have to be used, as discussed in section 2.

We shall now present and discuss the scaling theory of localization by Abrahams, Anderson, Licciardello and Ramakrishnan (1979) (see also Wegner 1976, 1979). This theory is really a clever guess based on an *interpolation* between the limits of a good conductor, $g_L \gg 1$, and a localized insulator, $g_L \ll 1$. It is consistent with the first correction to the good (weak scattering) conductor, with most of the presently available numerical work (e.g., Kramer et al. 1990; many of those numerical results that contradicted it seem to have been superseded by more reliable ones), with the 1D and thin wire cases, and with some analytical approximations (Vollhardt and Wölfle 1980, 1982). However, there is as yet no truly compelling theoretical argument for it and there have been many unfounded criticisms and some serious theoretical queries as well, some of which having been given good answers. Being an interpolation picture, it should be qualitatively correct (except possibly for the details insider the interpolation range, some of which can be quite important). While the behavior of real systems may be sensitive to other effects too (notably, electron–electron interactions), the scaling theory does explain better than qualitatively a large amount of data on any systems, and has made surprising (at the time) predictions that have been confirmed by experiment. Moreover, concerning reports about disagreements with the scaling theory predictions, one should make sure that real predictions of the scaling theory (e.g., eq. 2.28 is *not* valid in the insulator) are tested and that electron–electron interactions do not play a role.

In the limit of a good conductor one expects the usual Ohm's law to hold, i.e., $\sigma(L) = \text{const.} \Rightarrow g(L) \propto L^{d-2}$. In the opposite limit, one expects (see section 2) both σ and g to decrease exponentially with L. Thus, in these two limits, the logarithmic derivative of g is given by

$$\beta \equiv \frac{d \ln g}{d \ln L} = \begin{cases} d-2, & g \gg 1 \\ \text{const} + \ln g, & g \ll 1, \end{cases} \tag{2.29}$$

where small corrections due to a possible power-law prefactor for $g \ll 1$ have been neglected. The important large-g corrections (Gorkov et al. 1979, Abrahams et al. 1979, Hikami et al. 1981, Fukuyama 1980, 1981a,b, Altshuler et al. 1982a,b,c) will be discussed later. These behaviors are independent of L and of the details of the system. Since β can be obtained from computations on finite systems, it must be analytic and it can be expected not to decrease with g. One assumes that β stays a function *only of* g, that is, using RG (renormalization group) language, that g is the only "relevant" variable and that L is large enough so that the "irrelevant" ones vanish already in the whole range $L \gg l, a$ (l being the elastic mean free path and a the microscopic length). The above is obviously a stronger assumption than g being the only relevant parameter only for $g \gg 1$. One then arrives at the picture of $\beta(g)$ changing monotonically and smoothly between the two limits of eq. 2.29, as shown in Fig. 2.2, where β is schematically given as function of $\ln g$ in various dimensions.

The Case $d \leq 2$

For $d \leq 2$, β is always negative (a notable exception is the case of spin–orbit interaction at 2D; see, e.g., Altshuler et al. 1982b). This means that if we know $g(L_0) = g_0$ for some small L_0, then, obtaining $g(L)$ for every L by solving $d \ln g / d \ln L = \beta(g)$ with $g(L_0) = g_0$, we find that *always* $g(L) \sim \exp(-\alpha L)$ as $L \to \infty$. This can be visualized by noting that g_0 is represented by some point on the $\beta(g)$ graph and that $g(L)$ will simply flow down that curve with increasing L, until it reaches the linear range at small g (large negative $\ln g$). The above procedure for this very simple case is called "solving the RG equations" and the motion of the point along the $\beta(g)$ curve, an "RG-flow," in the theoretical jargon. The localization length, ξ, is the L above which the flow has reached the linear range in $\ln g$, which is the macroscopic limit. Since $\beta(g)$ becomes very flat when $g \gg 1$, ξ increases when g_0 is increased. For $g \gg 1$ we expect, from analyticity in g^{-1},

$$\beta(g) \sim d - 2 - \frac{C}{g}. \tag{2.30}$$

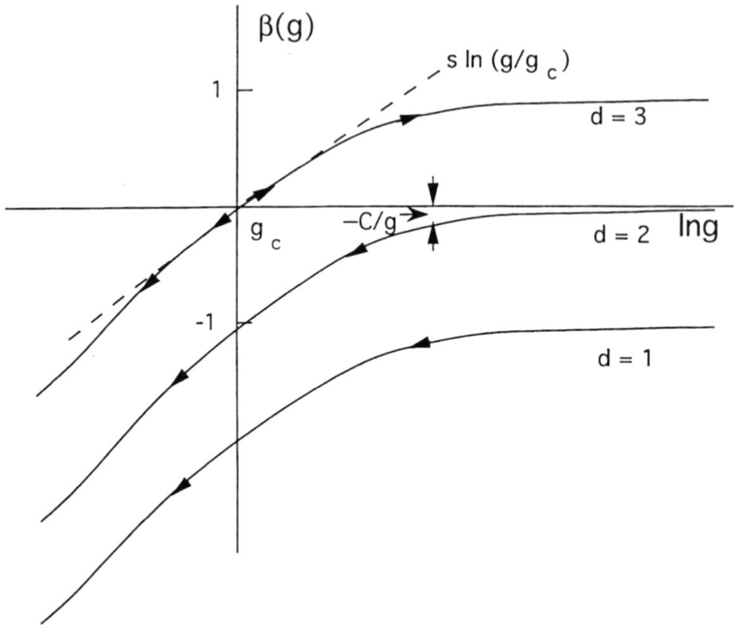

Figure 2.2 $\beta(g)$ at d =1, 2, 3 (schematic).

This has been confirmed (Gorkov et al. 1979, Abrahams et al. 1979) and the numerical constant C computed as a function of the number of dimensions, d, by perturbation theory. This is called the weak localization regime and will be discussed in section 6. ξ is roughly defined as the scale at which β decreases substantially from $d-2$ and approaches the linear range. This can be estimated using the rough approximation eq. 2.31 or even $\beta \sim d-2$, extrapolated to $g \sim 1$, and is consistent with taking

$$g(\xi) = \text{a numerical constant of order unity.} \tag{2.31}$$

We note that this agrees with eq. 2.23 for the effectively 1D case. In 2D, ξ will be exponentially large for large g_0 (i.e., larger than the distance to the sun for $R(L_0) \cong 10^{-3}$).

Note that eq. 2.29 yields a correction to the ohmic behavior, $\sigma(L) = \text{const}$, for $L \ll \xi$, that is,

$$\begin{aligned} \sigma(L) &= g_0 L_0 - C_1 L \qquad & d &= 1 \\ g(L) &= g_0 - C_2 \ln(L/L_0) \qquad & d &= 2. \end{aligned} \tag{2.32}$$

Thus, as a function of L, when L is increased, one should first obtain the weak

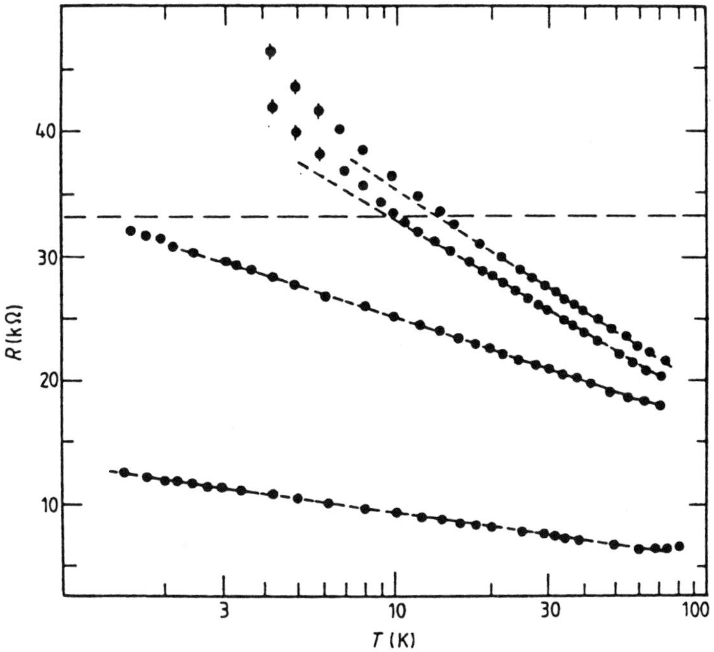

Figure 2.3 Resistance versus temperature for several effectively 2D samples ($d = 210$ Å). The broken horizontal curve marks the "critical" $R_{\square}$ above which the resistance of the samples increases faster than logarithmically.

localization correction eq. 2.32, until $L \sim \xi$, and then one gets into the (strongly) localized range with the behavior discussed before. Most interesting is the 2D behavior: The system is, in principle, never a metal. Nonclassical logarithmic corrections will be seen at relatively high temperatures, where $L_\phi \ll \xi$ (however, when the temperature is too high, so that L_ϕ is smaller than the smallest microscopic L_0 allowed ($L_0 \sim l_{el}$) this theory will break down). Upon *decreasing* T, g will decrease and R will *increase*. Once $R \sim h/e^2$ (the numerical constants yield $R_{critical} \sim 30$ kΩ), one crosses over into strong localization with an exponential increase of R with L. Thus "2D metals are not really metals." A necessary condition for 2D behavior is that the thickness of film $\ll L_\phi$, as before.

These surprising predictions are now confirmed in the weakly localized range by many experiments. It is also apparent that $\sigma(T \to 0) \to 0$ for 2D samples with $R > 30$ kΩ on an attainable scale. There is an experiment by Ovadyahu and Imry (1983) on thin InO films, shown in Fig. 2.3, where the *same* sample crosses over from weak ln T behavior to a stronger increase with decrease temperature around $R \cong 30$ kΩ, which is a remarkable confirmation of the above surprising prediction.

The Case $d > 2$, the Metal–Insulator (M–I) Transition

The new feature which appears for $d > 2$ is the occurrence of a metal–nonmetal transition, associated with the fact that $\beta(g)$ vanishes at some $g = g_c$. This zero of $\beta(g)$ follows because β is positive for $g \to \infty$ and negative for $g \to 0$. The value of g_c, as well as the slope, s, of β at g_c—which will have an important role to play—are numerical constants that can be obtained from approximations (Vollhardt and Wölfle 1980, 1982) or from simulations (Stein and Krey 1979, 1980, MacKinnon and Kramer 1981, Kramer et al. 1990). It is agreed that s is of order unity (but not exactly equal to unity) and g_c is perhaps 2–3 for $d = 3$, which is the case of most interest. We note that if g on any scale is $> g_c$, the conductance will "flow" to the Ohmic, conducting, limit as $L \to \infty$. Likewise, if g is sometimes $< g_c$, it will "flow" to the insulating range, $g \sim e^{-\alpha L}$, when $L \to \infty$. $g = g_c$ is a "fixed point" of the RG transformation, that is, if $g = g_c$ on some scale then $g = g_c$ on all scales, including $L \to \infty$ and thus $\sigma \sim g_c L^{2-d} \to 0$ in the macroscopic limit. According to this simple theory, all materials (where an important necessary condition for the applicability of the theory is their being homogeneous on the scales of interest) "sit" on the same universal $\beta(g)$ curve and can be distinguished from each other, for example, by their conductance, g_0, on some microscopic scale, L_0. Clearly, all materials with $g_0 > g_c$ are conductors and all those with $g_0 > g_c$ are insulators. It is interesting to find out what happens when the transition is approached by changing the "control parameter" $\epsilon \equiv |\ln g_0 - \ln g_c| \cong |g_0 - g_c|/g_c \ll 1$ ($\epsilon = 0$ at the transition). For a range of scales, L, from L_0 on, the behavior can be approximated by $\beta(g) \sim s \ln(g/g_c)$, until g changes enough to get into the *macroscopic* range (where the limiting forms of eq. 2.29 are valid). The scale beyond which the macroscopic laws apply is denoted by ξ, in analogy with the usual correlation length in other phase transitions. By integrating the linear approximation to $\beta(g)$ from L_0 to ξ, we find

$$\frac{\ln(g/g_c)}{\ln(g_0/g_c)} = \left(\frac{\xi}{L_0}\right)^s \tag{2.33}$$

so that

$$\xi \sim L_0 \frac{\text{const}}{\epsilon^{1/s}}. \tag{2.34}$$

Thus, the critical exponent of the diverging ξ is $\nu = 1/s$ (ν is a constant of order 1 at $d = 3$), employing the usual notation. In the "macroscopic," $L \gg \xi$, regime, $g \propto e^{-L/\xi}$ in the insulating and $g \propto L^{d-2}$ in the conducting phase, where the first correction to the latter can be obtained from eq. 2.30. In both cases, in the whole range $l_0 \leq L \leq \xi$, g should not change by much more than an order of magnitude within the crossover range from the "critical" (or "microscopic") regime to the macroscopic one. Thus, the macroscopic conductivity for the metal is given by

$$\sigma_\infty(L \to \infty) \sim \frac{\text{const}}{\xi^{d-2}} \sim \epsilon^{(d-2)\nu}. \tag{2.35}$$

Note that the M–I transition in this theory is a continuous second-order transition ($\sigma \to 0$ continuously as the transition is approached and $\xi \to \infty$) and there is no "minimum metallic conductivity." The "minimum metallic conductivity" value, σ_m, still gives one an estimate of when nontrivial things start to happen. The continuity of the localization transition for *noninteracting* electrons is now believed rather generally. The value of the exponent ν is still under active study; its numerical value appears to have converged to a value of 1.5 ± 0.1 (Ulloa et al. 1992, Kramer and MacKinnon 1993, Hofstetter and Schreiber 1993). The physical meaning of ξ in the insulating phase is obvious—it is the localization length. The above considerations suggest that ξ in the conducting phase is the length below which the behavior is roughly the same as in the insulating phase in the same regime ($L \ll \xi$), where the wavefunctions and various correlation functions and Green's functions behave similarly in the two phases. It is only for $L > \xi$ that the difference between exponential decay and a nonzero average value is apparent.

For $L \ll \xi$, the two phases (insulating and metallic) are qualitatively similar. In both of them g does not change by more than an order of magnitude in the range from L_0 to ξ, as mentioned above. Thus (apart from the variation in g, which is relatively unimportant for large ξ):

$$\sigma(L) \sim \sigma_\infty \left(\frac{\xi}{L}\right)^{d-2} \qquad (L \lesssim \xi), \tag{2.36}$$

that is, the conductivity (and therefore the diffusion constant) is *scale dependent* for $L \ll \xi$ (Imry 1981b, Shapiro and Abrahams 1981, Shapiro 1982, Imry and Ovadyahu 1982b).

The "anomalous" diffusion in the microscopic regime $L < \xi$, as given by eq. 2.36, has interesting consequences for the relationship between time and length scales in this regime. Instead of $L^2 \sim Dt$ for usual diffusion, here D is renormalized as the scale is changing, and $dL^2/dt = D_L$ where D_L is the diffusion constant on scale L. This implies

$$L^d \sim \xi^{d-2} D_\infty t, \tag{2.37}$$

where D_∞ is the macroscopic diffusion constant in the metallic phase for the given ξ. Thus, in terms of an inelastic scattering time, $\tau_\phi \propto T^{-p}$, as before, the appropriate length which is given by $L_\phi \sim \sqrt{D\tau_\phi} \sim T^{-p/2}$ (eq. 2.25) in the macroscopic regime, is given here by

$$L_\phi \sim (\xi^{d-2} D_\infty \tau_\phi)^{1/d} \propto T^{-p/3}, \qquad (2.38)$$

in the microscopic critical regime ($d = 3$ was taken in the last relationship).

In both the macroscopic metal and the "microscopic" regime of both metal and insulator the large-sample conductivity as a function of temperature is given by $g(L_\phi) \cdot L_\phi^{2-d}$ as discussed earlier. In this critical regime g stays a constant within about an order of magnitude and

$$\sigma(T) \sim L_\phi^{2-d} \sim T^{(d-2)p/3} \sim T^{p/3} \qquad (\text{at } d = 3). \qquad (2.39)$$

In the macroscopic conducting regime $\beta(g)$ is given by eq. 2.30, whose integration yields $g(L_\phi)$ and hence

$$\sigma(T) = \text{const} + CL_\phi^{2-d} = \text{const} + O(T^{p(d-2)/2}), \qquad (2.40)$$

where the correction ("weak localization") *increases* with T like an appropriate power (which becomes a log in 2D). There exists now experimental evidence for both eq. 2.39 and eq. 2.40 in the appropriate domains, where the former is the relevant correction to the normal behavior in the "microscopic regime." In the range where these considerations are valid ($L_\phi \gg l$ is an important condition necessitating not too clean samples and low temperatures) we are getting a negative temperature coefficient of resistivity (TCR), which is a universal attribute (Imry 1980a) of dirty conductors! Moreover, to get a negative TCR at temperatures around room temperature, where l_ϕ is very small, one needs an l comparable to the interelectron distance and σ just somewhat larger than σ_{min}. We believe that this constitutes a valid qualitative explanation for the Mooij correlations discussed in section 1.

The smallest permissible value of the short-distance cut-off length L_0 is usually taken as $\sim l$. However, there exist many cases where inhomogeneity of some sort exists in the system and it may be viewed as homogeneous *only* on scales larger than some homogenization length l_{ho}. For granular metals, l_{ho} should be on the order of the grain size, d. Near the percolation threshold, l_{ho} will be on the order of the percolation correlation length. l_{ho} values of 10^3 Å are not uncommon. It is well known experimentally that in these systems the "nominal" elastic mean free path near the M–I transition is very small (10^{-2}–10^{-1} Å is possible) and hence the associated conductivity is much smaller than any appropriate σ_{min}. In granular metals (for a review, see Abeles et al. 1975), these conductivity values characterizing the M–I transition are found to go like $1/d$ (d being here the grain size). These facts are easily understood noting that l_{ho} (or d in granular metals) is the appropriate scale L_0. Thus, from the scaling theory the conductivity around which localization is important should be (Imry 1980)

$$\sigma_0 \sim \frac{e^2}{\hbar l_{ho}}, \tag{2.41}$$

in good agreement with both the order of magnitude and grain size dependence (Adkins 1976) mentioned above. These facts are hard to understand using naive σ_{min} Yoffe–Regel considerations.

In the presence of a magnetic field B, the relevant length scale, called l_H (see eq. 2.46 below), depends on B, and it is especially relevant once $l_H \ll L_\phi$. This may yield a relatively large negative magnetoresistance (MR), in analogy to the weak localization case (section 5). This may be the explanation for many cases of "anomalously large" negative MR in dirty systems. It is easy to see from the scaling theory that a magnetic field yields delocalization (Efetov 1983, Lerner and Imry 1995). Further complications, such as spin–orbit scattering, may change the sign of the MR, and there is now a large body of work on these aspects, including also the changes of ξ with B and the possible effects of electron–electron interactions (Altshuler and Aronov 1979, 1985, Altshuler et al. 1980a,b).

A concise presentation of the results of the scaling theory for $\sigma(L)$ in the 3D metallic range not too far from the transition is provided by Fig. 2.4. The lower curve depicts $\sigma(L)$ at the transition, it simply goes as $\sigma_c L_0/L$ where $\sigma_c = g_c e^2/\hbar L_0$. The upper curve is for $\sigma(L)_0$ somewhat above the transition. Here $\sigma(L)$ goes like $1/L$ in the microscopic range $\xi > L > L_0$, and like $\sigma_m + e^2 C/\hbar L$ in the macroscopic regime $L \gg \xi$. The macroscopic σ_m is $Ae^2/\hbar\xi$ where A is the order of magnitude of g where $\beta(g)$ becomes close to unity, that is, $A \sim 10$. The correction to σ at $L = \xi$ is on the order of $e^2/\hbar\xi$. Thus σ_m is still important for $L \leq \xi$; this is relevant for the interpretation of experiments in the microscopic regime that have supported eq. 2.39 (Ovadyahu and Imry 1983).

As mentioned before, many experiments on quasi 1D and 2D, and on 3D systems are in very good qualitative and semiquantitative agreement with this picture, *provided* that some new ideas on the mechanisms for inelastic scattering are accepted (chapter 3). One can now approach a quantitative under-

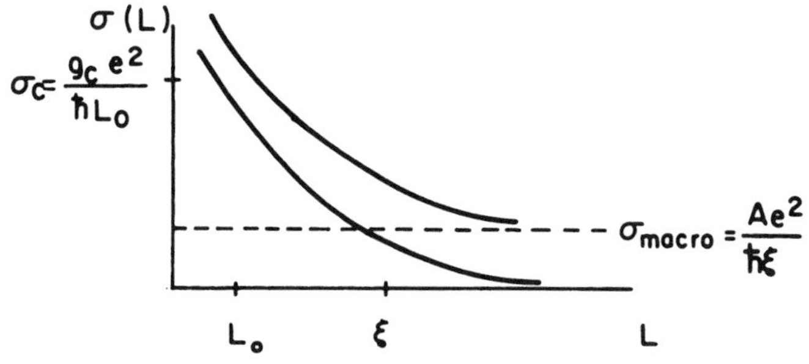

Figure 2.4 $\sigma(L)$ in 3D (schematic).

standing of the experiments, especially if the effects of electron–electron correlations are understood. The two general new features that the experiments have revealed are

1. τ_ϕ is typically shorter by a few orders of magnitude in dirty systems (Schmid 1074) than naively expected for otherwise similar pure ones.
2. The temperature dependence of τ_ϕ is weaker than in the pure case.

We are now in the process of gaining a quantitative understanding of these effects. This will be discussed in chapter 3.

We conclude this discussion of the scaling theory by briefly mentioning dielectric and optical properties. The ability to screen the long-range part of the Coulomb interaction is as important an attribute of the conducting state of matter as the finite conductivity itself. In addition to the static screening, the frequency dependences of σ and the dielectric constant ϵ determine the nontrivial optical properties (including the microwave and infrared ranges) of the conductor. Important anomalies in all these properties exist around the M–I transition and in the insulating phase near the transition. These were fully analyzed by Imry et al. (1982) and by Abrahams and Lee (1986). We just mention here that within the Thomas–Fermi screening picture, the *static* screening in the dirty metal is the same as in the pure one.

5. THE WEAKLY LOCALIZED REGIME

We shall very briefly discuss here the weak localization regime (for reviews see Fukuyama 1981b, Altshuler et al. 1982b, Bergmann 1984, Lee and Ramakrishnan 1985) where the quantum corrections to the classical conductivity are small, but quantitatively known. The constant C in the 2D case in eq. 2.32 is $1/\pi^2$. This means that the weak localization correction found by integrating eq. 2.29 is as in eq. 2.32:

$$\Delta G = -\frac{e^2}{\pi^2\hbar}\ln L, \qquad \Delta G(T) = +\frac{e^2 p}{2\pi^2\hbar}\ln T \tag{2.42}$$

In terms of the resistance per square, $R_\square$,

$$\Delta R_\square / R_\square = -\frac{e^2 p}{2\pi^2\hbar} R_\square \ln T \tag{2.43}$$

i.e., $R_\square$ decreases when T increases, the relative effect is increasing like $R_\square/(\hbar/e^2)$. In one dimension, a similar procedure leads to the "quantum correction" (see eq. 2.32)

$$\frac{\Delta\sigma}{\sigma} \sim \frac{\Delta R}{R} \propto -L \tag{2.44}$$

and in 3D

$$\frac{\Delta\sigma}{\sigma} \sim \text{const} + O\left(\frac{1}{L}\right), \tag{2.45}$$

where, as before, the appropriate scale is L_ϕ at finite temperatures. The effective dimension of the sample is determined by comparing the appropriate length with L_ϕ (see also Kaveh et al. 1981, Davies et al. 1983).

There is another correction to the conductivity due to electron–electron interactions (Altshuler and Aronov 1979), which has a temperature dependence that is not easy to separate from the above. It turns out that measurements of the magnetoresistance (MR) are extremely useful in this respect.

The easiest way to understand the weak localization MR is as follows. Assume a magnetic field B perpendicular, for example, to the 2D layer. Over a range l_H in space with

$$2\pi B l_H^2 = \Phi_0, \qquad \Phi_0 = hc/e = \text{the single-electron flux quantum} \tag{2.46}$$

a flux of the order of one flux quantum pierces the system. The usual gauge transformation (appendix C) on the $\hat{p} + eA/c$ term in the Hamiltonian yields that the electron acquires a phase ~ 1 by motion on a scale $l_H = \sqrt{\hbar c/eB}$ in this field. The length l_H is thus in this case a candidate for the physical length determining the scale on which the motion is as at $T = 0$, $H = 0$. (The physical understanding of the relevance of l_H is clearest using the semiclassical picture described later in this chapter.) Now we have two cases: (a) $l_H \ll l_\phi$ ("strong fields"), where the characteristic physical length is l_H, and, for example,

$$\Delta\sigma(B) \sim \frac{e^2}{e\pi^2\hbar} \ln B \tag{2.47}$$

in the weak localization regime in 2D; (b) $l_H \gg l_\phi$ ("weak fields"), here l_ϕ is roughly the relevant length, with small corrections proportional to B^2. Here the field causes a $O(B^2)$ magnetoconductance which can yield direct information on τ_ϕ. This "weak localization" MR is relatively large and often negative, but becomes positive for strong enough spin–orbit scattering.

There are now many cases in which the measured magnetoresistance agrees quantitatively with the detailed predictions in the weak localization regime (Bergmann 1984). Complications due to various spin and magnetic effects are rather well understood. In particular, a strong spin–orbit scattering multiplies the weak localization corrections by a universal *negative* constant. This is physically understood in terms of the semiclassical picture explained below. There exist several excellent reviews on this subject (Altshuler et al. 1982b, Bergman 1984, Lee and Ramakrishnan 1985, Aronov and Sharvin 1987).

A further measurement that can distinguish between the localization and interaction contributions is that of the Hall constant R_H. In the pure localization theory it should not have any ln T temperature term in 2D (this may be *very roughly* interpreted as confirming that the density of states (DOS) does not change for noninteracting electrons). On the other hand, the interaction terms yield a $\Delta R_H(T)/R_H$ which is *twice* $\Delta R(T)/R$. The overall behavior of the Hall effect is still a problem under active study, both experimentally and theoretically.

While the weak-localization results were first obtained using diagrammatic perturbation theory (a systematic expansion in $1/k_F l$), a very instructive interpretation based on semiclassical ideas was later developed (Larkin and Khmelnitskii 1982, Bergmann 1984, Chakravarty and Schmid 1986). The amplitude to go from one point to another is a sum over Feynman paths, which can be approximated by a sum on classical trajectories, j:

$$A_{1\to 2} = \sum_{j=1}^{\mathcal{N}} A_j e^{iS_j/\hbar} \tag{2.48}$$

where S_j is the action of the jth path from 1 to 2 and A_j an appropriate coefficient. The probability to go from 1 to 2 is a sum of the $\mathcal{N}$ classical terms ($\mathcal{N}$ is the "number of paths") $\sum_{j=1}^{\mathcal{N}} |A_j|^2$ and the interference terms $\sum_{i\neq j} A_i A_j e^{i(S_i - S_j)/\hbar}$. The latter sum has $O(\mathcal{N}^2)$ terms but it has strong cancellations and is usually argued to vanish upon "impurity ensemble averaging." By that we mean as usual an average over all microscopic realizations (e.g., different defect arrangements) of systems with the same macroscopic properties (e.g., average deffect concentrations). Such averaging is supposed to be automatically applicable in a large system. It is necessary in order to restore translational invariance to various correlation functions and Green's functions in theories of disordered systems. (We remark, however, that these interference terms are the source of the very important "mesoscopic fluctuations" to which we shall return later.) There exists a large class of trajectories whose contributions do *not* vanish upon ensemble averaging when time-reversal symmetry is obeyed ($B = 0$). These are the pairs of time-reversed paths that start at a point and return to it. Since the two members of such a pair have *the same phase*, they lead (without spin–orbit scattering) to an enhanced probability for return to the initial point, hence to a smaller probability for diffusion away. The corresponding negative correction to the conductivity can be shown to be given by, with τ_ϕ the long-time cut-off for these quantum effects,

$$\Delta\sigma = -\frac{2e^2}{\pi\hbar} D \int_{\tau_0}^{\tau_\phi} dt\, w(t) \tag{2.49}$$

where $w(t) = (4\pi Dt)^{-d/2}$ is the classical return probability and τ_0 a short-time cutoff. It is easy to see that this yields the weak localization corrections eqs.

2.32, 2.45. The magnetic field, once $l_H \ll L_\phi$, destroys the phase coherence of these conjugate paths and thus strongly reduces or eliminates the weak localization corrections, eventually restoring the classical conductivity. However, for (usually much) larger values of B, the large-field limit is obtained. There the physics is very different, and requires a separate study, which will be done in chapter 6.

Problems

1. Integrate the linear approximation to $\beta(g)$ around g_c to get eqs. 2.32, 2.45.
2. Prove that eq. 2.38 is indeed valid for anomalous diffusion, $D(L) \sim 1/L$.

3

Dephasing by Coupling with the Environment, Application to Coulomb Electron–Electron Interactions in Metals

1. INTRODUCTION AND REVIEW OF THE PRINCIPLES OF DEPHASING

Many of the interesting effects in mesoscopic systems are due to quantum interference. Among these are, for example, the weak localization corrections to the conductivity (section 5, chapter 2), the universal conductance fluctuations (chapter 5), persistent currents (chapter 4), and many others. These effects are known to be affected by the coupling of the interfering particle to its environment, for example, to a heat bath. The way such a coupling modifies quantum phenomena has been studied for a long time, both theoretically (Feynman and Vernon 1963, Caldeira and Leggett 1983), and experimentally. The effect of the coupling to the environment may be characterized by the "phase breaking" time, τ_ϕ, which is the characteristic time for the interfering particle to stay phase coherent as explained below.

Stern et al. (1980a,b) have studied the way the coupling of an interfering particle affects a two-wave interference experiment. This discussion will be based on their work. Two methods have been used to describe how the interaction of a quantum system with its environment might suppress quantum interference. The first regards the environment as measuring the path of the interfering particle. When the environment has the information on that path, no interference is seen. The second description answers the question naturally raised by the first: How does the interfering particle "know," when the interference is examined, that the environment has identified its path? This question is answered by the observation that the interaction of a partial wave with its environment can induce an uncertainty in this wave's phase (what counts physically is the uncertainty of the *relative* phases of the paths). This may be described as turning the interference pattern into a sum of many patterns, shifted relative to one another. The two descriptions were proved to be equivalent, and this has been applied to the dephasing by electromagnetic fluctuations in metals, and by photon modes in thermal and coherent states. Here we will review the two descriptions, and examine in sections 2–4 the dephasing by the electron–electron interaction in metals. We shall find it convenient to consider that problem from the first point of view mentioned above, rather than the second, that is, to find out where the information on the interfering electron path is hidden in the bath of electrons it interacts with. An early simple model for dephasing (Büttiker 1985b) considered the interfering electron going into a particle reservoir and an electron from the reservoir replacing it. For a reservoir with a continuous spectrum, this may yield a change in the state of the reservoir, which will cause dephasing.

As a guiding example, we consider an Aharonov–Bohm (A–B) interference experiment on a ring. The A–B effect has been proved to be a convenient way to observe interference patterns in mesoscopic samples, because it provides an experimentally straightforward way of shifting the interference pattern. This experiment starts with the construction of two electron wave packets, $l(x)$ and $r(x)$ (l, r stand for left, right), crossing the ring (see Fig.3.1) along its two opposite sides. We assume that the two wave packets follow well-defined classical paths, $x_l(t)$, $x_r(t)$ along the arms of the ring. The interference is examined after each of the two wave packets has traversed half of the ring's circumference. Therefore, the initial wavefunction of the electron (whose coordinate is x) and the environment (whose wavefunction and set of coordinates are respectively denoted by χ and η) is

$$\psi(t=0) = [l(x) + r(x)] \otimes \chi_0(\eta). \tag{3.1}$$

At time τ_0, when the interference is examined, the wavefunction is, in general,

$$\psi(\tau_0) = l(x, \tau_0) \otimes \chi_l(\eta, \tau_0) + r(x, \tau_0) \otimes \chi_r(\eta, \tau_0) \tag{3.2}$$

and the inference term is

$$2\,\mathrm{Re}\left[l^*(x, \tau_0) r(x, \tau_0) \int d\eta\, \chi_l^*(\eta, \tau_0) \chi_r(\eta, \tau_0)\right] \tag{3.3}$$

Had there been no environment present in the experiment, the interference term would have been just $2\,\mathrm{Re}[l^*(x, \tau_0) r(x, \tau_0)]$. So, the effect of the interaction is to multiply the interference term by $\int d\eta\, \chi_l^*(\eta) \chi_r(\eta)$ at τ_0. This is so since the environment is not observed in the interference experiment; its coordinate is therefore integrated upon; that is, the scalar product of the two

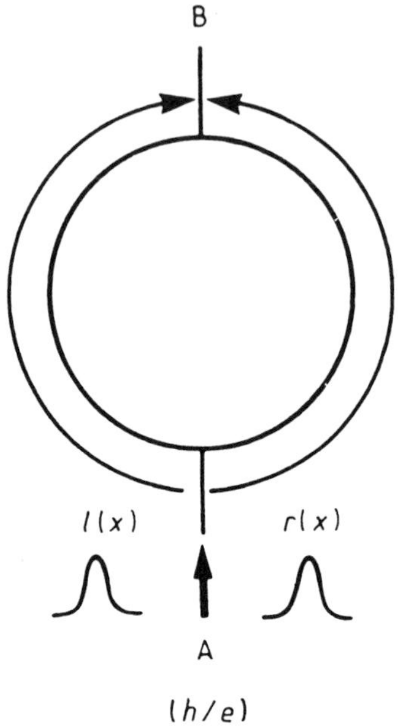

Figure 3.1 Schematics of interference experiments in A–B rings. Each partial wave traverses half the ring, and the interference is examined at the point *B*. This kind of interference gives rise to h/e oscillations of the conductance.

environmental states at τ_0 is taken. The first way to understand the dephasing is seen directly from this expression, which is the scalar product of the two environment states at τ_0, coupled to the two partial waves. At $t = 0$ these two states are identical. During the time of the experiment, each partial wave has its own interaction with the environment, and therefore the two states evolving in time become different. When the two states of the environment become orthogonal, the final state of the environment identifies the path the electron took. Quantum interference, which is the result of an uncertainty in this path, is then lost. Thus, the phase breaking time, τ_ϕ, is the time in which the two interfering partial waves shift the environment into states orthogonal to each other, that is, when the environment has the information on the path the electron takes.[1]

The second explanation for the loss of quantum interference regards it from the point of view of how the environment affects the partial waves, rather than how the waves affect the environment. It is well known that when a static potential $V(x)$ is exerted on one of the partial waves, this wave accumulates a phase

[1] We note that the question of whether somebody does or does not come in to observe the change of state of the environment, simply does *not* arise. The rather nebulous disussions of the importance and effect of that observation are best avoided.

$$\phi = -\int V(x(t))\, dt/\hbar \tag{3.4}$$

and the interference term is multiplied by $e^{i\phi}$. "A static potential" here is a potential which is a function of the particle's coordinate and momentum only, and does not involve any other degrees of freedom. For a given particle's path, the value of a static potential is well defined. When V is not static, but created by environmental degree(s) of freedom, V becomes an operator. Thus its value is no longer well defined. The uncertainty in this value results from the quantum uncertainty in the state of the environment. Therefore, ϕ is also not definite. In fact, ϕ becomes a statistical variable, described by a distribution function $P(\phi)$. (For the details of this description see Stern et al. 1990a,b.) The effect of the environment on the interference is then to multiply the interference term by the average value of $e^{i\phi}$, that is,

$$\langle e^{i\phi}\rangle = \int P(\phi)e^{i\phi}\, d\phi \tag{3.5}$$

The averaging is done on the interference "screen." Since $e^{i\phi}$ is periodic in ϕ, $\langle e^{i\phi}\rangle$ tends to zero when $P(\phi)$ is slowly varying over a region much larger than one period, of 2π. When this happens, one may say that the interference screen shows a superposition of many interference patterns, mutually canceling each other. Hence, the phase breaking time is also the time in which the uncertainty in the phase becomes of the order of the interference periodicity. In the Feynman–Vernon terminology, $\langle e^{i\phi}\rangle$ is the influence functional of the two paths taken by the two partial waves. This is, then, the second explanation for the loss of quantum interference.

The statement of equivalence between the two explanations is given by the equation

$$\langle e^{i\phi}\rangle = \int d\eta\, \chi_l^*(\eta)\chi_r(\chi) \tag{3.6}$$

When the environment measures the path taken by the particle (by χ_l becoming orthogonal to χ_r), it induces a phase shift whose uncertainty is of the order of 2π. The equivalence embodied in (3.6) is proved as follows.

We start considering dephasing of the right-hand path χ_r only. The generalization to two paths will be seen later. The Hamiltonian of the environment will be denoted by $H_{env}(\eta, p_\eta)$, while the interaction term is $V(\chi_r(t), \eta)$ (the left partial wave does not interact with the environment). Starting with the initial wavefunction (eq. 3.1) the wavefunction at time τ_0 is

$$\psi(\tau_0) = l(\tau_0)e^{-H_{env}\tau_0/\hbar}\chi_0(\eta) + r(\tau_0)\hat{T}\exp\left[-\frac{i}{\hbar}\int_0^{\tau_0} dt\,(H_{env}+V)\right]\chi_0(\eta), \tag{3.7}$$

where $\hat{T}$ is the time-ordering operator. It is useful at this point to write $\psi(\tau_0)$ in terms of $V_I(t) \equiv e^{iH_{env}t}V(\chi_r(t),\,\eta)e^{-iH_{env}t}$, that is, the potential V in the interaction picture. Using V_I, $\psi(t_0)$ can be written as

$$\psi(\tau_0) \equiv l(\tau_0)\otimes e^{-iH_{env}\tau_0/\hbar}\chi_0(\eta) + r(\tau_0)\otimes e^{-iH_{env}\tau_0/\hbar}\hat{T} \times \exp\left[-i\int_0^{\tau_0}\frac{dt}{\hbar}V_I(x_r(t),\,t)\right]\chi_0(\eta). \tag{3.8}$$

Hence the interference term is multiplied by

$$\langle\chi_0|e^{iH_{env}\tau_0}\hat{T}\exp\left[-i\int_0^{\tau_0}\frac{dt}{\hbar}(H_{env}+V)\right]|\chi_0(\eta)\rangle = \langle\chi_0|\hat{T}\exp\left[-\frac{i}{\hbar}\int_0^{\tau_0}dt\,V_I(x_r(t),\,t)\right]|\chi_0(\eta)\rangle. \tag{3.9}$$

The interpretation of this expression in terms of a scalar product of two environment states at time τ_0 is obvious. The interpretation in terms of phase uncertainty emerges from the observation that eq. 3.9 is the expectation value of a unitary operator which can be defined as the operator corresponding to $e^{i\phi}$. As all unitary operators, this operator can be expressed, if desired, as the exponential of a Hermitian operator ϕ, that is,

$$\langle\chi_0|\hat{T}\exp\left[-\frac{i}{\hbar}\int_0^{\tau_0}dt\,V_I(x_r(t),\,t)\right]|\chi_0\rangle = \langle\chi_0|e^{i\phi}|\chi_0\rangle. \tag{3.10}$$

Hence the effect of the interaction with the environment is to multiply the interference term by $\langle e^{i\phi}\rangle$, where the averaging is done with respect to the phase probability distribution, as determined by the environmental state χ_0.

The phase operator ϕ was introduced here by means of the mathematical properties of unitary transformations, so that it still deserves a physical interpretation. To obtain such an explanation, we first discuss the case where the potentials exerted by the environment at different points along the particle's path commute, that is,

$$[V_I(x_r(t),t),\;V_I(x_r(t'),t'] = 0. \tag{3.11}$$

Then,

$$\langle\chi_0|\hat{T}\exp\left[-\frac{i}{\hbar}\int_0^{T_0}dt\ V_I(x_r(t),t)\right]|\chi_0\rangle$$
$$=\langle\chi_0|\exp\left[-\frac{i}{\hbar}\int_0^{T_0}dt\ V_I(x_r(t),t)\right]|\chi_0\rangle, \quad (3.12)$$

and $\phi=-\frac{1}{\hbar}\int_0^{T_0}dt\ V_I(x_r(t),\ t)$. In this case $\dot{\phi}$, the rate of accumulation of the phase, is just the local potential acting on the interfering particle, independent of earlier interactions of the particle with the environment. One should distinguish here between two limits: for $\langle\delta\phi^2\rangle\ll 1$, eq. 3.12 yields

$$\langle e^{i\phi}\rangle\approx e^{i\langle\phi\rangle}(1-\tfrac{1}{2}\langle\delta\phi^2\rangle), \quad (3.13)$$

and the environment's potential can be approximated by a single-particle (possibly time-dependent) potential

$$\langle V_I(x_r(t),t)\rangle=\langle\chi_0|V_I(x_r(t),t)\chi_0\rangle. \quad (3.14)$$

For $\langle\delta\phi^2\rangle\gg 1$, on the other hand, the interference term tends to zero. The crossover between the two regimes is then at

$$\langle\delta\phi^2\rangle=\int_0^{T_0}\frac{dt}{\hbar}\int_0^{T_0}\frac{dt'}{\hbar}[\langle V_I(x_r(t),t)V_I(x_r(t'),t')\rangle$$
$$-\langle V_I(x_r(t),t)\rangle\langle V_I(x_r(t'),t')\rangle]\sim 1, \quad (3.15)$$

where $\chi_0(\eta,t)\equiv e^{-iH_{env}t}\eta_0(\eta)$ is the environment state as it evolves in time under H_{env}.

When is the condition in eq. 3.11 valid, and what happens when it is not? A typical case where the potentials at different points along the path are commutative is the case of an interfering electron interacting with a free electromagnetic field. In that case the interaction is

$$V_I(x_r(t),t)=-\frac{e}{c}\dot{x}_r(t)\cdot A(x_r(t),t), \quad (3.16)$$

where $A(X,t)$, the electromagnetic free field, is in obvious notation

$$A(x,t)=\sum_{k,\lambda}\epsilon_{k,\lambda}\left[\frac{2\pi c^2}{\omega_k}\right]^{1/2}(a_k\,e^{ik\cdot x-i\omega t}+a_k^\dagger\,e^{-ik\cdot x+i\omega t}), \quad (3.17)$$

and $[V_I(x,t),\ V_I(x't')]=0$ unless $|x-x'|=c|t-t'|$. Since $\dot{x}_r(t)<c$, the condition of eq. 3.11 is valid. Generally, this condition is valid when there is no amplitude for an environment excitation created at $(x_r(t),t)$ to be anihilated at $(x_r(t'),t')$, that is, when a change induced in the environment's state at

$(x_r(t), t)$ does not influence the potential the environment exerts on the interfering particle at $(x_r(t'), t')$. In the above example, a photon emitted by the electron at $(x_r(t), t)$ will *not* be at $(x_r(t'), t')$ when the electron gets there.

If instead of discussing the interaction with a photon field, we turn our attention to the interaction with phonons, the speed of light in eqs. 3.16 and 3.17 is replaced by the sound velocity in the analogous theory. Then, a phonon emitted by the electron at $(x_r(t), t)$ might be encountered again by the electron at $(x_r(t'), t')$. Hence lattice excitations created by the electron along its path may affect the potential it feels at a later stage of the path. The potential the electron feels at a given point of its path is now not a local function of that point, but depends on the path since it includes a "back reaction" of the environment to the potential exerted by the electron. Therefore this potential will be different from $V_I(x_r(t), t)$ and, consequently, the rate of phase accumulation will also differ from $V_I(x_r(t), t)$. However, in large many-body environments the potential exerted by the environment on the interfering particle is usually practically independent of the particle's history since the environment's memory time is very short. Therefore eq. 3.11 can be assumed to hold.

We thus see that the loss of interference due to an interaction with a dynamical environment can be understood in the two ways discussed. The interference is destroyed either when the state of the environment coupled to the right wave is orthogonal to that coupled to left wave, or, alternatively, when the width of the phase distribution function exceeds a magnitude of order unity. The interaction with the dynamical environment turns the phase into a statistical variable, and this, together with the fact that the phase is defined only over a range of 2π, determines the conditions for the phase to become completely uncertain. If the potential exerted by the environment on the interfering particle at a given point along its path is assumed to be independent of the path, the phase uncertainty is given by

$$\langle \delta\phi^2 \rangle = \int_0^{\tau_0} \frac{dt}{\hbar} \int_0^{\tau_0} \frac{dt'}{\hbar} [\langle V_I(x_r(t), t) V_I(x_r(t'), t') \rangle - \langle V_I(x_r(t), t) \rangle \times \langle V_I(x_r(t'), t') \rangle]. \tag{3.18}$$

The exact behavior of the interference term for $\langle \delta\phi^2 \rangle \gg 1$, that is, the value of $\langle e^{i\phi} \rangle$ for broad distribution functions, depends on the phase distribution, $P(\phi)$. However, the description of the phase as a statistical variable enables us, under appropriate conditions, to apply the central limit theorem, and conclude that $P(\phi)$ is a normal distribution. The central limit theorem is applicable, for example, when the phase is accumulated in a series of uncorrelated events (e.g., by a series of scattering events off different, noninteracting, scatterers), or, more generally, whenever the potential–potential correlation function decays to zero with a characteristic decay time much shorter than the duration of the experiment. In particular, the central limit theorem is usually applicable for coupling to a heat-bath. For a normal distribution,

$$\langle e^{i\phi} \rangle = e^{i\langle \phi \rangle - (1/2)\langle \delta\phi^2 \rangle}. \tag{3.19}$$

This expression is exact for the model of an environment composed of harmonic oscillators with a linear coupling to the interfering waves. The evaluation of $\langle e^{i\phi} \rangle$ by eq. 3.19 reproduces the result obtained by Feynman and Vernon for a rather similar model. Feynman and Vernon's result was obtained by integration of the environment's paths. This model was proved to be very useful in the investigation of the effect of the environment on quantum phenomena (e.g., Caldeira and Leggett 1983). Equation 3.19 is therefore a convenient way to calculate the influence functional for many-body environments, where the central limit theorem is usually applicable.

As seen from eq. 3.15, the phase uncertainty remains constant when the interfering wave does not interact with the environment. Thus, if a trace is left by a partial wave on its environment, this trace cannot be wiped out after the interaction is over. Neither internal interactions of the environment, nor a deliberate application of a classical force on it, can reduce back the phase uncertainty after the interaction with the environment is over. This statement can be proved also from the point of view of the change the interfering wave induces in its environment. This proof follows simply from unitarity. The scalar product of two states that evolve in time under the same Hamiltonian does not change in time. Therefore, if the state of the system (electron plus environment) after the electron–environment interaction has taken place is

$$|r(t)\rangle \otimes |\chi_{env}^{(1)}\rangle + |l(t)\rangle \otimes |\chi_{env}^{(2)}\rangle, \tag{3.20}$$

then the scalar product $\langle \chi_{env}^{(1)}(t) | \chi_{env}^{(2)}(t) \rangle$ does not change with time. The only way to change it is by another interaction of the electron with the same environment (see the discussion at the end of this chapter). Such an interaction keeps the product $\langle \chi_{env}^{(1)}(t) | \chi_{env}^{(2)}(t) \rangle \otimes \langle r(t) | l(t) \rangle$ constant, but changes $\langle \chi_{env}^{(1)}(t) | \chi_{env}^{(2)}(t) \rangle$. The interference will be retrieved only if the orthogonality is transferred from the environment wavefunction to the electronic wavefunctions which are not traced over in the experiment.

The above discussion was concerned with the phase $\phi = \phi_r$, accumulated by the right-hand path only. The left-hand path similarly accumulates a phase ϕ_l from the interaction with the environment. The interference pattern is governed by the *relative* phase $\phi_r - \phi_l$, and it is the uncertainty in *that* phase which determines the loss of quantum interference. This uncertainty is always smaller than, or equal to, the sum of uncertainties in the two partial waves' phases. The case of noncommuting phases will not be discussed here.

Often the same environment interacts with the two interfering waves. A typical example is the interaction of an interfering electron with the electromagnetic fluctuations in vacuum. In this case, if the two waves follow parallel paths with equal velocities, their dipole radiation, despite the energy it transfers to the field, does not dephase the interference. This radiation makes each of the partial waves' phases uncertain, but does not alter the relative phase. We shall encounter more examples later demonstrating that the environment excitations created must be able to distinguish l from r in order to dephase their interference. Another well-known example is that of "coherent inelastic neutron scattering" in crystals (see, e.g., Kittel 1963). This process follows from the coherent addition of the amplitudes for the processes in which the neutron exchanges *the same* phonon with *all* scatterers in the crystal.

The previous example demonstrates that an exchange of energy is not a sufficient condition for dephasing. It is also not a necessary condition for dephasing. What is important is that the two partial waves flip the environment to *orthogonal* states. It does not matter in principle that these states are degenerate. Simple examples were given by Stern et al. (1990a,b). Thus, it must be emphasized that, for example, long-wave excitations (phonons, photons) cannot dephase the interference. But that is *not* beause of their low energy but rather because they do not influence the *relative* phase of the paths.

We emphasize the dephasing may occur by coupling to a discrete or a continuous environment. In the former case the interfering particle is more likely to "reabsorb" the excitation and "reset" the phase. In the latter case, the excitation *may* move away to infinity and the loss of phase can usually be regarded as, practically speaking, irreversible. The latter case is that of an effective "bath" and there are no subtleties with the definition of ϕ since eq. 3.12 may be assumed. We point out that in special cases it is possible, even in the continuum case, to have a finite probability of reabsorption of the created excitation and thus retain coherence. This happens, for example, in a quantum interference model due to Holstein (1961) for the Hall effect in insulators (see also Entin-Wohlman et al. 1995a,b).

2. DEPHASING BY THE ELECTRON–ELECTRON INTERACTION

An interesting application of the above general principle is the dephasing of mesoscopic interference effects by electron–electron interaction in conducting samples. Stern et al. (1990a,b) have applied the phase uncertainty approach to dephasing by electron–electron interaction in metals in the diffusive regime. They have shown that this approach reproduces the results obtained in the pioneering work of Altshuler et al. (1981b, 1982a). Following Stern et al., we now consider the dephasing due to electron–electron interactions from the point of view of the changes induced in the state of the environment, using the response functions of the latter. In the original work of Altshuler, Aronov and Khmelnitskii, the phase uncertainty induced on the particle by the electromagnetic fluctuations of the environment has been considered. We will see that the fluctuation–dissipation theorem guarantees the equivalence of these two pictures.

The general picture is that of a test particle interacting with an environment. For definiteness, we consider an interfering "electron," whose paths are denoted by $x_{r,l}(t)$, interacting with a bath of environment electrons, whose coordinates are y_i. The identity of the interfering electron with those of the bath will be handled approximately later. The Coulomb interaction of the interfering electron with the rest of the electrons is, in the interaction picture,

$$\hat{V}_I(\boldsymbol{x}, t) = \int \frac{\hat{\rho}_I(\boldsymbol{r}', t)\, d^3r'}{|\boldsymbol{x} - \boldsymbol{r}'|}, \tag{3.21}$$

where $\hat{\rho}_I(r, t) = e\sum_i \delta(r - \hat{y}_I^i(t)) - \bar{\rho}$. For brevity of the following expressions, we first consider only the interaction of the electron bath with the right partial wave of the interfering electron, we omit the corresponding subscript, and we begin by assuming (this will be relaxed later) that the electron bath is initially in its ground state, $|0\rangle$. Assuming that the left partial wave does not interact with the electron bath, the intensity of the interference pattern is reduced by the probability that the bath's state coupled to the right wave becomes different from $|0\rangle$. Up to second order in the interaction, this probability is

$$P = \frac{1}{\hbar^2} \sum_{|n\rangle \neq |0\rangle} \int_0^{\tau_0} dt \int_0^{\tau_0} dt' \, \langle 0|V_I(\boldsymbol{x}(t), t)|n\rangle\langle n|V_I(\boldsymbol{x}(t'), t')|0\rangle. \tag{3.22}$$

For the ground state $\langle 0|\hat{\rho}|0\rangle = 0$, so that the summation in eq. 3.22 can be extended to include all states. We neglect the changes in the paths $x_{r,l}(t)$ due to the interaction; thus only the phase due to the latter is taken into account. The interpretation of eq. 3.22 as the variance of the phase given to the particle by the interaction with the environment is clear. We now express P in terms of the response of the environment. using the convolution theorem, $\int d^3r' f(\boldsymbol{r} - \boldsymbol{r}')g(\boldsymbol{r}') = (2\pi)^{-3} \int d^3q\, f_{\boldsymbol{q}} g_{\boldsymbol{q}} e^{-i\boldsymbol{q}\cdot\boldsymbol{r}}$ where $f_{\boldsymbol{q}}$ and $g_{\boldsymbol{q}}$ are the Fourier transforms of f and g, we write

$$P = \frac{1}{\hbar^2 (2\pi)^6} \int_0^{\tau_0} dt \int_0^{\tau_0} dt' \int d^3q \int d^3q' \frac{4\pi e}{q^2} \frac{4\pi e}{q'^2} \langle \rho_{\boldsymbol{q}}(t) \rho_{\boldsymbol{q}'}(t') \rangle e^{i_{\boldsymbol{q}} \cdot \boldsymbol{x}(t) - i\boldsymbol{q}' \cdot \boldsymbol{x}(t')}. \tag{3.23}$$

We assume translational invariance:

$$\langle \rho_q \rho_{q'} \rangle = \frac{(2\pi)^3}{\mathrm{Vol}} \delta(\boldsymbol{q} + \boldsymbol{q}') \langle \rho_{\boldsymbol{q}} \rho_{-\boldsymbol{q}} \rangle \tag{3.24}$$

(For a finite system the q's are discrete and one just has $\delta_{qq'}$. Going to the continuum, the Kronecker delta is replaced by $(2\pi)^3/\mathrm{Vol}$ times the Dirac delta.) By performing one q integration and inserting a complete set of intermediate states, we obtain:

$$P = \frac{1}{\mathrm{Vol}(2\pi)^3\hbar^2}\sum_{|n\rangle}\int_0^{\tau_0} dt \int_0^{\tau_0} dt' \int d^3q \frac{(4\pi e)^2}{q^4} \langle 0|\rho_I^{\boldsymbol{q}}(t)|n\rangle\langle n|\rho_I^{-\boldsymbol{q}}(t')|0\rangle e^{i\boldsymbol{q}\cdot(x(t)-x(t'))}. \tag{3.25}$$

By transforming into Schrödinger picture operators (see, e.g., appendix A) and inserting a dummy integration variable ω, P can be rewritten in the form

$$P = \frac{1}{\mathrm{Vol}(2\pi)^3\hbar^2}\sum_{|n\rangle}\int_0^{\tau_0} dt \int_0^{\tau_0} dt' \int d^3q \int d\omega \frac{(4\pi e)^2}{q^4} |\langle 0|\rho_S^{\boldsymbol{q}}|n\rangle|^2 \times \delta(\omega - \omega_{n0}) e^{i\boldsymbol{q}\cdot(x(t)-x(t'))-i\omega(t-t')}. \tag{3.26}$$

At first glance, eq. 3.26 looks useless, due to the practical impossibility of calculating the bath's eigenstates $|n\rangle$. However the usefulness of that expression stems from its relation to the linear response expression for dynamic structure factor (see appendix A, eq. A.11) and the imaginary part of the complex dielectric function where both are related by the fluctuation–dissipation theorem (see eq. A.13)

$$\mathrm{Im}\left(\frac{1}{\epsilon(\boldsymbol{q},\omega)}\right) = \frac{4\pi^2 e^2}{\mathrm{Vol}\ q^2\hbar}\sum_{|n\rangle}|\langle 0|\rho_S^{\boldsymbol{q}}|n\rangle|^2\ \delta(\omega-\omega_{n0}) = \frac{4\pi^2 e^2}{\mathrm{Vol}\ q^2\hbar} S(\boldsymbol{q},\omega). \tag{3.27}$$

Thus, eq. 3.27, becomes

$$P = \frac{1}{\hbar(2\pi)^3}\int_0^{\tau_0} dt \int_0^{\tau_0} dt' \int d^3q \int d\omega \frac{4e^2}{q^2} \mathrm{Im}\left(\frac{1}{\epsilon(\boldsymbol{q},\omega)}\right) e^{i\boldsymbol{q}\cdot(x(t)-x(t'))-i\omega(t-t')}. \tag{3.28}$$

Equation 3.28 is the powerful central result of this section. Before proceeding to a discussion of this result, we comment that the calculation can be generalized to treat an electron bath initially in a thermal state. The integrand in eq. 3.28 is then multiplied by $\coth(\omega/2k_BT)$. The probability that the state of the environment be changed during the time τ_0, is expressed through an integral over the *dissipative* part of the response, that is, the excitability of the system. The equivalent expression via the dynamic structure factor through the fluctuation–dissipation (F–D) theorem simply expresses this by integrals over the inelastic scattering probability. It is well known (see, e.g., Nozières 1963) that the energy loss to the system by inelastic scattering is given in a similar fashion by an integral over S, or $\mathrm{Im}(1/\epsilon)$. The new feature of our result is the appearance of the classical path $x(t)$, along which the excitation of the environment occurs. The phase in the exponentials in eq. 3.28 is the relative phase between two traversals of the path with the $(\boldsymbol{q},\omega)$ scattering occurring at t and t'. This is averaged using the (weak) scattering probability.

Equation 3.28 was obtained for the excitation of the environment caused by the electron along the path r only, which will lead to its phase uncertainty $\delta\phi_r^2$. The interaction of the path l will similarly lead to a $\delta\phi_l^2$ and one can similarly obtain the cross terms $\langle\delta\phi_r\,\delta\phi_l\rangle = \langle\delta\phi_l\,\delta\phi_r\rangle$. The total reduction of the interference will be governed by the fluctuation of the *relative* phase.

$$\langle(\delta(\phi_r - \phi_l)^2)\rangle = \langle\delta\phi_r^2\rangle = \langle\delta\phi_l^2\rangle - 2\langle\delta\phi_l\,\delta\phi_r\rangle, \tag{3.29}$$

where in $\langle\delta\phi_i\,\delta\phi_j\rangle$ $(i,\, j = l, r)$ the term $e^{i\boldsymbol{q}\cdot(x(t)-x(t'))}$ in eq. 3.28 is replaced by $e^{i\boldsymbol{q}\cdot(x_i(t)-x_j(t'))}$. The cancelation occurring among the terms in eq. 3.29 will be shown to be of decisive importance at and below two dimensions. We will now draw a few conclusions out of the above calculation.

1. For good conductors, $\mathrm{Im}(1/\epsilon(\boldsymbol{q},\omega)) = \omega/4\pi\sigma$, and the probability that the state of the electron's bath was changed in the path $x(t)$ is

$$P = \frac{1}{\hbar(2\pi)^3}\int_0^{\tau_0} dt \int_0^{\tau_0} dt' \int d^3q \int d\omega \frac{e^2\omega}{\pi q^2\sigma} e^{i\boldsymbol{q}\cdot(x(t)-x(t'))-i\omega(t-t')} \coth\frac{\omega}{2k_BT}. \tag{3.30}$$

   For $x(t) = x_r(t)$, this probability (as long as $P \ll 1$) is just one half of the uncertainty in the phase $\langle\delta\phi_r^2\rangle$ accumulated by the right partial wave. This result is equivalent to the AAK one, and we shall obtain below (section 3) the phase breaking time, τ_ϕ, out of it. The present derivation demonstrates that the origin of this dephasing is in the electrostatic electron–electron interaction, and establishes the connection with the linear response of the bath.
2. For poor conductors $\mathrm{Im}\ \epsilon(q,\omega) \ll \mathrm{Re}\ \epsilon(q,\omega)$. Then, the q, ω integrals in eq. 3.26 will have significant contributions only from those values of q, ω in which $\mathrm{Re}\ \epsilon = 0$. A typical example is $\omega = \omega_p$, the plasma frequency.
3. In both cases mentioned above, the rate of dephasing depends crucially on the imaginary part of the dielectric response function. This, in turn, determines the rate at which the electron bath is excited by the interfering electron. It should be emphasized here that the polarization of the electron bath by the interfering electron, reflected in the real part of the dielectric response, does not dephase the interference. This polarization disappears when the electron leaves the polarized region and therefore it does not identify the path taken by the electron. For a general discussion of the relation between dissipation, excitations and dephasing, the reader is referred to Stern et al. (1990a,b).
4. The above discussion of dephasing due to the Coulomb interaction can easily be generalized to any two-particle interaction $V(r - r')$. This is done by replacing $e^2/|r - r'|$ in eq. 3.21 with $V(r - r')$, and following the derivations in eqs. 3.22–3.30. In particular, it is interesting to consider the case of a short-range potential, which can be

approximated by $V(r-r') \propto \delta(r-r')$. For such a potential, the probability that the bath's state is changed is proportional to the density–density correlation function of the bath's electrons,

$$\propto \int_0^{\tau_0} dt \int_0^{\tau_0} dt' \langle \rho(x(t),t)\rho(x(t'),t')\rangle. \tag{3.31}$$

Thus, the intensity of the interference effects provides information on the density–density correlation function of the bath, which is again related to the dynamical structure factor and the dissipative part of the response.

5. As emphasized above, the interaction of the environment with the interfering partial waves changes the state of the environment, so that it acquires information on the path taken by the interfering particle. One might then consider the case of a very slow electron traversing a piece of metal and examine the change it induces in the state of the metal. At a first glance, it looks as if the state of the metal adiabatically follows the motion of the electron, so that when the electron leaves the metal, the metal is back in its initial state. However, since the excitation spectrum of the electron bath is continuous, the adiabatic argument is never applicable. It is true that the electron induces a polarization in the bath, polarization that follows its motion adiabatically and disappears when the electron leaves the metal. But, due to the bath's continuous spectrum, this polarization has to involve an *excitation* of the bath, and this excitation does not disappear when the interaction of the bath with the interfering electron is over.

   The situation is different, of course, for insulators. There, due to the gap in the excitations spectrum, a very slow electron can polarize the bath without exciting it, that is, without identifying its path.
6. We have chosen to express the dephasing in terms of dynamic correlations of densities. Using the continuity equation $\rho_{q\omega} = -(q/\omega)\cdot \boldsymbol{j}_{q\omega}$, one may express the latter in terms of dynamic current correlations. The longitudinal (parallel to q) components appear in the latter (due to having $\nabla \cdot \boldsymbol{j}$ in the charge conservation condition, which can in turn be expressed in terms of correlations of the longitudinal components of the vector potential $\boldsymbol{A}$) as in the original work of Altshuler, Aronov and Khmelnitskii. We believe that the presentation here makes the connection with the Coulomb e–e interaction very clear.
7. For $\omega \ll k_e T$, the last factor in eq. (3.30) may change the behavior when the small ω processes are dominant.

3. REVIEW OF RESULTS IN VARIOUS DIMENSIONS

The final expression (3.30) of the previous section defines the phase uncertainty accumulated by the right partial wave. The physically meaningful object is the uncertainty in the *phase difference* between any two paths, for example, the right and left ones. It may be felt that this difference should be of the same order as each of the uncertainties $\langle\delta\phi_l^2\rangle$ and $\langle\delta\phi_r^2\rangle$. This is in fact true for $d > 2$; however, for $d \leq 2$, each of the above single-path fluctuations diverges in the thermodynamic limit. This not-very-physical divergence is cancelled by its counterpart in the mixed $\langle\delta\phi_l\,\delta\phi_r\rangle$ terms. So, here the subtraction is crucial. The divergence is a typical "infrared," or low q effect and it is seen immediately from the $1/q^2$ in the denominator of eq. 3.30, which at $d \leq 2$ is not cured by the phase space factor q^{d-1} from the q integration. This divergence and its remedy are in exact analogy with, for example, the by now well-known anomaly in the fluctuations of the 1D and 2D lattices (e.g., Imry and Gunther 1971). It is also relevant, for example, for lower-dimension superconductors, see Chapter 7.

To evaluate eq. 3.30 and the three other terms discussed following eq. 3.29 we start with the ω-integration. The integrand will be seen in section 4 to vanish for energy transfers much larger than k_BT and to typically peak at much smaller energy transfers. Thus the integration produces a peak around $t' - t = 0$ whose width is $(k_BT)^{-1}$. We assume that the times of interest, such as τ_ϕ and the duration of the interference experiment, are much longer than $(k_BT)^{-1}$. The ω integral can then be approximated as proportional to $2\pi\delta(t - t')$. This is, in fact, the assumption that $k_BT\tau_\phi \gg 1$. This means that the width of the quasi-particle excitations is much smaller than their energies, which is a basic assumption of the Fermi liquid theory underlying much of our thinking about metals. The final results will indeed be consistent with this assumption. Summing together all the four terms of the phase uncertainty, we obtain

$$\langle\delta\phi^2\rangle = \frac{4}{\pi^2}\int_0^{\tau_0} dt \int d\boldsymbol{k}\,\frac{e^2 k_B T}{\sigma q^2}\sin^2\{\tfrac{1}{2}\,[\boldsymbol{k}\cdot(\boldsymbol{x}_1(t) - \boldsymbol{x}_2(t))]\}, \qquad (3.32)$$

and τ_ϕ, the phase breaking time, is the value of τ_0 for which the phase uncertainty is of order unity.

There are two important points that should be emphasized regarding this expression. The first is that $\langle\delta\phi^2\rangle$ is not necessarily a linear function of time. Since the intensity of the interference term is reduced by the factor $e^{-(1/2)\langle\delta\phi^2\rangle}$, this means that the reduction of the interference term does not have to be a simple exponential function of time. This result is important in the analysis of the conductance of a mesoscopic ring as a function of the magnetic flux inside the ring, discussed in Chapter 5.

The second point is the strong dimensionality dependence of the phase uncertainty. Equation 3.32 for $d = 1, 2$ where d is the dimensionality of the sample) can be approximated as follows. For $\boldsymbol{q}\cdot(\boldsymbol{x}_1 - \boldsymbol{x}_2) \ll 1$ the dangerous q^2 denominator is compensated by the $\sin^2$ and this contribution is easily seen to be small. For $\boldsymbol{q}\cdot(\boldsymbol{x}_1 - \boldsymbol{x}_2) \gg 1$ we have an oscillatory contribution which tends to cancel out, and we remain with the average, $\frac{1}{2}$, of the sine squared. This would diverge at small k, except that the integrand is cut off with $q(x_1 - x_2)$

becoming comparable with unity. Thus $q \sim 1/|\boldsymbol{x}_1 - \boldsymbol{x}_2|$ is the relevant "infrared" cut-off for a divergent $\int dq\, q^{d-3}$. This yields, with σ multiplied by the film thickness in thin films and by the wire cross-section in thin 1D wires,

$$\langle \delta\phi^2 \rangle \sim \frac{e^2 k_B T}{\sigma} \int_0^{\tau_0} dt\, |\boldsymbol{x}_1(t) - \boldsymbol{x}_2(t)|^{2-d}. \tag{3.33}$$

In other words, the main contribution to the $\boldsymbol{q}$ integral of eq. 3.32 comes from $q \sim |\boldsymbol{x}_1(t) - \boldsymbol{x}_2(t)|^{-1}$, and large values of $\boldsymbol{q}$ contribute much less. Since for typical paths in a diffusive medium $|\boldsymbol{x}_1(t) - \boldsymbol{x}_2(t)| \sim \sqrt{Dt}$, we obtain for these paths

$$\langle \delta\phi^2 \rangle \sim \frac{e^2 k_B T}{\sigma} D^{(2-d)/2} t^{(4-d)/2}, \tag{3.34}$$

and for the phase breaking time (at which $\langle \delta\phi^2 \rangle \sim 1$)

$$\tau_\phi \sim \left[\frac{1}{e^2 k_B T D^{(2-d)/2}} \right]^{2/(4-d)} \tag{3.35}$$

The case $d = 2$ is special in that logarithmic factors appear. A careful analysis (see appendix E) shows that there is a logarithmic correction to τ_ϕ, but it is of the nature of $\log(\sigma d\hbar/e^2)$, d being the thickness of the film, and *no* $\log T$ contribution appears. [The argument of the log is the dimensionless conductance per square, $g_\square$, of the system.]

For $d = 3$, the $\boldsymbol{q}$ integral of eq. 3.32 diverges at the upper limit. It is cut off by the condition $Dk^2 < \omega < k_B T$, that is, by $|q| = (k_B T/D)^{1/2}$ (see, e.g., Imry et al. 1982). Then

$$\langle \delta\phi^2 \rangle \sim \frac{e^2 k_B T}{\sigma} \left[\frac{k_B T}{D} \right]^{1/2} \cdot \tau_0,$$

where we assume that $(k_B T/D)^{1/2} |x_1(t) - x_2(t)| \gg 1$ for most values of t. Therefore, for $d = 3$,

$$\tau_\phi \sim \frac{\sigma D^{1/2}}{e^2 (k_B T)^{3/2}}. \tag{3.36}$$

These results lead often, especially at low T, to stronger dephasing than that due to the electron–phonon coupling. Thus, in disordered metals, interference is dephased mainly by the Coulomb interaction or, equivalently, by longitudinal fluctuations of the electromagnetic potential. Unlike the transverse fluctuations that originate in the photon modes and exist also for insulators, the longitudinal modes originate from electron–electron interactions, and are diminished when the metal becomes an insulator. The most effective fluctuations in one- and two-dimensional conductors are those of wavelengths comparable to the distance between the two interfering paths, that is, those where $q \sim l_\phi^{-1}$. Longer wavelengths contribute to the uncertainty in each wave's phase but keep the relative phase well-defined since they cannot resolve the two paths. Shorter wavelengths make the relative phase uncertain, but their magnitude is relatively small.

Above two dimensions the phase uncertainty increases linearly with time, but in a one-dimensional system, for example:

$$\langle \delta\phi^2 \rangle \sim \left[\frac{t}{\tau_\phi} \right]^{3/2} \tag{3.37}$$

with τ_ϕ given by eq. 3.35. The 2/3 power of eq. 3.35 has received convincing experimental confirmation by Wind et al. (1986), Pooke et al. (1989) and Echternach et al. (1993). The former results for τ_ϕ^{-1} as function of T are shown in Fig. 3.2. τ_ϕ was obtained from the weak-localization magnetoresistance.

We will now discuss how the effective dimensionality of the system is determined. Consider a slab of thickness d. Intuitively, one might argue that

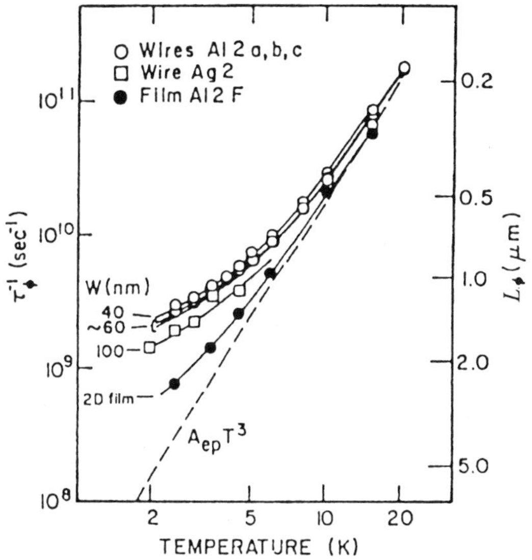

Figure 3.2 Phase breaking rate vs. temperature (from Wind et al. 1986). The solid lines for the wires are fits by eq. 3.35, $d = 1$. The data for wire Ag2 ($W = 100$ nm) from 2 to 4.5 K are normalized to the $R_\square$ and D of the Al samples to allow comparison with results for the Al wires. The solid line for the 2D Al film is a fit by the form$A'_{ee}T + A_{ep}T^3$, with $A'_{ee} = 3.9 \times 10^8$ K^{-1} s^{-1}. The dashed line gives the electron–phonon rate, ^{3}. The scale for L_ϕ applies for the Al samples only.

during the time τ_ϕ the electron propagates a characteristic distance L_ϕ. For $L_\phi \ll d$, it is clear that the propagating electron will typically dephase before feeling the finite thickness, and the dephasing process will be approximately 3D-like. On the other hand, for $L_\phi \gg d$, the diffusing electron cloud fills the whole film thickness and propagates in a 2D manner before dephasing has occurred. Thus, the condition for effectively low-dimensional behavior would appear to be

$$L_\phi \gg \text{appropriate length of the system.} \tag{3.38}$$

However, an analysis (e.g., Sivan et al. 1994b, mentioned below) of the $\boldsymbol{k}$ integration in eq. 3.32 shows that the length $L_T = (\hbar D/k_B T)$, and *not* L_ϕ is the relevant one in eq. 3.39. The relevance of the energy (i.e., $k_B T$)—dependent length to interactions has been noted before (e.g., Altshuler and Aronov 1979, 1985, Imry and Ovadyahu 1982a). Since $L_\phi \gtrsim L_T$, the correct condition is more restrictive.

In a thin wire, the 1D behavior will apply once L_T is larger than both transverse dimensions. When the wire is also of a finite length L, a further crossover will happen once $L_T \gg L$. The system will then become zero-dimensional (0D), where $1/\tau_\phi$ is proportional to T^2. Similar crossovers to 0D will happen for $d = 2$, 3 (Sivan et al. 1994b) as well. This result is of special interest due to the following circumstance. Equation 3.38 with L_T replacing L_ϕ implies that at the crossover to 0D, $k_B T \sim E_c$. We shall now demonstrate that the crossover to 0D prevents the Landau–Fermi-liquid theory from being violated in a narrow wire. Recall that a basic assumption of the Fermi-liquid theory is that the width of the quasiparticle excitation is much smaller than its energy. Since it is suggested physically that the above width is $\hbar/\tau_\phi$, this condition at temperature T is $k_B T \tau_\phi/\hbar \gg 1$. Since eq. 3.35 implies that $\tau_\phi \sim T^{-2/3}$ at $d = 1$, the above condition *appears* to be violated at low enough temperatures. using the Einstein relation and the fact that σ in the 1D result (eq. 3.35 with $d = 1$) is multiplied by the cross-section, we first write the 1D result in the particularly transparent fashion

$$\frac{\hbar}{\tau_\phi} \sim \left(\frac{k_B T \Delta}{\sqrt{E_c}}\right)^{2/3} \tag{3.39}$$

At the crossover $k_B T \sim E_c$. $k_B T$ is the characteristic energy of an excitation and its width satisfies $\hbar/\tau_\phi \sim E_c/g^{2/3} \ll k_B T$, since in the metallic regime $g \gg 1$. Thus at the crossover the assumption of a good Fermi liquid $(k_B T \tau_\phi/\hbar) \gg 1$ is well satisfied. Below the crossover, the rate $1/\tau_\phi$ decreases faster than $k_B T$. Thus, the condition for a valid Fermi-liquid picture is always satisfied. This is very gratifying both on general grounds and because we have assumed it in our derivation of τ_ϕ which is therefore self-consistent. Thus, although the 1D result *appears* to be very problematic in this respect, since

$\hbar/\tau_\phi$ decreases more slowly with temperature than the energy $k_B T$, the situation is saved by the 0D crossover. Another way to put the above argument is by saying that the condition for breakdown of the Fermi-liquid picture is (Altshuler and Aronov 1985) $L_T > \xi$, where ξ is the localization length. The transition to 0D occurs when $L_T > L$. Thus, for a wire which is not in the localized regime (i.e., its length satisfying $L < \xi$), the former will *not* happen. The 0D crossover occurs before $k_B T \tau_\phi/\hbar$ becomes smaller than unity, and rescues the Fermi-liquid theory. Since the the metallic limit the conductance per square of a thin film is large, the Fermi-liquid assumption is also valid in a 2D thin film (eq. E.3).

We now discuss briefly the evaluation by Sivan et al. (1994b) of the dephasing rate in a "quantum dot" (a small finite particle), including the zero-dimensional (0D limit). One has to calculate the relative phase fluctuation, eq. 3.29. It has four terms, each similar to eq. 3.28, and to evaluate them one needs averages of the type $\langle e^{i\mathbf{q}\cdot(\mathbf{x}_i(t)-\mathbf{x}_j(t'))}\rangle$ over the diffusive motion where $i, j = r, l$ are two diffusive paths. The calculation is done by expanding an initial wavepacket which is localized at the origin, in terms of eigenfunctions of the diffusion equation with the appropriate boundary conditions (zero current through the surface) for the dot. The time-dependence is then obtained by letting the wavepacket evolve with time according to the appropriate eigenvalues of the diffusion equation. It is important that the $q = 0$ mode is irrelevant for an *isolated* dot, due to charge neutrality. The calculation is done for an electron of energy ϵ above the Fermi energy at $T = 0$, and as shown in the next section the effect of the Pauli principle is to limit the ω integration to the interval $[0, \epsilon]$.

Each of the dot's dimensions L_i defines a Thouless energy in the diffusive regime,

$$E_c^i \equiv \frac{\hbar D}{L_i^2}. \tag{3.40}$$

For $\epsilon \gg$ all three E_c^i, the integration is three-dimensional and one obtains the 3D result eq. 3.36. Once $\epsilon \lesssim$ a given E_c^i, the effective dimension is reduced in that direction or those directions. In the quantitative calculation, important numerical factors appear (Sivan et al. 1994b). In the zero-dimensional limit $\epsilon \ll E_c \equiv \min_i(E_c^i)$, one finds

$$\tau_\phi^{-1} \sim \frac{\Delta}{\hbar} \cdot \left(\frac{\epsilon}{E_c}\right)^2. \tag{3.41}$$

This implies that, for energies $\epsilon \gtrsim E_c$, the inelastic broadening is enough to effectively smear the discrete spectrum $(\Delta\tau_\phi/\hbar \ll 1)$. This agrees with experiments (Sivan et al. 1994a).

4. DEPHASING TIME vs. ELECTRON–ELECTRON SCATTERING TIME

Sections 2 and 3 of this chapter were devoted to the calculation of the phase breaking time, τ_ϕ, due to the electron–electron interaction, using the general principles developed in section 1. In the present section we discuss the relation between this dephasing time and the electron–electron scattering time, τ_{ee}. The latter time-scale is defined in the following manner. An electron is put in a *single-electron eigenstate* in a disordered system with a given impurity configuration. This eigenstate is characterized by an excitation energy E. Due to the interaction of this electron with the Fermi sea, the eigenstate acquires a width. This width, averaged over all impurity configurations, is τ_{ee}^{-1}. This time-scale obviously depends on the state of the Fermi sea. The case in which the Fermi sea is at zero temperature was studied by Altshuler et al. (1981b), who obtained that $\tau_{ee}^{-1} \sim E^{\frac{d}{2}}$. The case of a Fermi sea at a finite temperature was discussed by Abrahams et al. (1981) for a two-dimensional system. Other cases were studied by Schmid (1974) and by Eiler (1985).

In this section we review the results of previous calculations of τ_{ee}, with two goals in mind. The first is to relate them to the τ_ϕ calculation, and to show whether, when, and why these two times are similar. The second is to discuss an important subtlety overlooked by our calculation of τ_ϕ, namely, the Pauli constraint of the energy loss of the interfering electron. Since, in the approach we have taken in sections 2 and 3, the interfering electron is taken to be distinguishable from the rest of the electrons, the state to which it is scattered is not restricted to be vacant. On top of that, the amplitude for an exchange of the interfering electron with an electron from the Fermi sea is neglected. We shall try to use the τ_{ee} calculation in order to study how significant the errors made in our present approach are, and how a relatively simple recipe can semiquantitatively correct them.

We start our review of the τ_{ee} calculations with the zero-temperature case, and then extend the results to finite temperatures. The golden rule expression for the scattering rate is, in this case (we take $\hbar = 1$ in the rest of this section),

$$\frac{1}{\tau_{ee}} = \frac{2\pi}{n(0)L^d}\int_0^E d\omega \int_{-\omega}^0 d\epsilon$$

$$\times \sum_{\alpha\beta\gamma\delta} |V_{\alpha\beta\gamma\delta}|^2 \delta(E-\epsilon_\alpha)\delta(E-\omega-\epsilon_\beta)\delta(\epsilon - E_\gamma)\delta(\epsilon+\omega-\epsilon_\delta), \qquad (3.42)$$

where $n(0)$ is the density of states at the Fermi energy; $\alpha, \beta, \gamma, \delta$ label exact single-particle states with energies $\epsilon_\alpha, \epsilon_\beta, \epsilon_\gamma, \epsilon_\delta$, and V is the Coulomb interaction. By diagrammatic methods Altshuler et al. obtained the following expression for the impurity-averaged τ_{ee}^{-1}:

$$\tau_{ee}^{-1} = \frac{2e^2}{\pi} \int_0^E d\omega \int \frac{d\boldsymbol{q}}{q^2} \frac{\sigma\omega}{\omega^2 + (Dq^2 + 4\pi\sigma)^2} \mathrm{Re}\left[\frac{1}{i\omega + Dq^2}\right]. \tag{3.43}$$

Before proceeding to study how this expression is related to the derivations of the previous sections, we consider its simple interpretation. The integration variable ω is the energy transfer in the scattering. It is limited by E because of the Pauli exclusion principle. We emphasize again that the electron sea is at zero temperature, so that the scattered electron cannot absorb energy from it. The integrand is composed of the imaginary part of the screened Coulomb potential,

$$\frac{e^2}{q^2} \frac{\sigma\omega}{\omega^2 + (Dq^2 + 4\pi\sigma)^2} = \frac{e^2}{q^2} \mathrm{Im}\left(\frac{1}{\epsilon(\boldsymbol{q}, \omega)}\right),$$

multiplied by the "diffusion pole" $\mathrm{Re}[1(i\omega + Dq^2)]$. As shown in appendix D:

$$\frac{1}{\pi\hbar N(0)} \mathrm{Re}\left[\frac{1}{i\omega + Dq^2}\right] = |\langle m|e^{i\boldsymbol{q}\cdot\boldsymbol{r}}|n\rangle|_{av}^2,$$

where the subscript *av* denotes ensemble-averaging over diffusive states. We note that eq. 3.43 resembles eq. F.6 of appendix F, except that the imaginary part of the screened Coulomb potential $(4\pi e^2/q^2)(1/\epsilon(q, \omega))$ replaces its real part (taken in the static approximation) $1/\pi n(0)$. Thus, eq. 3.43 is the first-order exchange contribution to the imaginary part of the electron's self energy, where the perturbation is the complex potential $4\pi e^2/q^2\epsilon(\boldsymbol{q}, \omega)$. The Pauli constraints on the electron–hole excitation (e.g., the limitation of its energy to be at the most ω below the Fermi energy) are all hidden in the dielectric function. Similar remarks apply to a related calculation by Giuliani and Quinn (1982) for 2D ballistic systems.

The relation of eq. 3.43 to the expressions discussed in the context of τ_ϕ becomes evident when we write the diffusion pole as

$$\begin{aligned}\frac{1}{i\omega + Dq^2} &= \int_0^\infty dt\, e^{-Dq^2 t - i\omega t} \\ &= \tilde{N} \int_0^\infty dt \int D[\boldsymbol{x}(t)] \exp\left[\int \frac{\dot{x}^2(t')dt'}{4D} + i\boldsymbol{q}\cdot(\boldsymbol{x}(t) - (\boldsymbol{x}(0)) - i\omega t\right],\end{aligned} \tag{3.44}$$

that is, as the Laplace–Fourier transform of the average over the diffusive probability distribution of $e^{i\boldsymbol{q}\cdot(\boldsymbol{x}(t)-\boldsymbol{x}(0))}$, which is $e^{-Dq^2 t}$. Here $\tilde{N}$ is a normalization factor. The electron–electron scattering time then becomes, defining $x(0) = 0$,

$$\tau_{ee}^{-1} = \tilde{N} \int_0^\infty dt \int D[\boldsymbol{x}(t)] e^{-\int \dot{x}^2(t')dt'/4D} \frac{2e^2}{\pi} \times \int_0^E d\omega \int \frac{d\boldsymbol{q}}{q^2} \mathrm{Im}\left(\frac{1}{\epsilon(\boldsymbol{q},\omega)}\right) \mathrm{Re}\, e^{i\boldsymbol{q}\cdot \boldsymbol{x}(t) - i\omega t}. \tag{3.45}$$

In the previous section we derived an expression for the time after which an electron traversing a path $\boldsymbol{x}(t)$ changes the quantum state of the electron sea with which it interacts (cf. eq. 3.28). Looking carefully at that expression and at eq. 3.45, we see that τ_{ee}^{-1} is "almost" the average over diffusive paths $\boldsymbol{x}(t)$ of eq. 3.28 with $P(\tau_0) = O(1)$. The main difference between the two is the limits of integration on the energy transfer variable ω. In the expression for τ_{ee}, ω is bounded by the excess energy above the Fermi level, E. In the expression for the dephasing time of the path $\boldsymbol{x}(t)$, it is unbounded. This difference should come as no surprise to us, in view of the "distinguishability" of the interfering electron from the Fermi sea in our calculation of the dephasing time. It obviously suggests a cure to this flaw of eq. 3.28 by limiting the ω integration from 0 to E. This demonstrates that at higher dimensions, $d > 2$, when τ_ϕ is determined by the typical time at which a single path excites the environmennt, it is of the same order of magnitude as τ_{ee}. At $d \leq 2$, we have the subtlety, discussed in the beginning of section 3.3, that the divergence in the $\langle \delta\phi^2 \rangle$ of a single path necessitates the subtraction of the two paths, which yields the *physically meaningful* τ_ϕ.

What happens at finite temperatures? The finite-temperature expression for τ_{ee} was derived by Abrahams et al. (1981) (note that we use a different notation). As one might expect, the sharp Pauli constraint on the energy transfer is replaced by a smooth function of ω, E, and the temperature T.

$$\coth \frac{\omega}{2k_B T} - \tanh \frac{\omega - E}{2k_B T}. \tag{3.46}$$

The second term originates from the Pauli principle constraints of the electron of energy E, as hinted by its E dependence. Our dephasing time calculation has yielded the $\coth(\omega/2k_B T)$ factor, but has failed to yield the $\tanh[(\omega - E)/2k_B T]$ term. Again, the reason for this failure is the neglect of the Pauli constraint on the energy loss of the interfering electron.

Our comparison of the τ_ϕ and τ_{ee} calculations brings us then to the following recipe for the calculation of the diffusive electron–electron scattering time τ_{ee} when it is meaningful, that is, for $d > 2$. First, calculate the time it takes an electron whose path is $\boldsymbol{x}(t)$ to change the state of the Fermi sea it couples to. Second, average overall diffusive paths $\boldsymbol{x}(t)$. Third, correct for the Pauli constraints by using the thermal factor eq. 3.46. At $d \leq 2$, only the time for relative dephasing, τ_ϕ, is physically meaningful.

Being aware now of the approximations that were made in our calculation of τ_ϕ, we may examine the conditions under which *this* calculation is valid, that is, that our neglect of the Pauli principle does not significantly affect the final result. The answer to this question is easily found by an examination of eq. 3.46. The second term of the thermal factor is negligible when $\omega \ll k_B T$. Thus, if the dephasing is denominated by energy transfer ω which is much smaller than the temperature, then the neglect of the Pauli principle is not significant. As seen in section 3, this is the case for low dimensions, $d = 1, 2$. At low dimensions the time it takes each of the paths $x_1(t), x_2(t)$ to change the quantum state of the Fermi sea diverges due to small momentum and energy transfers. It is only the strong overlap of the excitations induced by each of the two paths that makes the dephasing time finite. Thus, the Fermi constraints of the interfering electron are unimportant. For $d = 3$ the situation is different. The dephasing time is dominated by energy transfers of the order of the temperature. Now, if the energy $E > k_B T$, again the $\tanh[(\omega - E)/2k_B T]$ term does not significantly affect the final result. It is only for $E \ll k_B T$ that our calculation of the dephasing time becomes wrong. Since the interfering electrons we discuss typically have an energy $k_B T$ above the Fermi energy, our calculation of the dephasing time remains qualitatively valid, however.

4

Mesoscopic Effects in Equilibrium and Static Properties

1. INTRODUCTORY REMARKS, THERMODYNAMIC FLUCTUATION EFFECTS

This chapter will be devoted to mesoscopic effects which are not neccesarily related to nonequilibrium or transport phenomena. Most of our discussion in the rest of this chapter will be concerned with electronic effects, often with fluctuations from sample to sample. We shall start here, however, by briefly reviewing more general finite-size effects due to ordinary thermodynamic fluctuations.

Such fluctuations are usually negligible in the "thermodynamic limit," as far as their contributions to, for example, intensive properties, or to the $O(N)$ part of extensive ones are concerned. However, thermodynamic fluctuations may play an important role in special situations where large length scales exist in the system. We mention here two of these: the elimination of certain types of long-range order (and, therefore, phase transitions) in low-dimensional systems and the effects of finite sample sizes on regular phase transitions. There are several varieties of the former effects.

In systems with "discrete order parameters," for example a liquid–gas equilibrium, the energy of a "wall" between two phases, U_W, is finite in the quasi-1D (finite cross-section) case for short-range interactions. Since the entropy (Landau and Lifshitz 1959) of such a wall is of the order of $k_B \ln(L/L_0)$, where L is the length of the system and L_0 is some atomic length, walls will always be spontaneously generated in equilibrium, for a large enough

L. Thus, long-range order in the conventional sense, as well as phase-equilibrium with arbitrary long segments of the two phases, are impossible. However, one can make the obvious observations that real systems have a finite L/L_0 (for typical mesoscopic systems $L/L_0 \sim 10^3$–10^4). Thus, once $U_W > k_B T \ln(L/L_0)$, these walls will *not* be generated and effective long-range order will be restored (Imry 1969a; it is also shown there that modestly long-range interactions have a strong effect on the above).

More interesting situations occur in systems having a "continuous symmetry" (i.e., when the order parameter can be "turned" continuously, e.g., by changing the phase in a superconductor, the magnetization direction in a Heisenberg ferromagnet, or by continuously shifting the atoms in the lattice case) Here, one has the phenomenon of a "Bloch wall" (Bloch 1930), where to create a wall between two orientations of the order parameter in a system of length L, it pays to turn the order parameter continuously over the whole length, since this involves a wall energy of $U_W \sim 1/L$ only. This makes the "lower critical dimension," at and below which long-range order is impossible, to be $d_l = 2$. Again, at $d = 2$, order is restored below $T \propto 1/\ln(L/L_0)$. An analogy to this that was encountered in chapter 2 is the occurrence of metallic behavior for weakly disordered metallic films at sizes $L/l \lesssim \exp(\pi^2 g_\square)$, (i.e., for $\pi^2 g_\square \gtrsim \ln(L/l)$). Superconductivity at low dimensions will be considered in chapter 7.

In such continuous symmetry systems the situation is even more subtle since the long-range order is destroyed by long-wavelength fluctuations (Hohenberg 1967) which are only weakly felt at short scales (Alexander 1968, private communication). Thus, not only is short-range order not affected, but in special cases one may have a power-law decay of correlations (Imry and Gunther 1971), which persists until a more interesting vortex unbinding (Berezinskii 1971, Kosterlitz and Thouless 1973) transition occurs. It is still not known for sure whether the localization problem in 2D is of a related type.

In addition to the above points, further special effects are possible in a system where long-range order is broken by thermal fluctuations. One example is a thin ring made of a superconducting material, to be discussed in chapter 7. There is a range of temperatures where thermal fluctuations destroy the superconducting order. This happens by populating a number of quantized-phase states of the order parameter along the ring. This quantization is determined by the periodic boundary conditions. If an A–B flux is introduced, superconducting currents flow in each quantized state to shield the noninteger part of the flux. Thus, the flux is *quantized* in each state as in a strict superconductor, but would appear not to be quantized upon thermal averaging over long times. The characteristic times for jumps among the quantized states are typically much longer than any microscopic time, and can (Langer and Ambegaokar 1967) easily be made astronomically long. Thus (Gunther and Imry 1969, Imry 1969b,c) this system, which is not a superconductor in the conventional sense, has equilibrium *persistent currents* and approximate flux quantization. The last example provides a motivation for going further and considering such currents in normal systems, which will be done later in this chapter.

In the context of the characterization of whether a given wire is an insulator, a conductor or a superconductor, it is extremely instructive to review a fundamental, very general, observation of Kohn (1964). Kohn examined the sensitivity to boundary conditions of the total many-body ground-state energy, writing for it an equation analogous to eq. B.4, except that v_x is the total velocity $\sum_n v_{xn}$, i is the ground state and j is an excited state of the (possibly interacting) many-body system. Similarly, Kohn derived an exact formal Kubo-type expression (see appendices A and B) for $\sigma(\omega)$ in terms of the matrix elements of $\hat{v}_x$. The "diamagnetic" contribution, proportional to the total particle number and due to the term containing A^2 in the Hamiltonian (see problem 1 of this chapter), was also included. By comparing both of these expressions and without any further assumptions, Kohn found that

$$\lim_{\omega \to 0} \omega \, \mathrm{Im} \, \sigma(\omega) = -\frac{e^2 L^2}{\mathrm{Vol} \cdot \hbar^2} \frac{\partial^2 E_0}{\partial \phi^2} \tag{4.1}$$

where ϕ is the phase shift (due to the flux) which applies when an electron encircles the ring, as in appendix C. The dielectric response of the metal is thus related to this general sensitivity to boundary conditions.[1] If the system is superconducting and the ground state supports a well-defined phase, this sensitivity is strongly enhanced, as we shall see in chapter 7.

Let us now briefly consider finite-size scaling near a second-order transition. We note that the various thermodynamic quantities of a small system weakly coupled to the environment are well defined only *on the average*. However, their instantaneous values fluctuate. When properly defined (Landau and Lifschitz 1959), intensive quantities such as the temperature can also be considered as fluctuating.[2] The temperature of a finite system exhibits in equilibrium the following thermodynamic fluctuation:

$$\langle \Delta T^2 \rangle = \frac{k_B T^2}{N c_v}, \tag{4.2}$$

where N is the number of atoms and c_v the specific heat per atom. Equation 4.2 gives only the magnitude of these fluctuations for weak enough coupling to the environment; their time-dependence is determined by several factors including the strength of the coupling to, for example, an isothermal environment. For very strong coupling the system may essentially fluctuate with the much larger

[1] See also the discussion following eq. 2.22 and in section 2 below.

[2] For a system thermally connected to a heat bath, the temperature of the bath is fixed and the energy of the system fluctuates. These fluctuations may be regarded either (Landau and Lifschitz 1959) as just fluctuations in the energy (the degree of excitation) of the system, or of its effective temperature.

environment. For weak coupling the system's own fluctuations are slow and their rates may be determined by the appropriate coupling strength (see problem 1 of chapter 8 for a related example).

We emphasize that while (4.2) is not particularly popular among many theoreticians, it can and does easily yield profound results. For example, the *first* theory of finite-size scaling near second-order transitions (see Fisher 1971) was done by D. J. Bergman and the author (Imry and Bergman 1971) using eq. 4.2 and its analogues. The idea was that the fluctuations limit how close one can effectively approach the transition in a finite system—the transition as a function of temperature will be rounded by $(T - T_c)_{min} \sim (\langle \Delta T^2 \rangle)^{1/2}$. Thus, near this broadened transition the correlation length ξ should reach its maximal possible value, the linear size of the system, and other quantities (e.g., the appropriate susceptibility) will have finite peak values going typically like *some powers of N*. This can provide a rather full description of "finite size scaling," along with a qualitative understanding of how, when $N \to \infty$, the transition sharpens and the finite peaks become mathematical singularities.

This picture can also be used to derive (Imry and Bergman 1971) bulk scaling laws among different critical exponents. For example, when c_v diverges, "hyperscaling" is immediately obtained from eq. 4.2. When only a higher-order derivative of c_v diverges, one has to go to the appropriate higher-order fluctuation formula (Landau and Lifschitz 1959) in order to establish the scaling. This picture can also be used to derive the finite-size rounding of first-order transitions (Imry 1980b) and to understand how, for example, a thin film crosses over between 2D and 3D behaviors.[3] It was also used to obtain expressions (Imry et al. 1973) that connecct critical exponents in different dimensions. While these expressions are not exact, their accuracy is surprisingly good in many cases.

Finally, we emphasize again that in a finite system the physical correlation range ξ is obviously limited by the finite size. Thus, even when long-range order does not exist in the "thermodynamic limit," the finite system may be as ordered *as if* the order existed in the bulk (Imry 1969a). This is exactly the same mechanism as the more recent one explaining why thin films made from a metal appear metallic, while "strictly speaking" they are not (Abrahams et al. 1979). The relevant scale to be compared with the localization length, ξ, is the smallest of the system's size, L, the length L_ϕ, and so on.

An interesting and very speculative consequence of 4.2 emerges at low temperatures (Imry 1986b, Gunther and Ford 1985 unpublished, Gunther 1989, 1990). We write

$$\frac{\langle \Delta T \rangle^2}{T^2} = \frac{1}{N(c_v/k_B)}. \tag{4.3}$$

[3] Similar considerations apply also in the electronic transport problem, see Chapter 2.

Now consider a small system, very weakly coupled to an isothermal environment. Since c_v vanishes as T approaches zero, the r.h.s. of eq. 4.3 will exceed unity below a certain temperature T_m. T_m is on the order of the level spacing Δ, i.e. $\sim$10 mK for a (300 A)^3 metallic particle, and is much larger than that for an insulating system. Let us now attempt to cool the particle below T_m by reducing the temperature of the environment. The particle's effective temperature, measured by the degree of excitation of its levels, will fluctuate so much that such cooling appears problematic. Thus, *if* ordinary thermodynamic formulas are valid, it would appear that T_m is the minimum temperature that the particle can in some sense be cooled to. Below T_m, the fluctuations in thermodynamic quantities (e.g., the energy) become larger than their averages.

The subject which will be of most interest to us will be that of quantum effects on the static and equilibrium electronic properties of mesoscopic systems. When treating the bulk, one is accustomed to use the simplification of a quasicontinuum of states. For a metallic "particle" with, say 10^5 atoms, the typical separation, Δ, between single-particle states at the Fermi energy is on the order of a few tenths of a kelvin. Thus, clearly at temperatures of 1 K or less, this "graininess" of the levels becomes important and may influence, for example, the thermodynamic properties of the system, such as the specific heat or the magnetic susceptibility (Kubo 1962, Gor'kov and Eliashberg 1965, Mühlschlegel 1983). The *precise* spectrum of such a system will usually depend on many details such as the specific defect arrangement as well as on the shape or morphology of the surface of the grain, but for many applications, especially those concerned with an ensemble of grains, it is enough to have some statistical information on the level distribution. Powerful theories of these distributions in effectively random systems (due, e.g., to the high sensitivity to many uncorrelated details) exist and have been extremely successful in atomic and nuclear physics. These methods and their applications are known and thoroughly reviewed (Wigner 1951, 1955, Dyson 1962, Mehta and Dyson 1963, Mehta 1967, Brody et al. 1981). Thus we shall not discuss these issues here in much detail, except for certain aspects that will be directly useful. It is clear that they are very relevant for disordered (Efetov 1982) and granular (Mühlschlegel 1983) as well as for some aspects of strongly localized (Sivan and Imry 1978) condensed-matter systems. We must state already at this stage that Δ is not the only relevant energy scale in the problem and it turns out that the Thouless energy, E_c (see eqs. 2.20–22 and eq. B6), plays an important role as well (Altshuler and Shklovskii 1986).

The effects we shall mainly concentrate on in this chapter have to do with the interference of the electron waves and, in particular, its sensitivity to magnetic fields or fluxes (Aharonov and Bohm 1959). Such interference phenomena exist in principle when the temperature is low enough not to disturb the coherence of the wavefunctions over the relevant spatial scales. The scattering of electrons by impurities, defects, imperfect surfaces, and so on, will play an important role in determining the magnitude of the interference effects but will not eliminate them.

Several of the interference phenomena that we shall discuss have some analogy to effects (e.g., related to flux quantization as discussed above) that are well known and documented in superconductors, where they are brought about by the appropriate "off-diagonal" long-range order which exists there (Byers and Yang 1961, Yang 1962, Bloch 1970). However, we emphasize that we shall consider, to start with, only *normal conductors*, where the possibility of coherence is not related to electron correlation but simply to the (finite) size of the sample being smaller than the appropriate phase randomization length. Later, in chapter 7 we shall consider systems with superconducting components as well.

A further consideration which may limit the *magnitude* of the phenomena under discussion is the possibility that while these phenomena exist they may experience some averaging due, for example, to the electrons not being mono-energetic because of the finite temperature. It will turn out that smaller sizes and/or lower temperatures will increase the magnitude of the interference effects, to the point where they become observable.

2. QUANTUM INTERFERENCE IN EQUILIBRIUM PROPERTIES, PERSISTENT CURRENTS

Generalities, Simple Situations

It was noticed rather early by Pauling (1936), London (1937), Hund (1938), and Dingle (1952) that equilibrium properties, such as the average energy or magnetization of a small *free electron* system with a simple, *ideal*, geometry, for example, a perfect disk or ring (which is realized in an aromatic benzene-type molecule), are sensitive to a magnetic field.[4] Oscillatory behavior is obtained as a function of the field, where the scale is set by the magnetic flux through the system being on the order of a flux quantum, $\Phi_0 = hc/e$. In the particular case of a ring (see Fig. 4.1a) with an Aharonov–Bohm flux Φ through its opening, the thermodynamic functions are periodic in Φ with a period Φ_0 (appendix C). Very large fields, $\sim 10^5$ tesla or more, are needed in order to observe the periodicities in molecules. We shall, however, mainly have in mind man-made conducting rings. Going continuously to the microscopic molecular limit is one of the exciting future directions.

We shall from now on concentrate in this section on the ring geometry. Results of Dingle's type were obtained later by several researchers in different contexts (Gunther and Imry 1969, Kulik 1970, Brandt et al. 1976, 1982). However, an important difficulty was that electron scattering was almost universally expected (Kulik 1970, Altshuler et al. 1981a, 1982b) to eliminate these effects in any realistic system.

[4] We are going to concentrate here on the case of a ring geometry. The orbital response of a singly connected "quantum dot" (see, for example, van Ruitenbeck and van Leeuwen 1991) is also of much interest.

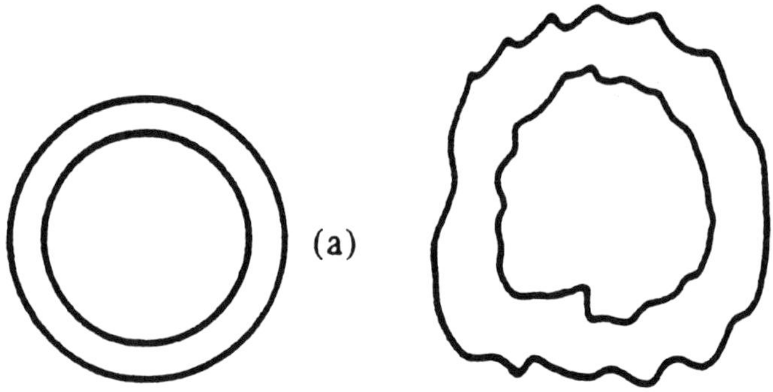

Figure 4.1 Schematic ring geometries: (a) ideal ring; (b) ring with very rough shape and surface.

The point is that it is very difficult to expect a real system to be not only impurity- and defect-free but also to have perfect surfaces. Some surface roughening (perhaps not as bad as in Fig. 4.1b) will practically always exist. Thus, the elastic mean free path, l, will be limited at best by the ring arms' width and thickness. Hence l will be typically much smaller than, say, the ring's circumference L, which is the distance over which the electron's wavefunctions experience A–B type interference. One's first intuition would be that the many scatterings the electron has to experience when traveling along the ring would completely eliminate any interference effect. This is, in fact, in agreement with common notions on electron beam diffraction experiments (including Aharonov–Bohm type: Chambers 1960, Merzbacher 1961, Tonomura et al. 1982) where care has sometimes to be taken to perform the experiment in a high enough vacuum to reduce the electron random scattering.

As should be clear from the discussion of chapter 3, this expectation is very seriously wrong and the analogy with beam experiments is misleading. The point is that there exists an important distinction between *elastic* scattering, due to some *static* potential in which wavefunctions with well-defined phases exist, and *inelastic* scattering. In the latter case, the electron may excite a phonon or alter the state of a "dust" particle, and so on. The electron will, as a result, not have a definite phase, as discussed in Chapter 3. The important distinction between the effects of elastic and inelastic scattering has become clear through the recent understanding (Thouless 1977, Bergmann 1984, Lee and Ramakrishnan 1985, Imry 1983c) of conduction in disordered systems via localization theory (Anderson 1981, Abrahams et al. 1979). Prior to that, in 1966 in an unpublished proposal, R. Landauer informally expressed similar insights (based on Landauer 1957) and Gunther and Imry (1969) found persistent "diamagnetic" currents in a system with a finite resistance as mentioned in section 1 (see also chapter 7). Before discussing this in more detail, we briefly review some simple, general results for "rings." There, the underlying insight is that for a general doubly-connected system with an Aharonov–Bohm flux Φ

through its opening (Byers and Yang 1961, Bloch 1970) all physical properties of this "ring" are periodic in Φ with a period Φ_0. The proof is given in appendix C and it consists of a gauge transformation which establishes an exact equivalence between Φ and a phase change of the transformed full wavefunction by

$$\phi = 2\pi\Phi/\Phi_0, \tag{4.4}$$

when one electronic coordinate is rotated once around the ring. Thus the fluxes Φ and $\Phi + n\Phi_0$ are *indistinguishable*. In addition to establishing the exact claimed periodicity of any physical property (energy levels, matrix elements etc.), eq. 4.4 also shows that a noninteger flux is mathematically equivalent to a change in boundary conditions. This concept will prove extremely useful throughout this book and we shall discuss an even more elementary way to establish it later.

In any system obeying the *classical* laws, which will happen if phase coherence is lost or averaged over, the Aharonov–Bohm flux Φ is clearly relevant. All physical properties do *not* depend on it. This is, of course, trivially consistent with the above theorem (a constant is also a periodic, but not a very interesting function). The real issue is whether there is a sizable sensitivity of, say, the energy levels or the transition probabilities to Φ. Periodicity is always guaranteed! Here, one sees the connection with the sensitivity of the energies of the system to changes in the boundary condition (Kohn 1964, Edwards and Thouless 1972, Thouless 1977). This is fundamentally related to whether the system is an insulator, a metal, or a superconductor.

A flux sensitivity of the ground-state energy E_0 at $T = 0$, or the free energy F at $T \neq 0$, yields a circulating current around the ring. This current is given by

$$I = -c\frac{\partial F}{\partial \Phi} \xrightarrow[T\to 0]{} -c\frac{\partial E_0}{\partial \Phi}. \tag{4.5}$$

This can be demonstrated microscopically since Φ is proportional to the azimuthal vector potential A, and the derivative $\partial F/\partial A$ is proportional to the average of the current operator. For a thermodynamic derivation (Bloch 1970) one may note that if Φ is changed very slowly with time, an EMF is induced given by $V = -(1/c)\dot{\Phi}$. The product $I \cdot V$ is then the rate at which free energy has to be supplied to the system at constant T. From this, eq. 4.5 is easily obtained. These currents are an equilibrium phenomenon, they thus "never decay" as long as Φ is kept on, hence the name "persistent currents." It is, of course, well known that such currents exist in superconductors. There, they exist both in equilibrium and in metastable states (e.g., flux-quantized states of a ring or cylinder, see chapter 7). Here we consider only equilibrium persistent currents, at finite Φ, in a realistic ring. The existence of such currents has been greeted with some skepticism until recently. This has been based on the wrong notion that impurity scattering destroys phase coherence, or on the insistence on taking the thermodynamic limit.

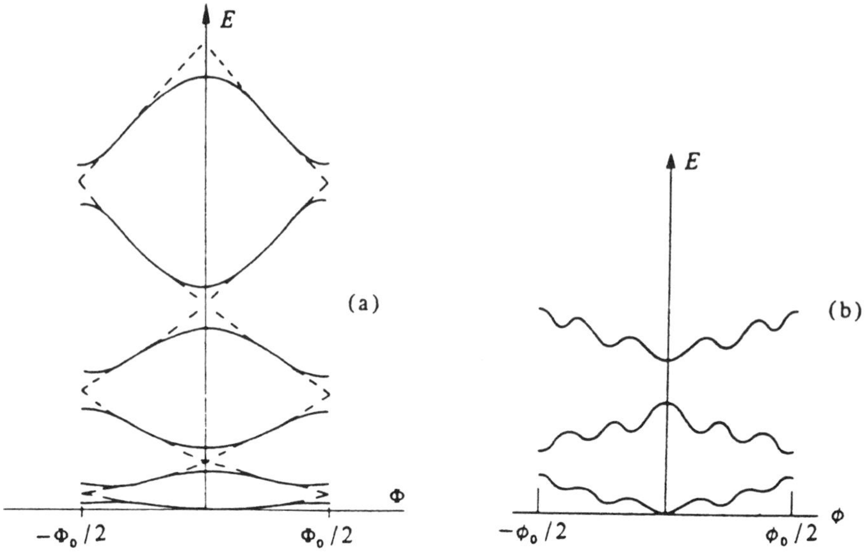

Figure 4.2 Energy levels as functions of Φ (schematic) for: (a) 1D ring, (b) non-1D ring.

Büttiker et al. (1983a) were the first to understand the effect of elastic scattering, using the simple model of a one-dimensional (1D) ring with disorder. They noted that the boundary condition (eq. 4.4) is similar to that satisfied by the Bloch function ψ_k in a periodic potential, across the unit cell of size L. Thus, identifying ϕ with kL establishes a one-to-one correspondence between the two problems. In fact, the condition (4.4) can then be understood since the electron experiences the same potential by moving again and again around the ring, that is, an effectively periodic situation, where the whole circumference of the ring plays the role of the unit cell. The electronic energy levels of the ring as functions of Φ are like those of 1D Bloch electrons. This is schematically depicted in Fig. 4.2. Note that the schematic form of Fig. 4.2a is applicable for an arbitrary random potential along the ring, since it can be shown (Peierls 1955) that in 1D the only extrema of $E(k)$ are at $k = 0, \pm\pi/a$. For nearly free electrons, one gets the usual wide bands and narrow gaps (i.e., $V \sim \Delta$) where V and Δ are the typical bandwidth and band gap, respectively, while for a strong potential (small transmission along the ring) the opposite tight-binding situation (narrow bands, large gaps, i.e., $V \ll \Delta$) is obtained. The latter case corresponds to strong localization.

It is possible to estimate the flux dependence of the total energy, E_0, at low temperatures ($k_B T \lesssim \Delta$) since it is the sum of all occupied levels. Due to the alternating signs of $\partial E_j/\partial\Phi$ for consecutive levels, one has a strong cancelation and the sum is on the order of the last term around E_F. Thus, for N electrons, assuming $V \sim \Delta$

$$I = -c\frac{dE}{d\Phi} \sim \frac{eE_F}{Nh} \sim \frac{ev_F}{L}. \tag{4.6}$$

For a strictly 1D ring made from metal with $E_F \sim 2$ eV with a circumference of 1 μm, $I \sim 10^{-8}$ amp. Since $\Delta \sim 10$ K, having $k_B T < \Delta$ is quite feasible. A further condition, necessary to observe the oscillation, is a long enough inelastic time, that is,

$$\hbar/\tau_\phi \ll \Delta,\ V, \tag{4.7}$$

namely, that the level width is much smaller than the level separation or the bandwidth, whichever is smaller. The condition $\hbar/\tau_\phi \ll V \sim E_c$ can be shown via the Thouless criterion to be equivalent to $\sqrt{D\tau_\phi} \ll L$. This is physically clear—the electron has to stay coherent along the whole ring. An unfavorable case is the limit $V \ll \Delta$, where the effects are very small—localized states are not sensitive to boundary conditions.

Note that in the presence of a finite nonintegral Φ the small diamagnetic type (however, see below for a discussion of the sign) currents flowing around the ring are *persistent* once the temperature is low enough so that eq 4.7 is satisfied. The currents do *not* decay if the inelastic scattering is weak enough—this condition just establishes, within a few τ_ϕ, an equilibrium population among the states. The current is given by the appropriate average, but there is no way for the persistent currents to decay. This result had seemed quite surprising to many but it is obviously correct, similarly to the persistence of ordinary diamagnetic currents in metals. The decrease of this equilibrium current amplitude with decreasing τ_ϕ and increasing dissipation has been discussed by Landauer and Büttiker (1985). These currents yield an *orbital* magnetic moment, M (sometimes referred to as "diamagnetic," although M may be parallel or antiparallel to H) and a magnetic susceptibility χ, oscillating as functions of Φ, with a period Φ_0.

So far we have discussed the case where all the magnetic fields were pure Aharonov–Bohm type. In the case where there are also some nonzero magnetic fields inside the metal itself, leading to a flux Φ_M, their (spin, as well as orbital) effects have to be added. There will be no exact periodicity in the total magnetic flux, only in the dependence on the Aharonov–Bohm part Φ. This dependence is added to the effects due to the magnetic fields inside the material. If the ratio of the area of the hole to that of the material is large enough ("good aspect ratio") one may expect the fast periodic dependence on Φ to be still visible on top of the slower variation due to Φ_M. Thus, aperiodic orbital effects analogous to conductance fluctuations (chapter 5) should occur too. An interesting case where the effective ring is provided by edge states was considered by Sivan and Imry (1988) and Sivan et al. (1989).

Many interesting things occur when the flux Φ changes with time to yield an e.m.f. $V = (-1/c)\,d\Phi/dt$. For the case where V is pure d.c., the resulting current will oscillate (Bloch 1968, 1970) with a Josephson-type frequency (e being the charge of the effective carrier)

$$\omega = eV/\hbar. \tag{4.8}$$

When the change of Φ is not slow enough (for a.c. voltage or for a finite d.c. one), Zener-type transitions (Zener 1930) may occur among the bands. This necessitates a dynamical treatment which we shall not do here.

Up to now we have discussed only the pure 1D case. However, in most conceivable experiments, the wires making the ring have a finite cross-section, A. Thus, the number of transverse states (across the wire) below E_F is on the order of

$$N_\perp \sim k_F^2 A \tag{4.9}$$

and the total number of electrons is

$$N \sim k_F L N_\perp. \tag{4.10}$$

Here, the levels as functions of Φ display a much more complicated structure than in the 1D case. The functions $E(\Phi)$ have many maxima and minima. Schematically, the structure is as in Fig. 4.2b. It is nontrivial to estimate even the order of magnitude of these currents, and we shall return to this later.

Before briefly reviewing the free-electron case, we repeat that persistent currents do exist not only in atoms and molecules but also in ordinary metals, leading to Landau diamagnetism. They cancel in the bulk, for homogeneous samples, but a nonzero surface contribution remains. An interesting case is that of a nonuniform system, for example, a mixture of a metal and an insulator, where nonzero diamagnetic currents may exist in the bulk—along the metal–insulator interfaces, for example. There are important differences, however, between the mesoscopic problem and the bulk case, where large-scale coherence is not necessary, for example.

Orbital moments can exist in equilibrium whenever the appropriate thermodynamic potential, J (e.g., E at zero temperatures and F at finite temperatures for canonical systems, $\Omega = F - \mu N$ for grand canonical ones), depends on the magnetic field, H. The system has then an equilibrium magnetization, M, given by $M = -\partial J/\partial H$ which is the counterpart of eq. 4.5 in the ring geometry. This magnetization can be regarded as due to some nonzero circulating currents. In the thick ring geometry, these currents flow along the inner and outer surfaces, and the novel property is the periodicity of the total circulating current as a function of the A–B flux Φ.

Since elastic scattering by itself is not detrimental to the persistent currents, the very simple case of free electrons on a multichannel ring ($N_\perp \gg 1$) is instructive.[5] The interesting result is, essentially, that both the total persistent

[5] For simplicity we consider only a narrow, short height ring; the long cylinder can also be straightforwardly treated (Gunther and Imry 1969, Cheung et al. 1988).

current and its temperature dependence are determined, in this case, by the energy level separation for motion *around the ring*,

$$E_1 \equiv \frac{\hbar v_F}{L} \sim N_\perp \Delta. \tag{4.11}$$

This is valid in the "just ballistic" regime $l \gtrsim L$. There is a possible enhancement (Cheung et al. 1988, 1989) by a factor $\sqrt{N_\perp}$ in the exra-pure case, where the disorder is a small perturbation (Sivan and Imry 1987) and $l \gg N_\perp L$. The subtle crossover between this perturbative regime and the usual ballistic one (where eq. 4.11 holds) is discussed by Altland and Gefen (1995).

Independent Electrons in Disordered Systems

Since for most physical properties the "geometric" channel number $N_\perp$ in the ballistic regime is replaced in the diffusive one by the effective channel number, which is of the order of g (Imry 1986a; see chapter 5), one is motivated to expect that the appropriate energy scale replacing $N_\perp \Delta$ in eq. 4.11 for the diffusive regime should be $g\Delta \sim E_c$. It turns out that this is exactly what happens, as we shall see.

The order of magnitude of the typical, sample-specific persistent current flowing in a disordered ring is therefore E_c/Φ_0, Over the impurity-ensemble, this current tends to cancel out, since it has, for example, a random slope at the origin ($\Phi = 0$). Thus, ensemble averaging may be expected to cancel out, or at least strongly reduce, the magnitude of the persistent currents. In fact, calculations for weak disorder have shown (Entin-Wohlman and Gefen 1989, Cheung et al. 1989) that for noninteracting electrons in the grand-canonical ensemble, the persistent currents indeed practically vanish upon ensemble averaging. "Grand canonical" means in this connection that the rings are taken to be connected to a particle bath having a chemical potential μ, and the electron number N may change with Φ in each ring.

We are now in a position to review the experimental situation regarding the persistent currents. There are now published results from three independent groups. The first experiment by Levy et al. (1990) was performed on an ensemble (about 10^7) of small (about 0.5 μm in diameter) copper rings. The second experiment was performed on a single gold ring or a small number of such rings, with somewhat larger size (Chandrasekhar et al. 1991). The third was performed on still larger (but with appropriately longer L_ϕ's) single GaAs rings entering the ballistic regime ($l \sim L$) (Mailly et al. 1993), with the advantage that the ring could be connected and disconnected by suitable "gates" to outside leads. This enabled transport h/e Aharonov–Bohm measurements (chapter 5) to be performed to confirm the soundness of the sample. All three experiments used sensitive SQUID magnetometry to measure the magnetic signals, which are periodic functions of Φ, due to the currents in the rings. The latter two experiments, which did not have the increased signal due to the

Table 4.1 Summary of Results

	Theory, Noninteracting Electrons	Theory with Interaction	Experiment
One ring	$L \lesssim l$, $\sim \frac{v_F}{L\phi_0}$	Same as without interactions	$\sim \frac{v_F}{L\phi_0}$
	$L \gtrsim l$, $\sim \frac{v_F l}{L^2\phi_0}$		$\gtrsim 10\text{–}50 \frac{E_c}{\phi_0}$
Ensemble-average	0, grand canonical $\sim \frac{\Delta}{\phi_0}$, constant N	$\frac{\tilde{V} E_c/\phi_0}{1+\ln(E_>/E_<)} \sim \frac{E_c/\phi_0}{5\text{–}10}$	$\frac{E_c}{\Phi_0} \sim 10^2 \frac{\Delta}{\phi_0}$

many samples, necessitated an even more remarkable sensitivity and noise reduction.

These experiments are extremely difficult. The combined reesults (those of the first two experiments are given in Figs. 4.3 and 4.4) seem to show the *existence* of persistent currents that do not decay on time-scales of almost a second (which is effectively "infinite" on any microscopic scale). The magnitudes of the measured signals are relatively large, in the many-ring experiment larger by two orders of magnitude than theories for noninteracting electrons and by about one order of magnitude than perturbative interaction theories (to be discussed later). The gold-ring experiments yielded results larger by more than an order of magnitude than any existing theory, while the results on the ballistic samples of Mailly et al. (1993) agree with theory (4.11) in this simpler regime. If confirmed, the two former results pose a serious challenge to theory. For convenience, the results are summarized in order of magnitude in Table 4.1. Recently, Mohanty et al. (1995) found that the h/e component in gold rings may be smaller than previously obtained. This needs further experimentation.

To understand the theory of the persistent current magnitudes, we start with the "typical," sample-specific case. The Thouless energy immediately gives the typical current response of any level to the A-B flux ϕ at small ϕ. As long as $E_n(\phi)$ can be approximated by its small ϕ quadratic, $\sim E_c\phi^2$, dependence we find for the typical first derivative at small Φ

$$|I_{n,typ}| = -c\left|\frac{\partial E_n}{\partial \Phi}\right|_c \sim \frac{e}{h} E_c \frac{\Phi}{\Phi_0^2} \qquad (\phi \ll 1). \tag{4.12}$$

What is the typical total current in the whole interval $|\phi| \leq \pi$? Cheung et al. (1989) calculated it perturbatively. Their result is

$$I_{tot,typ} \sim cE_c/\Phi_0. \tag{4.13}$$

(*a*)

(*b*)

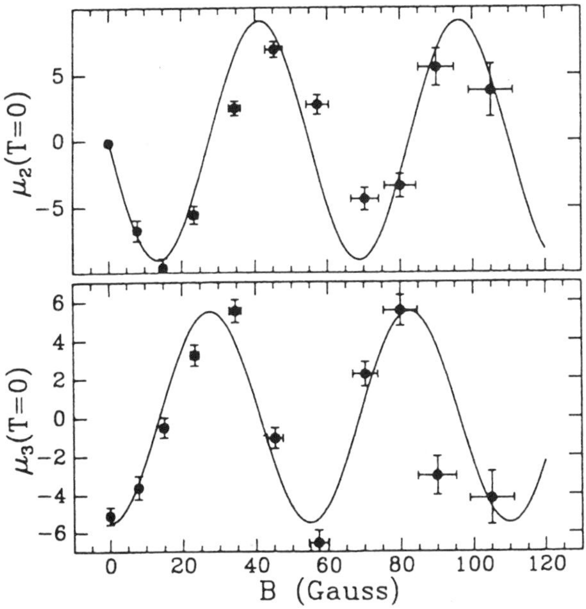

Figure 4.3 From Levy et al. (1990) who measured the nonlinear response of the ensemble of 10^7 copper rings a small part of which is shown is shown in (*a*), as function of flux (a value of B of 130 G corresponds to h/e). An a.c. signal at a low ($\sim$ 1 Hz) frequency and an amplitude of 15 G was employed and the second (μ_2) and third (μ_3) harmonics (double and triple the a.c. frequency) measured and shown in the figure. The nonlinearity arises from the periodic flux dependence and these results imply a persistent current with a period $h/2e$.

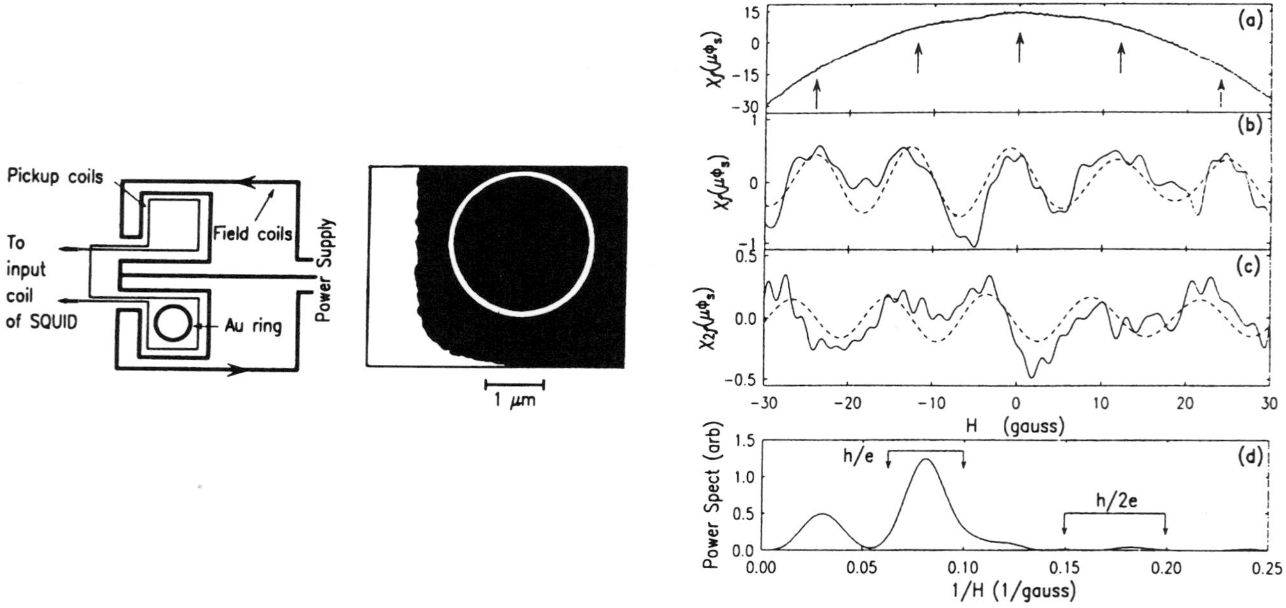

Figure 4.4 From Chandrasekhar et al. (1991). The experimental arrangement for obtaining the response of the single ring (shown) minus that of an empty substrate is depicted on the left. The results are shown on the right: the magnetic-field dependence of the amplitude of the f and $2f$ signals at 7.6 mK for the 1.4 μm × 2.6 μm gold loop. (a) f response with no signal processing. The arrows point to the maxima of the h/e periodic signal. (b) Data of (a), with the quadratic background subtracted. (c) $2f$ response, after subtraction of a linear background signal. The amplitude of the 4 Hz a.c. drive field was 4.12 G. (d) Power spectrum for the data displayed in (b). The h/e arrows show the region, centered about the expected frequency for h/e oscillations, over which the data in (b) and (c) were bandpassed to produce the dashed curves. The region where an $h/2e$ signal is expected to appear based upon the inside and outside area of the sample is also shown. The data in (b) and (c) has been digitally filtered to eliminate high-frequency contributions above 0.50 G^{-1}.

Such a result was also obtained by Altshuler and Spivak (1987) on the related model of an SNS (superconducting–normal–superconducting) junction, see chapter 7. (A related model of a ring with an A–B flux inducing a phase shift has been treated by Büttiker and Klapwijk (1986) in a study of the h/e vs. the $h/2e$ periodicity of the conductance). The similarity of the SNS model to our wholly normal A–B ring stems from the following observation to which we shall return at a later stage in chapter 7: an electron hitting the N–S boundary from the normal side is Andreev-reflected (Andreev 1964) as a hole into its time-reversed path, acquiring an additional phase of $-\chi_2$ (χ_2 is the phase of the superconducting order parameter in the reflecting S part). The hole in turn gets to the other N–S boundary and is Andreev-reflected there as an electron with an additional phase of χ_1 (χ_1 is the phase of the superconducting order parameter in the second S region). Thus, the net result of the two Andreev reflections is that the electron comes back to the same path it started on, with an additional phase of $\chi_1 - \chi_2$. This is equivalent to our A–B ring with $\phi = \chi_1 - \chi_2$. Thus, the supercurrent as a function of $\chi_1 - \chi_2$ in the SNS problem is similar to our persistent current as a function of ϕ. The above discussion ignores normal reflections in the boundaries, which vanish (Blonder et al. 1982) in ideal situations.

Still another way to understand the result (4.13), which includes some many-body aspects, is provided by Kohn's (1964) consideration of the flux-sensitivity of the *total* ground-state energy E_0. Adding the "diamagnetic" term to the equation replacing eq. B.4 for E_0, Kohn found (see eq. 4.1)

$$\left.\frac{\partial^2 E_0}{\partial \phi^2}\right|_{\phi=0} = \frac{\hbar^2 N}{mL^2} - \frac{2\hbar^2}{m^2 L^2} \sum_{j \neq 0} \frac{|\hat{p}_{0j}|^2}{E_j - E_0}, \qquad (4.14)$$

where p is the total momentum in the x direction. (Using the Kubo formula, it is straightforward to obtain eq. 4.1 from 4.14.) As Edwards and Thouless (1972) did, one assumes that the two terms in the r.h.s. of eq. 4.14 almost cancel out. The result is given by the (small) difference between the two terms on the r.h.s. of eq. 4.14, which is of the order of the term with the smallest $E_j - E_0$ (of order Δ) in the sum in that equation. Using eq. D.1 with $q \to 0$ for the semiclassical matrix elements of $\hat{x}$ between the ground state and the first excited state and going from x_{0j} to p_{0j} in the usual way or by determining $|p_{0j}|^2$ from the conductivity (eq. B.2), we obtain eq. 4.13 in order of magnitude quite generally.

The Semiclassical Picture

It turns out that much insight into the persistent-current problem can be obtained by employing a semiclassical approximation to the path-integral formulation of the problem (Imry 1991, Argaman et al. 1993). This is also helpful in order to understand under what conditions an ensemble-averaged persistent current may arise.

Let us start from the Green's function $G(\mathbf{r}, \mathbf{r}', E)$. The density of states (DOS) is given by

$$n(E) = \sum_j \delta(E - E_j) = (1/\pi) \int d^d r \, \mathrm{Im} \, G(\mathbf{r}, \mathbf{r}, E). \tag{4.15}$$

In the following, we shall leave the space integration understood and use in this section a volume of unity. We also use the approximation for $G(\mathbf{r}, \mathbf{r}, E)$ as a sum over all classical paths of given energy which start at and come back to $\mathbf{r}$. Let us denote by A_n the contribution to $G(E)$ from paths winding n times in the clockwise direction around the hole of the ring. The magnitude of A_n will be discussed later. We denote at zero magnetic flux $A_n = |A_n| e^{i\phi_n}$ and for systems with time-reversal symmetry $A_n = A_{-n}$. The presence of the A–B flux, Φ, will multiply A_n by $e^{in\phi}$ $(\phi = 2\pi\Phi/\Phi_0)$. Thus, in the presence of the flux the modified DOS at zero temperature is given by

$$\begin{aligned} n_\Phi(E) &= n_0(E) + 2|A_1| \sin\phi_1 \cos\phi = 2|A_2| \sin\phi_2 \cos 2\phi + \cdots \\ &\equiv n_0(E) + \delta n_1 \cos\phi + \delta n_2 \cos 2\phi + \cdots \\ &\equiv n_0(E) + n_\Phi(E), \end{aligned} \tag{4.16}$$

$n_0(E)$ being the (usually dominant) flux-independent part and $\delta n_\Phi(E)$ the flux-dependent correction. A_n and ϕ_n are functions of E. The typical phase of a diffusive path covering a length L is $k_F L^2/l \sim E_F L^2/\hbar D$; thus we expect ϕ_1 to generally increase with E at a rate of $\sim\pi/E_c$, and this rate for ϕ_n will increase strongly with n.[6] Equation 4.16 is a very convenient presentation of the flux-dependence of $n_\phi(E)$ as a series of harmonics. Several important and nontrivial insights can be gained by simple inspection using this presentation:

1. Given a dephasing time τ_ϕ, all paths requiring a time much longer than τ_ϕ to go around the ring, will be exponentially cut off. Thus, all details of the DOS on scales finer than $\hbar/\tau_\phi$ will be smeared out. The A_1 contribution typically requires a time $t_L \sim \hbar/E_c$. Thus, the condition for survival of the first harmonic of the flux-dependence is

$$\tau_\phi \gtrsim t_L \Leftrightarrow L_\phi \gtrsim L \tag{4.17}$$

and higher harmonics will be cut off by dephasing progressively faster. The energy scale Δ is irrelevant for this issue (Stern et al. 1990a,b).
2. In addition to dephasing, the flux-dependent harmonics will be progressively damped with increasing temperatures by energy-averaging. Here too, the relevant energy scale to be compared with $k_B T$ (disregarding the relativity unimportant energy dependence of $|A_i|$) is E_c[7]

[6] The resulting almost periodic fluctuation of $n_\Phi(E)$ is in agreement with numerical results (Bouchiat and Montambaux 1989 and Divincenzo 1990, unpublished).

[7] Notice that $E_c/k_B T \sim (L_T/L)^2$.

for $n = 1$ and smaller and smaller scales for increasing n. It can be shown that this averaging also cuts off the flux dependence in an exponential manner.

3. When a quantity such as $\sin\phi_n$ is averaged over the usual impurity-ensemble the phase ϕ_n will fluctuate so much that the ensemble average $\langle\sin\phi_n\rangle_{av}$ vanishes. It is this impurity-ensemble averaging which destroys the h/e-periodic A–B oscillation and the universal conductance fluctuations (UCF) in rings and wires (chapter 5). This is expected to happen once the system is large enough in comparison with the impurity scattering scales. It was indeed found (Entin-Wohlman and Gefen 1989, Cheng et al. 1989) that under appropriate conditions, ensemble-averaged persistent currents vanish exponentially with L/l.
4. The representation of δn by sums over closed paths (see in this connection Berry (1985)) is related to the weak-localization corrections to the conductivity, σ (Larkin and Khmelnitskii 1982, Khmelnitskii 1984b, Bergmann 1984, Chakravarty and Schmid 1986, Argaman 1993 unpublished). However, these quantities are determined respectively by Im G and $|G|^2$. The correction to g is of the order of unity. Thus, we expect

$$\delta n_1 \sim 1/\Delta, \qquad \text{etc.} \tag{4.18}$$

It is interesting that the flux-dependent contributions to the DOS are related to those of the conductivity! This analogy can be used to confirm the result in eq. 4.13.

A strong result that follows from the observations (2) and (3) above is that the *ensemble-averaged DOS is flux-independent for* $L \gg l$. The same statement is valid for a given sample with $k_BT \gg E_c$. This is the basis for the belief commonly expressed in the literature (see, e.g., Altshuler et al. 1981a, 1982b) that equilibrium properties should *not* show flux-dependence after ensemble-averaging. This result follows immediately from the observation that, for example, the partition function is an integral (Laplace transform) over the DOS, so that its flux-sensitivity is destroyed by ensemble averaging!

The above is indeed valid for the single-particle partition function. The situation is different, however, for a many-particle system, even without the effect of interactions. For a grand-canonical system, that is, one where the chemical potential rather than the particle number is given, all equilibrium properties are again given by integrals of $n(E)$ times the Fermi function and/or its derivatives. Thus, in this case too, ensemble averages of equilibrium properties should have *no* flux dependence for a large system. This is in agreement with the results of Entin-Wohlman and Gefen (1989) and Cheung et al. (1989). In order to get flux-dependence in ensemble-averaged equilibrium properties for noninteracting quasi-particles, one clearly needs a nonlinear dependence on $\delta n_\phi(E)$. It turns out that this is exactly what happens in an ensemble which is actually more appropriate for many experimental configurations (probably including that of Levy et al.). Consider an ensemble of systems whose macroscopic parameters are identical but in which the specific arrangement of the impurities in each of them is different. Suppose that each of these rings has a fixed number of electrons which *cannot change when* Φ *is varied.* This should happen, for example, when the metallic rings are deposited on a neutral insulating substrate. We shall refer to this as a canonical ensemble, still allowing the number of electrons, N, to vary among members of the ensemble (for discussions of the various ensembles, see Kamenev and Gefen 1993, 1994, Kamenev et al. 1994). The crucial point is that N is *constant in each member,*[8] but we believe it is not crucial for the diffusive regime. This variation of N from system to system is, of course possible, but it does not change the results. Moreover, the systematic dependence of the results on the parity of N is lost in the diffusive regime. Differences between the usual grand-canonical case and that of constant N have been discussed by Landauer and Büttiker (1985). They also appear in Cheung et al. (1988).

For clarity, we present this part of the treatment at zero temperature, where the N lowest levels are filled up to some E_F, and since N is constant, E_F may vary with the flux ϕ. Writing, following the discussion by Imry (1991),

$$E_F(\phi) = E_F^0 + \Delta(\phi), \tag{4.19}$$

we find from the constancy of N that to second order in the flux-dependent terms δn_ϕ

$$\Delta(\phi) + \frac{\delta N_\phi(E_F^0)}{n_0} = -\frac{1}{2}\frac{n_0'}{n_0}\Delta^2(\phi). \tag{4.20}$$

Here n_0 stands for $n_0(E_F^0)$, n_0' for the derivative at (E_F^0), and $\delta N_\phi(E)$ is defined by

$$\delta N_\phi(E) \equiv \int_0^E \delta n_\phi(\epsilon)\, d\epsilon. \tag{4.21}$$

Next we evaluate the canonical average of the energy for a specific system, subtracting the value, E_0, of the energy at $\phi = 0$:

$$E - E_0 = \int_{E_F^0}^{E_F^0+\Delta(\phi)} n_0(\epsilon)\epsilon\, d\epsilon + \int_0^{E_F} \delta n_\phi(\epsilon)\epsilon\, d\epsilon. \tag{4.22}$$

We expand the first term in powers of $\Delta(\phi)$ (or δn) and evaluate the second one by parts. The first-order term in the former cancels and the total first-order contribution to $E - E_0$ becomes

[8] The variability of N among ensemble members was emphasized by Bouchiat and Montambaux (1989).

$$\Delta E^{(1)}(\phi) = -\int_0^{E_F^0} \delta N_\phi(\epsilon)\, d\epsilon. \tag{4.23}$$

This is a sample-specific contribution which varies randomly from sample to sample (but retains the symmetries—periodicity in ϕ and evenness in ϕ for the systems obeying time-reversal symmetry considered here). It is immediately seen that a sample-specific h/e-periodic term similar to eq. 4.23 exists in the grand-canonical case too.

One can use eqs. 4.18 and 4.23 to make estimates for the magnitude of $\Delta E^{(1)}(\phi)$ (and hence of the persistent currents). To do that, one needs more information on the delicate cancelation in integrals such as eqs. 4.21 and 4.23. All we can do, at this stage, is make reasonable but *ad hoc* and *tentative* assumptions on this issue. A better understanding necessitates a more complete treatment of the appropriate spectral correlations, to be discussed later. Let us thus tentatively assume that the oscillating integrands in eqs. 4.21 and 4.22 cancel except for the last shell of width E_c (as discussed before, this is the average "period" of $\delta n(E)$). The number of levels in this interval is of order g. Assuming that the contribution of that interval add randomly, we obtain

$$|\delta N(E_F)|_{typ} \sim n_0 \Delta \sqrt{g} \sim 0(\sqrt{g}). \tag{4.24}$$

Assuming again that the integral (4.23) is similarly determined by the "last shell," one obtains $|\delta E|_{typ} \sim E_c$, in agreement with Cheung et al. (1989), Altshuler and Spivak (1987) and Montambaux et al. (1990).

Upon ensemble-averaging, the contribution $\Delta E^{(1)}(\phi)$ will, of course, vanish, as all other direct higher-order terms do. However, in the second order (in δN and $\delta(\phi)$), the terms of eq. 4.22 are

$$\Delta E^{(2)}(\phi) = \left[\frac{1}{2}\frac{\partial}{\partial \epsilon}(n_0 \epsilon)\right]_{E_F^0} \Delta^2(\phi) + \Delta(\phi)\delta N_\phi(E_F^0)$$
$$- \frac{1}{2} E_F^0 \left.\frac{\partial n_0}{\partial \epsilon}\right|_{E_F^0} \Delta^2(\phi) - \Delta\phi \delta N_\phi(E_F^0) = \frac{n_0}{2}\Delta^2(\phi), \tag{4.25}$$

where we have used eq. 4.20. Upon ensemble averaging we find, using eq. 4.18 and the first-order part of eq. 4.20,

$$\langle \Delta E^{(2)}(\phi)\rangle_{av} = \frac{1}{2n_0}\langle \Delta N_\phi^2(E_F^0)\rangle = \frac{1}{2}\frac{\partial \mu}{\partial N}\langle \Delta N_F^2 \rangle\rangle$$
$$= \text{const} \times \cos^2 \phi. \tag{4.26}$$

At small ϕ, we obtain

$$\Delta E^{(2)}(\phi) = \text{const} - \text{const}' \times \phi^2. \tag{4.27}$$

The small-ϕ persistent current is linear and odd in ϕ and "paramagnetic" in sign! We are going to see later that this sign follows from a very fundamental property, the *decrease* with ϕ of the fluctuation appearing in eq. 4.26. This decrease is due to the tendency of the magnetic flux to break time reversal symmetry, which leads to an effective weakening of the level repulsion.

The ensemble-averaged persistent current (related to results by Bouchiat and Montambaux 1989) is in general odd in ϕ, $h/2e$-periodic, and "paramagnetic" for small ϕ in the simplest case. The survival of the $h/2e$ harmonic is essentially (and very simply) related to the constancy of N as a function of ϕ. The "period halving" (see also Landauer 1990) is related to but not identical with the analogous effect in the conductivity σ (chapter 5). The latter is obtained from $|G|^2$ using terms such as $|A_1|^2$ in which ϕ_1 has totally canceled out.

General Results on Ensemble-averaged Persistent Currents for Constant *N*

The final relation of eq. 4.26 for the contribution to the ensemble-averaged "canonical" persistent current is actually quite general and follows (Altshuler et al. 1991, Schmid 1991, Montambaux et al. 1990) from a rather simple general thermodynamic consideration. This is based on taking into account the variation of the chemical potential μ with Φ, needed in order to keep the electron number in the ring, N, constant. One starts with the general thermodynamic relationship (interesting corrections to which, for $k_B T \gtrsim \Delta$, were found by Kamenev and Gefen 1996)

$$\left.\frac{\partial F}{\partial \Phi}\right|_N = \left.\frac{\partial \Omega}{\partial \Phi}\right|_\mu, \tag{4.28}$$

where F is the free energy (calculated in the canonical ensemble), and $\Omega = F - \mu N$ is the thermodynamic potential (in the grand-canonical one) and the derivatives are taken with the quantities remaining constant at their equilibrium, "equation of state" values, e.g., $\mu = \left.\frac{\partial F}{\partial N}\right|_\Phi$. When Φ is varying, μ is changing as explained above. We write

$$\mu(\Phi) = \langle\mu\rangle + \delta\mu(\Phi), \tag{4.29}$$

where $\langle\mu\rangle$ is an average over a fluxed period.

Expanding the r.h.s. of eq. 4.28 in the small quantity $\Delta\mu$ we find

$$\left.\frac{\partial \Omega}{\partial \Phi}\right|_\mu = \left.\frac{\partial \Omega}{\partial \Phi}\right|_{\langle\mu\rangle} + \delta\mu(\Phi) \left.\frac{\partial}{\partial \mu}\frac{\partial \Omega}{\partial \Phi}\right|_{\langle\mu\rangle}. \tag{4.30}$$

Changing the order of the derivatives in the last term ("Maxwell-relation"), noting that $\left.\frac{\partial \Omega}{\partial \mu}\right|_{\phi} = -N$, and applying ensemble averaging, we find

$$\overline{\left.\frac{\partial F}{\partial \Phi}\right|_N} = \overline{\left.\frac{\partial \Omega}{\partial \Phi}\right|_{\langle\mu\rangle}} - \overline{\delta\mu \frac{\partial}{\partial \Phi} N(\langle\mu\rangle)} = \overline{\left.\frac{\partial \Omega}{\partial \Phi}\right|_{\langle\mu\rangle}} + \left.\frac{\partial \mu}{\partial N}\right|_{\Phi} \overline{\delta N \left.\frac{\partial N}{\partial \Phi}\right|_{\langle\mu\rangle}}, \tag{4.31}$$

where an implicit function derivative was used:

$$-\left.\frac{\partial \mu}{\partial \Phi}\right|_N = \left.\frac{\partial N}{\partial \Phi}\right|_{\mu} \left.\frac{\partial \mu}{\partial N}\right|_{\Phi}.$$

Finally, we use the result that the ensemble-averaged persistent current is exponentially small at a constant $\mu = \langle\mu\rangle$, and the fact that for noninteracting quasiparticles $\partial N/\partial \mu$, the DOS is to lowest order just the inverse of the typical level-spacing, Δ, to write

$$\overline{\left.\frac{\partial F}{\partial \Phi}\right|_N} = \frac{\Delta}{2} \frac{\partial}{\partial \Phi} \langle \Delta N_{\mu}^2 \rangle. \tag{4.32}$$

Here we have expressed the average canonical persistent current as the level spacing times the flux derivative of a grand-canonical fluctuation, as in eq. 4.26. The latter is much easier to calculate, as grand-canonical quantities usually are.

To evaluate the r.h.s. of eq. 4.32 one needs the spectral correlation function

$$K(E,\epsilon) \equiv \left\langle \delta n\left(E - \frac{\epsilon}{2}\right) \delta n\left(E + \frac{\epsilon}{2}\right) \right\rangle \tag{4.33}$$

of the fluctuations of the level density (DOS) between two energies separated by ϵ around an energy E (later we take $E \cong E_F$). The relevant fluctuation in the r.h.s. of eq. 4.32 is given at $T = 0$ by

$$\begin{aligned} \langle \Delta N_{E_F}^2 \rangle &= \int_0^{E_F} \int_0^{E_F} \langle \delta n(E_1) \delta n(E_2) \rangle \, dE_1 \, dE_2 \\ &= \int_0^{E_F} dE \int_{-E}^{E} d\epsilon \, K(E,\epsilon). \end{aligned} \tag{4.34}$$

This was evaluated by Schmid (1991) and by Altshuler et al. (1991) using the results of Altshuler and Shkovskii (1986) on the spectral correlation function, K. The Φ-dependence of those was due to the "double Cooperon" diagram. The results could be written as

$$\bar{I}(\phi) = \sum_m I_m \, e^{2im\phi} \tag{4.35}$$

with

$$I_m = \frac{4i\Delta}{\pi\Phi_0} e^{-2m/\sqrt{E_c\tau_\phi}} \operatorname{sgn}(m), \tag{4.36}$$

where at finite temperatures $\hbar/k_BT$ replaces τ_ϕ in the exponential reduction. As expected, this has a paramagnetic sign (due to the strengthening of "spectral rigidity" with increasing Φ, reflected in the correlation $K(\epsilon)$). The current amplitude is only on the order of Δ/ϕ_0 for each "harmonic," although the energy scale for the reduction due to τ_ϕ or k_BT is the Thouless energy E_c.

Semiclassical Theory of Spectral Correlations, Applications to Rings

Equation 4.33 is an example for the physical relevance of the spectral correlation function $K(E, \epsilon)$ which contains the information on the spectral properties of the system and how universal they are. It is by now firmly established that the energy-level spectra of a finite disordered "mesoscopic" quantum dot, including the case of a ring, obeys (Efetov 1982, 1983) random matrix theory (RMT; see, e.g., Mehta 1967) statistics in certain parameter ranges. In particular, the level density autocorrelation function has been shown by Altshuler and Shklovskii (1986) to lead to RMT-type spectral rigidity for energy differences in the range between the average level spacing and the Thouless energy (see below). At energies above the latter (but below the inverse elastic scattering time), a new universality class was found in that work. Earlier, Efetov (1982) proved that RMT statistics apply to such spectra at low energies using a different method. To obtain physical understanding of how the statistical properties which are relevant to many physical phenomena are generated, it is advantageous to use a quasiclassical picture. In the related field of "quantum chaology" the observation (Bohigas et al. 1984) that the quasiclassical spectrum of a system whose classical motion is chaotic obeys RMT rules has been substantiated by Berry (1985). Recently, Argaman et al. (1993) used the Berry method to demonstrate that RMT correlations apply to the quasiclassical spectrum of a particle whose classical motion is diffusive, for low energies. All the results of Altshuler and Shklovsii (1986) were reproduced in detail. This, with some generalizations, will be reviewed in appendix G. Here we shall summarize the results and indicate the application to persistent currents.

Argaman et al. (see also Doron et al. 1992) expressed Berry's results in the following way. The "spectral form factor" (Fourier transform of $K(K, \epsilon)$ from ϵ to t), is given in the semiclassical approximation by

$$\tilde{K}(t) = \frac{t}{\hbar^2} \frac{d\Omega}{dE} P_{cl}(t), \tag{4.37}$$

where $\Omega(E)$ is the (purely classical) phase-space volume for given energy E and $P_{cl}(t)$ is the classical probability density to return to the origin after a time t for a diffusing particle in a given finite volume. This can be applied in the metallic limit

$$\Delta \ll E_c \ll \hbar/\tau_{el} \ll E_F. \tag{4.38}$$

Both the RMT results for $E < E_c$ ($t > L^2/D$, so that the diffusing particle has filled the volume L^d) and the novel Altshuler–Shklovskii ones for $E > E_c$ (diffusing partile behaving as in an infinite volume) were recovered.

It is straightforward to present the quasiclassical argument for the ring with an A–B flux. One has to group, as in the section on the semiclassical picture, all orbits with winding number m and A–B phase $e^{im\phi}$ with all their time-reversed counterparts having A–B phase factors $e^{-im\phi}$. One can in fact rederive in this way (Argaman et al. 1993) the flux-dependent $K(\epsilon, \phi)$ used for both sample-specific and ensemble-averaged persistent currents.

To obtain the former, one writes

$$\overline{I^2} = \overline{c^2 \frac{\partial}{\partial \Phi} E \frac{\partial}{\partial \Phi'} E}$$

where at $T = 0$, E is expressed as $\int_{-\infty}^{E_F} \epsilon n(\epsilon)\, d\epsilon$. We write the flux-dependent fluctuations as in eq. 4.16 and employ a diagonal approximation as in eq. G.3 to the sum over paths. Again using eq. G.3 and eq. G.4 we see that the classical probability that appears in the term with n revolutions around the ring (again retained with the time-reversed counterpart, but here these have phases of $e^{\pm in\phi}$) is the one to *come back to the origin after n revolutions.* That probability is proportional to $e^{-n^2L^2/4D|t|}$. Finally one obtains (Argaman et al. 1993)

$$\langle I^2 \rangle \simeq 4c^2 \int_{-\infty}^{0} d\epsilon\, \epsilon \int_{-\infty}^{0} d\epsilon' \epsilon' \int_{-\infty}^{0} dt\, e^{-i(\epsilon-\epsilon')t/\hbar} \frac{|t|}{h^2} \frac{L}{\sqrt{4\pi D|t|}}$$
$$\times \sum_{n=1}^{\infty} \left(\frac{2\pi n}{\Phi_0}\right)^2 \exp\left(-\frac{n^2L^2}{4D|t|}\right) \sin^2 \frac{2\pi n \Phi}{\Phi_0}. \tag{4.39}$$

Evaluating this, one finds that at $T = 0$ the nth harmonic of the typical current has the expected order of magitude of $eE_c/\hbar$ for small n and it is given by

$$\overline{I_n(\Phi)^2} = \frac{24}{\pi^2 n^3} \left(\frac{eD}{L^2} \sin \frac{2\pi n \Phi}{\Phi_0}\right)^2. \tag{4.40}$$

At finite temperature and τ_ϕ, the time integral in eq. 4.39 is cut off exponentially at $(\hbar)/k_B T$ or τ_ϕ, leading to a gradual decay of the harmonics, strongest for the high ones as explained in the text around eq. 4.17.

Using a similar line of reasoning and eqs. 4.32 and 4.34, one can find (Argaman et al. 1993) the ensemble-averaged persistent current. Results equivalent to eqs. 4.35 and 4.36 are obtained in agreement with the calculation based on the Altshuler–Shklovskii results for $K(E, \epsilon)$.

Since, as explained before, the results for the ensemble-averaged currents are smaller than experiment by at least two orders of magnitude, it is clear that the model of noninteracting electrons does not capture the necessary physics. One thus has to treat electron–electron interactions seriously.

The effect of the magnetic field is also of interest for a singly-connected system (Quantum dot). The magnetic field breaks time-reversal symmetry and, within RMT, changes the spectral correlations to "unitary" ones. Within the semiclassical picture, for the case of the quantum dot, the magnetic field simply removes the constructive interference between time-reversed orbits and that is how the reduction factor of $\frac{1}{2}$ is obtained in $K(\epsilon)$. It is in full analogy with the removal of the "Cooperon" weak-localization-type contribution by Altshuler and Shklovskii (1986). The details of this "orthogonal-unitary" crossover are discussed by Argaman et al. (1993) and form the basis for a new, nonlinear, paramagnetic orbital susceptibility in a system of quantum dots, which can be quite substantial (Altshuler et al. 1991, 1993, Oh et al. 1991, Raveh and Shapiro 1992).

Interaction Effects on the Persistent Currents

The first microscopic calculations on the orbital response of interacting electrons were done by Aslamazov and Larkin (1974) in the context of superconducting fluctuations (see also Aslamazov et al. 1969). Earlier, Schmid (1969) considered that problem in the G–L approach. The result is actually more general and applies also when the interactions do not lead to superconductivity. The response of the current to a vector potential is proportional to the interaction when the latter is weak enough. The signs are such that the current, or the orbital magnetic response, is diamagnetic for an attractive effective interaction (e.g., a superconductor above the critical temperature T_c) and paramagnetic in the more usual case of a normal metal with an effective repulsive interaction. The former sign is consistent, in the case of a mesoscopic ring made of a superconducting material but rendered nonsuperconducting by thermal fluctuations, with the notion that for $\Phi \ll \Phi_0$ the circulating current will flow in order to shield the noninteger flux (Gunther and Imry 1969). Thus, the fact that a positive interaction yields a paramagnetic response is *natural within perturbation theory* which is first-order in the interaction (Halperin 1991, private communication). Technically, this contribution is obtained by adding an "interaction line" to the "two-Cooperon" diagrams used, for example, by Altshuler and Shklovskii (1986) for $K(e)$ and by Altshuler et al. (1991) and Schmid (1991) for the ensemble-averaged persistent current for noninteracting electrons. The calculation for the ensemble-averaged persistent current $\bar{I}$, was performed for the first time in this connection by Ambegaokar and Eckern (1990, 1991) and Eckern (1991). To first-order in the interaction, the level spacing Δ in the noninteracting case (eq. 4.36) is replaced by $E_c \cdot \tilde{V}$:

$$\Delta \rightarrow E_c \tilde{V} \qquad \text{to 1st order in } \tilde{V}. \tag{4.41}$$

Being the dimensionless coupling constant at low energies, $\tilde{V}$ should include screening in order to incorporate a partial summation of higher-order terms. A result similar to eq. 4.41 will be obtained below using the charge neutrality concept (Schmid 1991, Argaman and Imry 1993). It is strictly valid for $\tilde{V} \ll 1$. If used with relatively large values of $\tilde{V}$ ($\simeq$0.3–10), appropriate for a real metal like copper, the result agrees within perhaps a factor of 2 with the experiment of Levy et al.

However, for such values of $\tilde{V}$, high-order corrections become important. These can be calculated and in fact were obtained at a rather early stage by Spivak and Khmelnitskii (1982) and Altshuler et al. (1983), following early work by Aslamazov and Larkin (1974). The result is that the relevant interaction, which is termed the one in the "Cooper channel," is renormalized in the following schematic manner

$$\tilde{V}_{eff} = \frac{\tilde{V}}{1 + \tilde{V}\ln(E_>/E_<)} \rightarrow \begin{cases} \tilde{V} & \text{for } \tilde{V} \ll \dfrac{1}{\ln(E_>/E_<)} \\ \dfrac{1}{\ln(E_>/E_<)} & \text{for } \tilde{V} \gtrsim \dfrac{1}{\ln(E_>/E_<)}, \end{cases} \tag{4.42}$$

where $E_>$ and $E_<$ are some large and small energy scales in the problem, depending upon conditions. For realistic values of $\tilde{V}$, however, the interaction is reduced by roughly an order of magnitude. This renormalization is of the same nature as the similar decrease obtained by Bogoliubov and Tolmachev (see, e.g., DeGennes 1966) for the Coulomb interaction in the theory of superconductivity. This decrease of the repulsion, essential for the existence of superconductivity, is obtained by considering how the interaction, V_{eff}, at low energies (near E_F), is related to that at high energies. It is easily evaluated by integrating out the higher-energy transfers. Unfortunately, it reduces the theoretical results for $\bar{I}$ to ones that are smaller by about an order of magnitude than experiment. It appears that the perturbative theory in $\tilde{V}$ is *not* able to account for the present discrepancy between theory and experiment.

It is advantageous at this stage to have a physical picture for the effects of the interactions on $\bar{I}$. Such a picture was first given by Schmid (1991). Let us remember that an important consequence of the Coulomb interactions in a metal is to enforce local charge neutrality in any volume element larger than screening length Λ. Schmid argued that this strong constraint of local charge neutrality gives a more substantial contribution to the persistent current than just constraining the *total* electron number N. *This* reproduces the first order result of the interaction, but may be reduced as well by the "Cooper channel" renormalization mentioned above.

To present a simplified version of Schmid's argument, we employ a spatially varying (Argaman and Imry 1993) effective potential, which preserves local charge neutrality (to the extent specified by an appropriate Thomas–Fermi-type screening erquation) when the flux is varied. The result for the disorder-ensemble-averaged persistent current includes a term proportional to the (integral over space of the) product of the flux sensitivities of the effective potential and the noninteracting local electron density (see eqs. 4.43–4.45 below). It is a direct generalization of the canonical-ensemble result (eqs. 4.31 and 4.32), in which only the global chemical potential and (grand-canonical) number of electrons are taken as flux-sensitive. the expression based on eq. 4.45 below (Argaman and Imry 1993) still resembles the first-order result of Ambegaokar and Eckern. The strong renormalization of the interaction does not emerge on this level in an obvious manner, and the question of its relevance requires further work. For small $\tilde{V}$, this corroborates and gives additional insights to the argument of Schmid.

Following Schmid (1991), and as a generalization of the picture of Altshuler et al. (1991), we divide the sample into cells, i, larger than the screening length but smaller than all other characteristic lengths. All the effects of the Coulomb interaction are taken to yield a potential energy v_i for the electrons in the cell i, apart from the fluctuating impurity potential, which may vary on a finer scale. The free energy of our effective noninteracting electrons is thus given by $\Omega_{eff}(\{v\}, \Phi)$, where $\{v\}$ denotes the set of v_i for all i, and Φ is the Aharonov–Bohm flux (Ω_{eff} also depends implicitly on the temperature, the chemical potential, and the impurity potential). The persistent current is given by $I(\{v\}, \Phi) = -c\partial\Omega_{eff}/\partial\Phi$, and the number of electrons in the cell i is given by $N_i(\{v\}, \Phi) = \partial\Omega_{eff}/\partial v_i$. When Φ is varied, the v_i are adjusted to maintain *local* charge neutrality, so that N_i is kept equal to the ionic charge in the cell i (to get eq. 4.30, only *global* charge neutrality was maintained by allowing variation of the chemical potential, which may be regarded as an i-independent contribution to v_i).

Defining v_i^o as the average of v_i, for a *given* sample, over a period of Φ, and writing v_i as $v_i^o + \Delta v_i(\Phi)$, we have

$$I(\{v\}, \Phi) = I(\{v^0\}, \Phi) + \sum_i \Delta v_i \left.\frac{\partial I}{\partial v_i}\right|_{\{v^0\}}$$

$$= I(\{v^0\}, \Phi) - c\sum_i \Delta v_i \left.\frac{\partial N_i}{\partial \Phi}\right|_{\{v^0\}}. \qquad (4.43)$$

In the last line we have used the thermodynamic (Maxwell-type) relationship

$$\frac{\partial I}{\partial v_i} = -c\frac{\partial N_I}{\partial \Phi}.$$

Higher-order corrections in Δv_i are ignored, as the flux dependence of all physical properties of the rings is assumed to be weak. Δv_i is determined using a similar equation for $N_i(\{v\}, \Phi)$, which has been assumed to remain constant:

$$\begin{aligned} N_i(\{v\}, \Phi) &= N_i(\{v^0\}, \Phi) + \sum_i \Delta v_j \left.\frac{\partial N_i}{\partial v_j}\right|_{\{v^0\}} \\ &\simeq N_i(\{v^0\}, \Phi) - N(0)\Delta v_i, \end{aligned} \tag{4.44}$$

where we have introduced the local approximation $\partial N_i/\partial v_j \simeq -N(0)\delta_{i,j}$, and taken the density of states of each cell $N(0) = \partial N_i/\partial \mu$ to be independent of i. Defining $\Delta_\Phi N_i = N_i(\{v^0\}, \Phi) - N_i$, we have

$$\begin{aligned} \langle I \rangle &= \langle I(\{v^0\}, \Phi)\rangle - c\left\langle \sum_i \frac{\delta_\Phi N_i}{N(0)} \frac{\partial N_i}{\partial \Phi} \right\rangle \\ &= -\frac{c}{2N(0)}\frac{\partial}{\partial \Phi}\left\langle \sum_i (\delta_\Phi N_i)^2 \right\rangle, \end{aligned} \tag{4.45}$$

where we have employed the impurity ensemble averaging, and neglected $\langle I(\{v^0\}, \Phi)\rangle$, which is similar to the (grand canonical) averaged noninteracting current.

The last term is the (flux derivative of the) sum of the local number fluctuations when the local potential is kept constant, that is, the fluctuations of noninteracting electrons in the flux-averaged disordered potential v_i^0. Taking this potential to vary randomly (e.g., white noise) over the disorder ensemble, one can evaluate the fluctuations using established noninteracting-electron theories. This contribution appears similar to the first-order results of eq. 4.41 (for the case of a repulsive electron–electron interaction): it is large, paramagnetic at small Φ, periodic in Φ with a period $\Phi_0/2 = hc/2e$, and is in rough agreement with experiment *on the most naive level which disregards the possible reduction due to renormalization effects.* Argaman and Imry (1993) have also obtained the above results from a systematic application of density functional theory (Kohn and Sham (1965), a good review can be found in Kohn and Vashishta (1985)). To what extent will the "Cooper-channel renormalization" decrease this result as well, as one might expect, is presently an open issue.

To summarize the situation with regard to persistent currents at the time of writing: It appears that their existence is confirmed. But, according to the present experiments, theory gives results that are much too small for the magnitude of the currents for the ensemble-averaged and, possibly, sample-specific situations. Unless fresh ideas, such as invoking surface or interface phenomena, are generated, a nonperturbative theory for the Coulomb interactions is needed. The two existing speculations are the charge neutrality argument above and the very interesting observation by Altland et al. 1992, Müller-Groeling et al. 1993 and Müller-Groeling and Weidenmuler 1994) that interactions may strongly reduce the effect of disorder. This is supported by recent numerical results of Berkovits and Avishai (1995a,b). It has to be demonstrated, however, that this reduction is stronger for the equilibrium properties than for the transport. Both of the above ideas need further work.

Problem

1. (This is an exercise on the "subtleties" mentioned at the beginning of appendix B). Prove, using $[\hat{p}, \hat{x}] = \hbar/i$, that two terms on the r.h.s. of eq. B.4 cancel exactly. You had to employ $v_{ij} = i\omega_{ij}x_{ij}$. Why and when is that problematic? When is it justified? Convince yourself that the above cancelation would mean that there is no sensitivity to boundary conditions and no orbital magnetic response for our system!

5

Quantum Interference Effects in Transport Properties, the Landauer Formulation and Applications

1. GENERALITIES, REMARKS ON THE KUBO CONDUCTIVITY FOR FINITE SYSTEMS

The transport properties of mesoscopic systems display a wealth of interesting phenomena that are quite novel in respect to the usual macroscopic systems. Among these one can mention the nonadditivity of series resistances and parallel conductances, the periodic sample-specific oscillations (Gefen et al. 1984a,b, Webb et al. 1985a,b) in the magnetoresistance of a ring (i.e., two parallel resistors) as a function of the Aharonov–Bohm flux through its opening, and the analogous aperiodic conductance fluctuations (Blonder 1984, Umbach et al. 1984, Altshuler 1985, Lee and Stone 1985, Licini et al. 1985a, Stone 1985, Skocpol et al. 1986), in a fine singly-connected wire. These fluctuations have interesting universal aspects. As in the previous chapters, there is an important distinction here between the effects of elastic and inelastic scattering. Also, since the system is so small, its measured resistance may depend on the existence, type, and structure of contacts made onto it. Various other effects, familiar to varying extents from waveguiding systems, may also occur. For example, an open-ended branch can greatly change the resistance of the system (Gefen et al. 1984a,b); the resistance may be nonlocal in the sense that what is measured between a given pair of points may depend on things connected further away (Anderson et al. 1980, Engquist and Anderson 1981). Contact and spreading resistances, which are not always negligible, may play a role (Imry 1986b, see section 2).

From the theoretical point of view, this problem is also interesting due to the electrons being possibly further removed from equilibrium (due to the scarcity of inelastic scattering) than in ordinary transport situations. In some cases, one has to develop special methods to handle such aspects. We shall first briefly review some subtleties of the Kubo linear response formalism (Kubo 1957, 1962) for our case, reflecting on the Thouless picture of conductance in disordered systems. Then we shall develop in section 2 the Landauer-type scattering formulation for the conductance of a segment of a disordered system between two ideal leads, as well as the generalization to more terminals. The similarities and differences among these approaches will be discussed. Various applications will be reviewed in section 3.

For an infinite system, the Kubo-type conductivity at frequency ω may be obtained as in appendices A and B (see eq. B.1), or by calculating, using the golden rule, the power absorbed by the system from a classical e.m. field (we shall consider here the σ_{xx} component). An additional contribution, the Debye relaxation absorption, will be discussed later. The field used in this formulation is the actual one, containing the self-consistent field provided by the polarization of charges inside the system (see, e.g., Landauer 1978):

$$\sigma(\omega) = -\frac{1}{\text{Vol}}\frac{\pi}{\omega}e^2\sum_{k,l}|\langle k|\hat{v}_x|l\rangle|^2\delta(E_l - E_k\hbar\omega)(f_k - f_l). \qquad (5.1)$$

For simplicity, we consider noninteracting (or Hartree–Fock) electrons. Corrections for the self-consistent field, as mentioned above, are included. Vol is the volume of the system, $|k\rangle, |l\rangle$ are the free- (or self-consistent single-) electron states and f_k, f_l their populations. $\hat{v}_x$ is the velocity of the electron in the x direction. The assumption of an infinite system is crucial here, in order to have a continuum of states. Otherwise, the field does not induce real transitions. An *isolated* finite system with a truly discrete spectrum does *not* in fact really absorb energy from the monochromatic field. In order to obtain a finite conductivity, the small system has to be (and to some extent is in real situations) coupled to a very large heat bath—for example, to an assembly of thermal phonons. This enables energy to be transferred from the e.m. field into the bath via the small electronic system. For a weak enough interaction with the bath, one may say that the discrete levels of the system have just acquired finite widths. It then makes sense to write down eq. 5.1 with E_k having a finite width or with an imaginary part $i\eta$ to the frequency ω. Thus, for d.c. (Re $\omega \to 0$), Thouless and Kirkpatrick (1981), following Czycholl and Kramer (1979), suggest the following expression for σ_{dc} for finite systems based on eq. 5.1:

$$\sigma(i\eta) = \frac{1}{\pi}\int_{-\infty}^{\infty}\frac{\sigma(\omega')\eta}{\omega'^2+\eta^2}d\omega'$$

$$= \frac{2e^2\hbar}{\Omega}\sum_{k,l}\frac{|v_{kl}|^2}{E_k - E_l}\cdot\frac{\hbar\eta(f_k - f_l)}{(E_k - E_l)^2 + (\hbar\eta)^2}. \qquad (5.2)$$

While this procedure certainly makes sense in smoothing out the δ-functions of eq. 5.1, it does need a more rigorous justification in terms of the combined electron–bath system. Van Vleck and Weisskopf (1945) obtained similar results using a semiclassical picture with collision broadening, which is discussed further by Imry and Shiren (1986). The following discussion will be based on eq. 5.2

It is possible to show that once $\hbar\eta$ is much larger than the energy level separation, Δ, of the electrons at E_F (though much smaller than all other relevant energy scales in the problem), eq. 5.2 goes over, as it should, to the appropriate "bulk" expression. This condition is always very well satisfied for macroscopic systems where $\Delta/k_B \sim 10^{-18}$ K and $\hbar/(\tau_{in}k_B)$ is rarely smaller than $\sim 10^{-4}$–10^{-5} K (and it usually attains such values only at millikelvin temperatures). To get the usual expression for σ from eq. 5.1, one straightforwardly obtains the low-temperature d.c. conductivity by replacing the sums by integrals and assuming that $|\langle l|\hat{v}|k\rangle|^2$ has some typical value denoted by $|\langle v\rangle|^2$ near E_F (see appendix B),

$$\sigma_{KG} = \pi e^2 \,\mathrm{Vol}\, \hbar |\langle v\rangle|^2 [n(0)]^2. \tag{5.3}$$

Here $n(0)$ is the density of states per unit volume, at E_F. This is the Kubo–Greenwood conductivity (Kubo 1957, Greenwood 1958).

However, for the typical small metallic systems that are of interest to us here, Δ can become of the order of a few millikelvins. Thus, at temperatures below $\sim$0.1 K, one may encounter an interesting and novel range where

$$\hbar\eta \lesssim \Delta. \tag{5.4}$$

In the limit $\hbar\eta \ll \Delta$ the average Kubo-type conductivity is easily estimated (using $|E_k - E_l| \sim \Delta \sim [n(0)\Omega]^{-1}$) to be on the order of

$$\sigma \sim \sigma_{KG} \frac{\hbar\eta}{\Delta}. \tag{5.5}$$

This has the interesting feature that the $\omega \to 0$ conductivity which is defined by energy absorption from the e.m. field vanishes (Landauer and Büttiker 1985, Büttiker 1985b, Imry and Shiren 1986) when $\hbar\eta/\Delta \to 0$. In this limit, we have discrete states with no real energy absorption. Thus, the Kubo d.c. conductivity as a function of η looks schematically like Fig. 5.1. We emphasize that this discussion is concerned with a particular definition of the conductivity, as measured, for example, by putting the sample, with no contacts, in an electromagnetic cavity and measuring the extra absorption due to the sample at low frequencies. It will turn out that this definition is *not necessarily identical* to others. For example, we shall find in section 2 that the same sample may display a well-defined, finite resistance which will be independent of η for small enough η, if measured by connecting to it two appropriate contacts. In

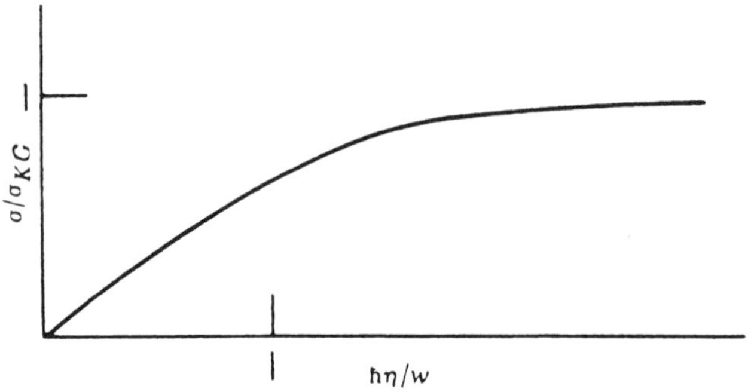

Figure 5.1 The d.c. Kubo conductivity as function of the ratio $\hbar\eta/\Delta$. It approaches the Kubo–Greenwood conductivity σ_{KG} for $\hbar\eta \gg \Delta$, and is proportional to $\sigma_{KG}\hbar\eta/\Delta$ for $\hbar\eta \ll \Delta$.

that case the Joule energy dissipation will take place inside the thermal baths that have to be assumed to be associated with the contacts.

The coupling of a discrete-levels system to a heat bath at finite temperatures also leads to well-known "Debye-relaxation" absorption (e.g., Gorter 1936, Gorter and Kronig 1936, Kittel 1986) which has to be *added* to the Kubo-type terms. This absorption vanishes both when $\omega/\eta \to 0$ and $\omega \gg \eta, \omega_0$. As emphasized by Landauer and Büttiker (1985), this absorption is due to the oscillation of the level separation with the applied field. The level populations attempt to relax to equilibrium with a time constant η^{-1}, but lag behind the field. This effect vanishes both when $\omega \ll \eta$ and the system follows the field, and when $\omega \gg \eta$ and the relaxation becomes negligible. Such absorption was obtained for rings with a.c. A–B fluxes by Landauer and Büttiker (1985) and by Trivedi and Browne (1988). $T \neq 0$ is necessary, since at $T = 0$ the populations do not depend on the level separation.

The Kubo–Greenwood formulation, which can be conveniently cast in terms of time-dependent correlation functions, has proved to be an extremely useful formulation of transport theory. It is also the basis for a systematic diagrammatic expansion in the strength of the disorder, characterized by the small parameter $(k_F l)^{-1}$, l being the elastic mean free path. The first correction (Langer and Neal 1996, Gorkov et al. 1979, Abrahams et al. 1979, Hikami et al. 1981) to classical Boltzmann transport yields the weak localization contributions discussed in chapter 2.

The Thouless expression for conductance was originally derived (appendix B) by employing the Kubo formulation. It thus appears, for example, that the condition $\hbar\eta \gg \Delta$ should be necessary to validate the Thouless expression. This is also consistent with the golden-rule picture of eq. 2.4. However, we shall see that the Landauer formulation described in the next section, which indeed provides a finite conductance for a finite segment with $\eta \to 0$, is intimately related to the Thouless conductance. The latter will be equivalent to the Kubo expression only when $\hbar\eta \gg \Delta$ applies. It remains to be discussed whether the inherent coupling of the system to reservoirs in the Landauer formulation might play a similar role to the parameter η above.

2. THE LANDAUER-TYPE FORMULATION FOR CONDUCTANCE IN A MESOSCOPIC SYSTEM AND SOME OF ITS GENERALIZATIONS

Introduction: the "Single-Channel" Case

The Landauer formulation (Landauer 1957, 1970, 1975, 1985), which expresses the conductance in terms of the scattering properties of the system, is especially suited to treating the conductance of a segment of a (possibly disordered) system to which two appropriate contacts are made. It has been extremely useful not only as a computational tool but also, perhaps more importantly, as a picture from which physical insights on new phenomena can be obtained. It is concerned with a given system, and no ensemble-averaging is necessary. Thus, mesoscopic fluctuation effects emerge very naturally. In 1957 Landauer first introduced the 1D version, which consisted of a given barrier connected through ideal 1D wires (flat potentials) to some external source (i.e., a pair of electron reservoirs with different chemical potentials) which drives a current I through the 1D system. The barrier is characterized by its transmission coefficient T and its reflection coefficient $R = 1 - T$ (for linear transport and at zero temperature, we need T and R at the Fermi energy only). As emphasized by Landauer (1970, 1975), it is very important to take the waves coming from the two reservoirs to be "incoherent" with each other (i.e., having no definite phase relationship), otherwise nonphysical results follow for the time-reversed situation.

Landauer first considered neutral particles and obtained the density difference across the barrier, and thence the appropriate diffusion coefficient. Then the Einstein relation was invoked to obtain the conductance. For charged particles, self-consistent screening (Landauer 1957) yields the same result. The conductance *due to the barrier*, including spin degeneracy, is given (Landauer 1957, 1970) by

$$G = \frac{e^2}{\pi\hbar}\frac{T}{R}. \tag{5.6}$$

We emphasize that this is the conductance of the barrier *itself*, defined as the ratio of the current, I, which runs through it, and the electrochemical potential difference which develops between its two sides. Some confusion has been generated (and later clarified, see below) in the literature, due to the following circumstance: A common way to drive a current through the system is to connect the ideal wires on its two sides to particle reservoirs of chemical potentials μ_1 and μ_2 ($\mu_1 > \mu_2$) as in Fig. 5.2. If now now computes a conductance G_c using the ratio of I and $\mu_1 - \mu_2$ one obtains (for the derivation of eqs. 5.6 and 5.7 see the discussion of eqs. 5.16 and 5.19 below)

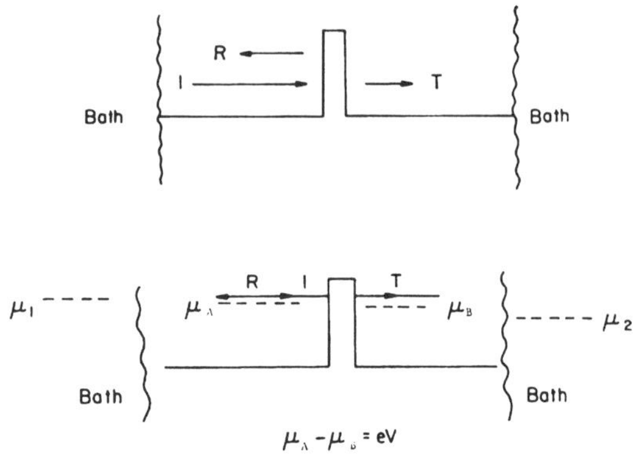

Figure 5.2 The Landauer geometry: μ_1, μ_2 are the chemical potentials of the baths; μ_A and μ_B are those of the ideal conductors on the two sides of the barrier.

$$G_c = \frac{I}{\mu_1 - \mu_2} = \frac{e^2}{\pi\hbar} T, \tag{5.7}$$

while the previously defined G (eq. 5.6) is given by $G = I/(\mu_A - \mu_B)$, where μ_A and μ_B are the chemical potentials on the ideal wires on the l.h.s. and r.h.s. of the barrier (see Fig. 5.2). Thus, G_c (which turns out to be smaller than G) is the conductance *measured between the two outside reservoirs.* Even for $T = 1$, $G_c = (e^2/\pi\hbar)$ is the finite conductance (Imry 1986b) due to the narrow channel between the two large reservoirs (Sharvin 1965, Jansen et al. 1983). This resistance can be thought of as due to two contact resistances, $(\pi\hbar/2e)^2$ each, between the wires and the corresponding reservoirs. In fact, $G_c^{-1} = G^{-1} + (\pi\hbar/e^2)$, that is, the total resistance between the reservoirs is the sum of the barrier resistance and the two contact resistances. Derivations (for example, Economou and Soukoulis 1981a,b) of the conductance from the Kubo formulation yielded the two-terminal conductance G_c *only*. This started a long controversy on "which of the Landau formulas is correct?" The answer (Imry 1986b) was that they pertain to *different* physical quantities.

The contact resistances are due to the geometry of a narrow channel feeding into a large reservoir, and to the electrons thermalizing in the baths by inelastic scattering. The corresponding contact resistance per channel will turn out to be of the same order of magnitude in the multichannel case as well, as will be discussed later. It thus becomes an interesting issue whether this order of magnitude is a universal quantity (as one would tend to suspect) or whether it depends strongly on the details of the connection of the thin wire to the reservoir. When the transmission between the "wire" and the reservoirs is perfect, $T = 1$ and the channel leads to an inter-reservoir conductance of $e^2/(\pi\hbar)$. This result is "universal"; it is generalizable to many channels and will be discussed later.

The conceptually correct way (which is still, however, subject to some questions) to measure the chemical potential difference $\mu_A - \mu_B$, was suggested by Engquist and Anderson (1981) and further discussed by Büttiker et al. (1985) and Sivan and Imry (1986a,b). It will be discussed in some detail later.

The "barrier" may be *any object fed by the two 1D wires*, for example, a segment of a linear chain. Thus, the above can be used to consider the conductance of any problem which is 1D in the above sense. In fact, at a very early stage, Landauer (1970) obtained the addition law (see eqs. 5.36, 5.39, 5.40) of two such barriers or "quantum resistors" in series, and thence, by induction, the resistance of a linear chain with n randomly placed barriers. The exponential increase of this resistance with n, for n larger than some characteristic size (the localization length, in modern terminology) was obtained. This was the first demonstration of 1D localization as manifested in the resistance, and the basis for the scaling theory for localization in 1D, which was presented by Anderson et al. (1980), after identifying the appropriate variables to be averaged. One should also note that the resistance of two such resistors in series is typically larger than that given by the usual $R_1 + R_2$ law.

The case of two parallel 1D resistors using the Landauer formulation was first solved by Gefen et al. (1984a), who also found that the addition law is *different* from the classical one. By introducing an Aharonov–Bohn type flux Φ in the space inside the loop formed by the two resistors, they obtained oscillations in the transmission coefficient. Hence the resistance between the two 1D leads oscillates as a function of Φ with a fundamental, usually dominant (Gefen et al. 1984b) period of Φ_0, in agreement with the general expectations discussed in the previous chapters. Further discussion of both the series and parallel cases will be given later with other applications of the Landauer formulation. The formulation is general enough to include some aspects of electron–electron interactions in the system, superconducting components, resonant states, and other complications. The above has been the first clear case of a sample-specific transport phenomenon, which is really a "conductance fluctuation." Such fluctuations had been discovered experimentally in the quest for the Φ_0-periodic oscillation. Their "universality" (Altshuler 1985, Lee and Stone 1985; the second and third subsections of section 3 and appendix I) is one of the fundamentally interesting mesoscopic phenomena.

The generalization of the Landauer approach to the multichannel case is of interest in order, for example, to consider the scaling theory for localization in more than one dimension (Anderson 1981). Here we shall be mainly interested in the application of this formulation to mesoscopic situations, for example, to the resistance of a small piece of wire or a small ring-type structure. In particular, the sensitivity of those structures to magnetic fields will be of major interest to us. We shall thus start by reviewing the general multichannel conductance formulation, emphasizing the two-terminal vs. four-terminal aspects (Anderson et al. 1980, Azbel 1981, Anderson 1981, Fisher and Lee 1981, Langreth and Abrahams 1981, Büttiker et al. 1985). In the following subsection we shall review the generalization due to Büttiker of the two- to several-terminal case, with emphasis on the Onsager-type reciprocity relationships. The rest of the chapter will be devoted to applications.

We also point out that this formulation is applicable to many other problems such as the scanning-tunneling microscope (Binning et al. 1982) as well as various interface resistances (Castaing and Nozières 1980, Uwaha and Nozières 1985).

Many generalizations are possible beyond straightforward ones such as that to phonon transport: for example, thermal and thermoelectric transport (Sivan and Imry 1986), the inclusion of inelastic processes in the system itself (Büttiker 1985a, 1986a), the Hall effect (Entin-Wohlman et al. 1986, Büttiker 1988), and various types of noises. The generalization to finite frequencies (Büttiker 1993) and that including Coulomb interactions are important problems as well.

The Multichannel Landauer Formulation

We now consider the multichannel and finite-temperature generalizations of the Landauer formulation, depicted in Fig. 5.3. The leads feeding into the general elastic scattering system S are now ideal wires with a finite cross-section A. Due to the quantization in the transverse direction leading to discrete transverse energies E_i, we now have $N_\perp$ conducting channels at the fermi energy E_F, each characterized at zero temperature by a longitudinal wave vector k_i (and velocity $\hbar k_i/m = v_i$), so that

$$E_i + \frac{\hbar^2 k_i^2}{2m} = E_F, \qquad i = 1, \ldots, N_\perp. \tag{5.8}$$

$N_\perp = Ak_F^2/2\pi$ for a 2D cross-section, and $N_\perp = 2Wk_F/\pi$ for a 1D cross-section of width W, both including spin. At finite temperatures the values of k_i acquire a finite thermal width. The incoming channels (right-going on the l.h.s. and left-going on the r.h.s.) are fed from electron baths with chemical potentials μ_1, μ_2, and the overall temperature is T (the case with $T_1 \neq T_2$ is discussed

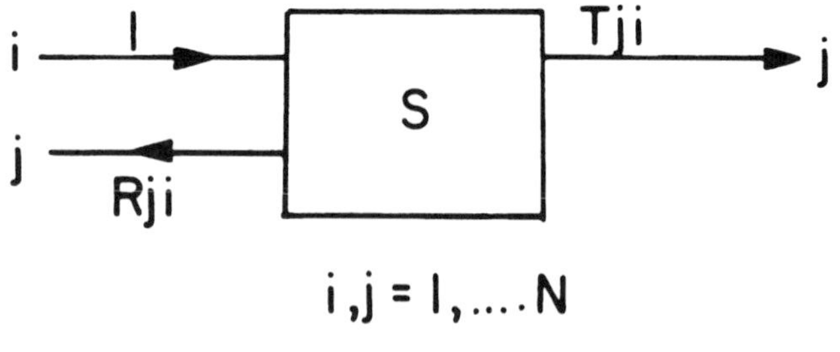

Figure 5.3 A multichannel scatterer S. An incoming wave in channel j from the left with amplitude 1 has probabilities R_{ij} and T_{ij} to be reflected or transmitted into the ith l.h.s. or r.h.s. channel, respectively.

by Sivan and Imry (1986) and will be briefly mentioned later). We assume that the *outgoing* channels from each reservoir are fed up to a *thermal equilibrium population.* As long as scattering has not occurred, this assumption can be justified (Sivan and Imry 1986) along the same lines as the proof (Landau and Lifschitz 1959) that the *outgoing* radiation in a given solid angle from a black body has a thermal equilibrium distribution. This proof utilizes the Liouville theorem. It is *convenient* to assume that particles *reaching* the "sink" reservoir via the ideal lead are totally absorbed there. This is not obviously correct, since electrons reaching the reservoir much below its Fermi energy do not have vacant states to go into. However, if they are reflected, they will just contribute to the outgoing streams from the "sink" reservoir and reduce the rate of particles emanating from it. Therefore this assumption is effectively justified and it is irrelevant whether these electrons "really" enter the sink reservoir or not. In equilibrium, the above assumptions yield a vanishing *net* current, and this is really a dynamic equilibrium situation, in which fluctuating "noise" currents flow. Those will be seen in chapter 8 to be consistent with the known thermal equilibrium noise. We also assume that there are no phase relationships among electrons in different channels (i.e., that the channels behave like "incoherent sources") and consider the case where $\mu_1 - \mu_2$ is small enough to insure a linear transport regime. The system S scatters in the following fashion: An incoming wave (see Fig. 5.3) from the left jth channel has probabilities $T_{ij} = |t_{ij}|^2$ and $R_{ij} = |r_{ij}|^2$ for transmission into the r.h.s. ith channel and reflection into the l.h.s. ith channel, respectively. The analogous matrices for incoming waves from the r.h.s. are denoted by primes. The $2N_\perp \times 2N_\perp$ matrix S given by

$$S = \begin{pmatrix} r & t' \\ t & r' \end{pmatrix} \tag{5.9}$$

is unitary due to current conservation, because the T_{ij}, R_{ij} matrices transform the lead *currents.* Furthermore, when time-reversal symmetry holds,

$$SS^* = I, \qquad S = \tilde{S}, \tag{5.10}$$

where the star denotes complex conjugation, the tilde the matrix transpose, and I is the unit matrix. When a magnetic field is present, the second relationship in eq. 5.10 becomes $S(H) = \tilde{S}(-H)$.

We define the total transmission and reflection probability into the ith channel by

$$T_i = \sum_j T_{ij}, \qquad R_i = \sum_j R_{ij}, \tag{5.11}$$

with similar definitions for the primed quantities. The unitarity conditions imply

$$\sum_i T_i = \sum_i (1 - R_i), \tag{5.12}$$

with a similar condition for the primed quantities.

We also note, for completeness, that the more detailed equalities

$$R_i' + T_i = 1, \qquad R_i + T_i' = 1 \tag{5.13}$$

are valid only between transmission *to* the right (left) and reflection *from* the right (left). They also state that if all incident channels on both sides of the barrier are fully occupied, all outgoing channels will also be fully occupied. Further unitarity conditions follow from the columns of S. There are additional conditions (see eq. 5.10) that exist when time-reversal symmetry holds.

As long as they satisfy the above constraints, the elements of S can otherwise be completely arbitrary and depend, in principle, on energy (although this is a small effect at low temperatures). We note that our assumptions on the incoming channels plus the known matrix S completely determine all the distributions of the outgoing channels. These are rather out of equilibrium, since there are no processes that give the usual "shifted Fermi distribution," or even transfer electrons among the channels to equalize their chemical potentials. Nevertheless, these distributions are precisely known. One can straightforwardly calculate all currents, electron densities, energy densities, entropy densities, and so on. Since there are good methods for computing S for given models (such as a tight-binding Anderson one), this formulation is very suitable for numerical computation.

Before briefly presenting the derivation of the conductance, we point out that these assumptions are not the only possible ones. For example, Langreth and Abrahams (1981) had assumed that the various channels on the leads reach a common chemical potential on each side, presumably via electron–electron interaction, and then dropped the assumption of Fermi distributions in the input channels. While we think that our assumption stated above is quite reasonable and is, in fact, analogous, as mentioned above, to the known distribution of photons coming out of a photon bath (Landau and Lifschitz 1959), we certainly cannot rule out different physical situations, where the assumptions of Langreth and Abrahams might apply.

Since the densities of states in the channels are 1D-like, and given, including spin, by

$$n_i(E) = (\pi \hbar v_i)^{-1}, \tag{5.14}$$

we write the current on the r.h.s. as

$$\begin{aligned} I &= \frac{e}{\pi\hbar} \sum_i \int dE [f_1(E) T_i(E) + f_2(E) R_i'(E) - f_2(E)] \\ &= \frac{(\mu_1 - \mu_2) e}{\pi\hbar} \int dE \left(-\frac{\partial f}{\partial E} \right) \sum_i T_i(E), \end{aligned} \tag{5.15}$$

where the velocities canceled with the density of states factors, and in order to get the last equality we used eq. 5.12, staying in the linear transport regime. It is straightforwardly checked that the current on the l.h.s. is also equal to I (conservation of current). The conductance measured between the *the outside reservoirs*, similarly to eq. 5.7, is thus given by

$$\begin{aligned} G_c &\equiv \frac{I}{\mu_1 - \mu_2} \\ &= \frac{e^2}{\pi\hbar}\int dE\left(-\frac{\partial f}{\partial E}\right)\sum_i T_i(E) \underset{T\to 0}{\longrightarrow} = \frac{e^2}{\pi\hbar}\sum_{ij} T_{ij} \\ &= \frac{e^2}{\pi\hbar}\mathrm{tr}\, tt^{\dagger}, \end{aligned} \tag{5.16}$$

where at the zero-temperature limit everything is evaluated at E_F.

Equation 5.16 is the *correct* expression for the two-terminal conductance (Imry 1986b). Much of the earlier discussion in the literature on "What is the correct Landauer formula" was based on misunderstanding this. Equation 5.16 implies that one is looking at the flowing current divided by the *electrochemical potential difference between the "source" and "sink" reservoirs.* G_c thus includes the contact resistances between the reservoirs and the system. It will therefore be finite, and *at most* equal to $N_\perp e^2/\pi\hbar$, obtained when the system and its contacts with the reservoir are ideal (Imry 1985, 1986b). After the beautiful experimental observations (Van Wees et al. 1988, Wharam et al. 1988) of the "quantized" conductance of narrow links ("quantum point contacts") between two 2D gases, a lot of work went into elucidating what the conditions are for this ideal behavior. It seems that an essential practical condition for observing good apparent "steps" in the conductance, due to opening of channels (having a larger number of transverse states below E_F, with increasing electron density in the narrow wire), is a gradual opening of the lead to the reservoir. This gradual opening should guarantee an "adiabatic" condition for the early stages of the motion in the widening channel. This would cause the reflections, due to breaking of this adiabaticity, to be small (Yacoby and Imry 1990). For observing the conductance steps (Imry 1985, 1986b) the temperature should be lower than the separation of the quantized thresholds of the channels. For the original GaAs samples, this necessitated temperatures in the subkelvin range, but very quickly such effects could be seen at tens of kelvins. For atomic-sized constructions (provided they have good transmission) the temperature range could exceed room temperatures. In fact, very recently Costa-Krämer et al. (1995), in a remarkable experiment, observed such effects at room temperature with naturally formed atomic-sized contacts (see also Lang 1987).

There are various ways to consider the more complex issue of four-terminal resistances, which should characterize the sample *itself*. The simplest way to arrive at an idealized expression for the conductance of the sample *itself*, defined by $I/(\mu_A - \mu_B)$, is to define μ_a and μ_B as follows. The electron density on the l.h.s. is

$$n_l = \frac{1}{2\pi\hbar}\int dE \sum_i \frac{1}{v_i}[(1+R_i)f_1 + T_i f_2]. \tag{5.17}$$

If the electron gas on the l.h.s. were in equilibrium, with a chemical potential μ_A and Fermi function f_A, its density would be

$$n_A = \frac{1}{\pi\hbar}\int dE \sum_i f_A(E)/v_i. \tag{5.18}$$

We define μ_A so that $n_l = n_A$, which is a very reasonable definition. It also has the advantage that the Einstein relation will be automatically satisfied (actually, this definition of $(\mu_A - \mu_B)$ is equivalent to the Einstein relation). Using the value of $(\mu_A - \mu_B)$ thus obtained, one arrives at the results

$$G = \frac{I}{\mu_A - \mu_B} = \frac{2e^2}{\pi\hbar} \frac{\int dE \frac{\partial f}{\partial E} \sum_i T_i(E)}{\int dE \sum_i \frac{\partial f}{\partial E}[1 + R_i(E) - T_i(E)]/v_i}$$

$$\xrightarrow[T\to 0]{} \frac{2e^2}{\pi\hbar} \frac{\sum T_i \sum 1/v_i}{\sum(1 + R_i - T_i)/v_i}. \tag{5.19}$$

The above definition of μ_A and μ_B applies for noninteracting electrons. For real electrons in a real conductor with Coulomb interactions, we know that *charge neutrality must prevail* over length scales larger than the screening length λ (see also section 4.2). Assuming that the ideal conductors on the two sides of the sample are large enough, this will happen via self-consistent potentials δV_A and δV_B that *must* be generated to keep the electron density equal to its equilibrium (charge neutrality with the positive ions) value $\bar{n}$. Thus, $-e\delta V_A = \frac{\partial\mu}{\partial n}(n_A - \bar{n})$. Obviously, this amounts to a change in the *electrochemical* potential, bringing it up to μ_A. Similar remarks apply to V_B and μ_B. The difference in electrostatic potential $V_A - V_B$ *can be measured by capacitative methods*. The measurement of $\mu_A - \mu_B$ is more subtle and will be discussed later. We emphasize, however, that the above discussion provides a satisfactory justification for the fundamental physical validity of eq. 5.19 as the basic four-terminal conductance. The measurability of $V_A - V_B$ by capacitative methods has repeatedly been emphasized by Landauer 1989c, private communication, 1990b); see also Payne (1989).

In the important case where all the T_i's are small, $T_i \ll 1, 1 + R_i \simeq 2$, and the difference between G and G_c is small. This should be applicable for large

$N_\perp$ whenever $G \ll e^2 N_\perp/\hbar$, or sample length $L \gg l$. This clearly happens with a wide margin near the localization transition where $G \sim e^2/\pi\hbar$. The zero-temperature limit of eq. 5.19 was first obtained by Azbel (1981) from apparently similar assumptions. The derivation was later clarified and substantiated by Büttiker et al. (1985). The generalization to finite temperatures was considered by Büttiker (1985a,b and personal communication) and by Sivan and Imry (1986 and unpublished results), who also discussed thermoelectric transport. Even the zero-temperature limit in eq. 5.19 does not agree (except in the important case where the transmissions are small enough for G_c to be a good approximation to G) with all the other multichannel results which existed previously in the literature (Abrahams et al. 1979, Anderson et al. 1980, Anderson 1981, Fisher and Lee 1981, Langreth and Abrahams 1981). This can easily be appreciated by noting that even in the simple independent channel case, eq. 5.1 does not reduce to the parallel addition form (which will look like $\sum_i T_i(1 + R_i - T_i)^{-1}$). The reason is, of course, that the latter also assumes a common electrochemical potential difference for all channels. The disagreement with Langreth and Abrahams (1981) has already been discussed above. Discussions overlapping the one presented here and by Azbel (1981) existed previously in the context of the problem of Kapitza resistance (Castaing and Nozières 1980, Uwaha and Nozières 1985). It is also important to note that eq. 5.19 is similar to the finite-temperature single-channel result of Engquist and Anderson (1981). The latter can be obtained with a small modification, even from the zero-temperature version of eq. 5.19, by regarding the different energies as (a continuum of) independent channels. It is indeed not of the parallel addition form (for a caveat, see Sivan and Imry 1986).

Engquist and Anderson (1981) also introduced the conceptual method by which the chemical potentials μ_A and μ_B could be measured. This is accomplished in principle by bringing in two "measurement reservoirs" with temperature T and with adjustable chemical potentials μ'_A and μ'_B. These are now allowed to exchange electrons with the l.h.s. and r.h.s. wires, respectively. This exchange must be weak, in order not to disturb the system! Now, one adjusts, say, μ'_A until no *net* current flows between the appropriate measurement reservoir and the l.h.s. wire. Then, one applies the same procedure to μ'_B. By very general principles, it is at this point where the chemical potentials of the measurement reservoirs are equal to those of the measured systems

$$\mu'_A = \mu_A, \qquad \mu'_B = \mu_B.$$

This clearly accomplishes a four-terminal measurement. The voltmeter (measuring $\mu_A - \mu_B$) does not draw any current. Such a measurement would second the capacitative one mentioned above. Although it is not easy to achieve, it yields the correctly defined voltage across the sample.

Three subtleties have to be handled here. First, as already mentioned, the coupling of the measurement reservoir to the system has to be weak enough. Otherwise, although it draws no *total* current, it might affect some interchannel electron transfer by drawing current from some channels and sending it into others (Castaing and Nozières 1985, private communication). This may include transferring electrons from right-moving to left-moving channels. Note that this effect (which has to be avoided in the conceptually correct experiment, but *might well exist* in a given experimental system) contributes toward inter-channel equilibration. (This will be, in principle, an undesirable "invasive measurement" limit where the coupling is strong enough for the measurement process to affect the measured property, i.e., the chemical potential of the system.) Such effects exist in general in the four-terminal generalization by Büttiker (1986b), to be discussed later, which does not require a weak coupling to the bath. It is an interesting and an unavoidable characteristic of our mesoscopic system that normally irrelevant details of the measurement process may thus affect the results.

Second, we have to understand the requirement of the energy dependence of the coupling between the measurement reservoir and the system. Obviously, we do not want the measurement to distinguish between right- and left-moving electrons, different channels, and energies. In order to obtain the detailed condition on the coupling, we take the reservoir to have a density of states $n_r(E)$ and to be coupled to the jth channel of the system with matrix elements $V_j(E)$. The condition of zero net current from the reservoir μ_A to the system, using the Fermi golden rule, reads

$$\int dE \sum_i n_r(E)|V_i(E)|^2 f_A(E)[2 - f_1(1 + R_i(E)) - f_2 T_i'(E)]/v_i$$

$$= \int dE \sum_i \frac{1}{v_i} [f_1(1 + R_i) + f_2 T_i'] n_r(E)(1 - f_A(E))|V_i(E)|^2. \quad (5.20)$$

The l.h.s. gives the current from the reservoir to the system as the integral of the density of available electrons times the absolute values squared of the matrix elements, times the final density of states, times the density of available holes in the system. The r.h.s. similarly yields the current from the system to the reservoir. Using the relationship in eq. 5.13 and comparing the relationship in eq. 5.20 with the one obtained by equating eqs. 5.17 and 5.18, we find after some algebra that the two definitions of μ_A are equivalent, provided $n_r(E)|V_i(E)|^2$ is *independent of E and of the channel number i* (Sivan and Imry 1986 and unpublished results). It might be argued that this is a nontrivial condition to satisfy in a strict fashion. For a large number of channels, however, the important requirement is that there be no *systematic* variation of $n_r(E)|V_i(E)|^2$ with either E or i. The effects of such variations when they are *random* will tend to average out. Thus, it stands to reason that a real measurement may qualify in this respect and give an unbiased determination of $\mu_A - \mu_B$. However, at this stage there is no proof of this and it may well need some further averaging. In particular, the measurement should be done on a spatial scale much larger than both the wavelength and the screening length, in order to avoid obvious oscillations on the scale of the wavelength and to agree with the charge neutrality consideration (and with the electrostatic measurement) discussed above. For further discussion of G_c vs. G and the issue of probes, see Landauer (1989c).

Third, the quantity G evaluated here is an *effective conductance* in the sense of being the ratio of the measured current to the measured voltage. At finite temperatures, however, it was shown (Sivan and Imry 1986) that the above G may also have a (usually small) thermoelectric component. The reason for this is that even when the temperatures of the outside reservoirs, T_1 and T_2 are equal, the temperatures that are measured on the two sides of the sample, T_A and T_B (analogous to μ_A and μ_B) will, in general, be different, at a finite overall temperature. The current I may thus have a component due to the nonvanishing of $T_A - T_B$. This observation is relevant only when the denominator of the conductance formula (eq. 5.19) is important (i.e., when the approximation $G \simeq G_c$ is not valid. The condition for G_c to be a good approximation to G is (as mentioned before) that $G \ll N_\perp e^2/\hbar$, which is equivalent to the sample length L being much longer than the elastic mean free path l.

The Onsager-type Relationship in a Magnetic Field: Generalized Multiterminal Conductance Formulas

One of the interesting aspects of the multichannel four-terminal formula (eq. 5.19) is that the various constraints and symmetries in the general case do *not* guarantee the validity of the Onsager-type relationships among transport coefficients (Onsager 1931). For example, the relation $G(H) = G(-H)$ is not guaranteed to hold (although for the two-terminal conductance, $G_c(H) = G_c(-H)$ *is valid*, see below). Büttiker and Imry (1985) constructed specific examples with small numbers of channels where

$$G(H) \neq G(-H). \tag{5.21}$$

Numerical calculations by Stone (1985) on larger-size disordered models also produced this asymmetry. Such "“asymmetry” has also appeared in experiments on mesoscopic systems and there were indications that it might be related, if no magnetic impurities (see Shtrikman and Thomas 1965) are present, to sample inhomogeneity and the four-terminal nature of the experiment (Von Klitzing 1985, personal communication[1]).

The naive Onsager relationship (eq. 5.21) holds, in fact, for G_c, the "two-terminal" conductance between the outside reservoirs. To obtain this, one

[1] Von Klitzing has pointed out to the author that the Onsager relation implies $\sigma(H) = \sigma(-H)$, which impies $G(H) = G(-H)$ only for homogeneous systems. otherwise, the (antisymmetric in H) Hall part of the conductance tensor might come in and contribute to the effective G. The author is indebted to von Klitzing for pointing this out to him.

notes that by generalizing the time-reversal symmetry relation eq. 5.10 to a finite magnetic field, one finds

$$T_{ij}(H) = T_{ji}(-H). \tag{5.22}$$

It follows (Büttiker and Imry 1985) that indeed

$$G_c(H) = G_c(-H). \tag{5.23}$$

Thus the apparent "deviation" from the Onsager relation for G simply means that G, being a four-terminal object, should *not* have the naive symmetry $G(H) = G(-H)$. This reflects the need to consider the correct predictions of the Onsager symmetry for a four-probe measurement. A similar need arises for the four-terminal thermoelectric case (Sivan and Imry 1986).

This was considered very early by Casimir (1945) and the results were also proved, assuming the conductivity to be local (which is not necessarily the case in our situation), by Sample et al. (1987). For a review of the four-probe technique see van der Pauw (1958). Büttiker (1986) considered a generalized n-terminal Landauer-type conductance and confirmed the existence of the correct Onsager symmetries. Let us consider $n = 4$. Recall that the current between two reservoirs, 1 and 2, coupled to the system via the ideal leads is given in obvious notation by

$$I_{1\to 2} = \sum_{ij} T_{ij}^{1,2}(\mu_1 - \mu_2) = G_c^{12}(\mu_1 - \mu_2), \tag{5.24}$$

where eq. 5.23 for G_c^{12} is a statement of time-reversal symmetry. Suppose now that four probes with chemical potentials $\mu_1, \ldots, \mu_4$, are connected to the system. Since we have linear transport and carriers from different reservoirs are incoherent, the total current *from* the ith reservoir is found by adding all three contributions:

$$I_i = \sum_{j \neq i} G_c^{ij}(\mu_i - \mu_j). \tag{5.25}$$

This is a set of four linear equations relating the four currents $I_1, \ldots, I_4$ to the four chemical potentials. It may conveniently be expressed in matrix form, in obvious notation:

$$I = \hat{G}\mu \tag{5.26}$$

where $\hat{G}$ is a 4×4 matrix. This matrix is singular—its determinant vanishes. This is clear, since the vector having four equal μ components is an eigenvector of $\hat{G}$ with zero eigenvalue—no current flows in equilibrium. This implies that the sum of the elements in each row of G vanishes. Thus, eq. 5.26 does *not* have a solution for an arbitrary vector $\boldsymbol{I}$. Physically, charge conservation demands that the sum of all four components of $\boldsymbol{I}$ should vanish. Mathematically this follows because the sum of all elements in a column of $\hat{G}$ is equal to zero, by unitarity. Thus, only vectors $\boldsymbol{I}$, having a vanishing sum of all of their components, are allowed. Since time-reversal symmetry implies eq. 5.23, the set of circuit equations is *identical* to that of a four-probe classical conductor. The fact that the G_c^{ij} themselves may include coherence effects is *irrelevant* as far as symmetries are concerned. The Onsager symmetries for such a situation were generally obtained and analyzed by Casimir (1945). Büttiker (1986) followed Casimir's formulation and confirmed that the Onsager symmetries are valid, as they *must* be, according to the above. The four- (or in general n-) probe formulation has proved to be extremely useful for many situations.

It is advantageous (Casimir 1945) to take the following current configuration: Take $\{k\ l\ m\ n\}$ to be a permutation of $\{1\ 2\ 3\ 4\}$ and $I_k = -I_l \equiv J_1$, $I_m = -I_n \equiv J_2$. Solving eq. 5.26, J_1 and J_2 can be expressed in terms of the voltages $eV_1 = \mu_k - \mu_l$ and $eV_2 \equiv \mu_m - \mu_n$ as follows:

$$\begin{pmatrix} J_1 \\ J_2 \end{pmatrix} = \begin{pmatrix} \alpha_{11} & -\alpha_{12} \\ -\alpha_{21} & \alpha_{22} \end{pmatrix} \begin{pmatrix} V_1 \\ V_2 \end{pmatrix}, \tag{5.27}$$

where, as found by Büttiker (see appendix H for the case $k = 1$, $l = 3$, $m = 2$, $n = 4$) the matrix elements α_{ij} are given by

$$\begin{aligned}
\frac{h}{e^2}\alpha_{11} &= -\sum_{p\neq k} T_{kp} - (T_{kn} + T_{km})(T_{nk} + T_{mk})/S, \\
-\frac{h}{e^2}\alpha_{12} &= (T_{km}T_{ln} - T_{kn}T_{lm})/S, \\
\frac{h}{e^2}\alpha_{21} &= (T_{mk}T_{nl} - T_{ml}T_{nk})/S, \\
\frac{h}{e^2}\alpha_{22} &= -\sum_{p\neq l} T_{mp} - (T_{mk} + T_{ml})(T_{lm} + T_{km})/S,
\end{aligned} \tag{5.28}$$

$$S \equiv T_{km} + T_{kn} + T_{lm} + T_{ln} = T_{mk} + T_{nk} + T_{ml} + T_{nl}, \tag{5.29}$$

where $T_{rs} = G_c^{rs} = \mathrm{tr}\ t_{rs}t_{rs}^{\dagger}$. It follows immediately that

$$\alpha_{ij}(H) = \alpha_{ji}(-H), \tag{5.30}$$

which is the *appropriate Onsager symmetry* for this four-terminal conductance matrix. Equation 5.27 can be inverted to yield

$$\begin{pmatrix} V_1 \\ V_2 \end{pmatrix} = \frac{1}{\det} \begin{pmatrix} \alpha_{22} & \alpha_{12} \\ \alpha_{21} & \alpha_{11} \end{pmatrix} \begin{pmatrix} J_1 \\ J_2 \end{pmatrix}, \tag{5.31}$$

where $\det = \alpha_{11}\alpha_{22} - \alpha_{12}\alpha_{21}$. One is now in a position to discuss four-probe resistance measurements. Such a measurement, with k, l being the current terminals and m, n the voltage ones, is done by taking $J_1 = J_{kl}, J_2 = 0$ (no current in the voltmeter circuit, to avoid irrelevant contact resistances) and then $V_2 = V_{mn}$,

$$V_{mn} = R_{kl,mn} I_{kl}. \tag{5.32}$$

Thus

$$R_{kl,mn} = \frac{\alpha_{21}}{\det} = -\frac{T_{lk}T_{nm} - T_{lm}T_{nk}}{(\alpha_{11}\alpha_{22} - \alpha_{12}\alpha_{21})S} \tag{5.33}$$

is the proper four-terminal resistance, with k, l being the current terminals and m, n the voltage ones. It is seen at once that (see van der Pauw 1958)

$$R_{kl,mn}(H) = R_{mn,kl}(-H), \tag{5.34}$$

which is the *proper Onsager symmetry* for the four-terminal resistances. Obviously $R_{kl,mn}(H) \neq R_{kl,mn}(-H)$ in general. Thus it is not surprising that the simple Landauer four-terminal expression eq. 5.19 does not have this symmetry!

The above is obviously a generalization of the Engquist–Anderson type approach to eq. 5.19, to allow an arbitrary coupling strength of the voltage probes to the system. While it has the disadvantage that it cannot be regarded as a "noninvasive" measurement of the "resistance of the system itself," since the contacts play a role, it has two advantages. (a) It is extremely useful, since most of the transport experiments are currently done with voltage probes coupled to the system via lithographically made contacts, which are definitely not "noninvasive." (One must appreciate, however, that this is not a law of nature and that hopefully less invasive measurements are already possible, in principle, using, for example, STM-type contacts. It remains to be seen whether the required uniformity of the coupling to the different channels can be achieved. Alternativelly, as discussed before, capacitative measurements of the *electrostatic* potential are possible). (b) This formulation treats current and voltage probes on the same footing and it is thus possible to treat the proper Onsager symmetry (eq. 5.34), unlike the treatment of eq. 5.19 in which the voltage probes must be special.

It is possible to play more games with the above formulation by, for example, writing each four-terminal resistance as a sum of two parts, even and odd in H, and combining those between dual probe configurations. This is reviewed by Büttiker (1988) and Benoit et al. (1987a,b). The multiprobe formulation is also extremely convenient for describing nonlocal effects. For example, a current between k and l produces a voltage between another pair of contacts, m and n, a distance L away. As one might expect, such effects persist for $L \lesssim L_\phi$.

3. APPLICATIONS OF THE LANDAUER FORMULATION

Series Addition of Quantum Resistors, 1D Localization

We now consider, following Landauer (1970), the effect of two obstacles in series (see also problem 4). We denote the phase change of the wave passing through the constant potential (ideal conductor) between the obstacles, by ϕ. The waves in the region between the obstacles are the sums of all the multiply scattered waves yielding a total left-going wave and a right-going one. An amplitude A is reflected and D transmitted through the whole device. A, B, C, D are complex numbers. The wave just emerging from obstacle 1 is $Be^{i(kx-\omega t)}$, and it acquires a phase ϕ upon impinging on obstacle 2. The wave C suffers a similar phase change from 2 to 1. The barrier equations are

$$\begin{aligned} A &= r_1 + Ct_1, \qquad & B &= t_1 + Cr_1', \\ Ce^{-i\phi} &= Be^{i\phi}r_2, \qquad & D &= Be^{i\phi}t_2. \end{aligned} \tag{5.35}$$

Solving these equations, we find

$$D = \frac{e^{i\phi}t_1 t_2}{1 - e^{2i\phi}r_2 r_1'},$$

which yields the transmittance T_{12} of the device:

$$T_{12} = \frac{T_1 T_2}{1 + R_1 R_2 - 2\sqrt{R_1 R_2}\cos\theta}, \tag{5.36}$$

where $\theta = 2\phi + \arg(r_2 r_1')$, and, denoting $T_{12} = T, R_{12} = R$,

$$\frac{R}{T} = \frac{R_1 + R_2 - 2\sqrt{R_1 R_2}\cos\theta}{T_1 T_2}. \tag{5.37}$$

Assume now that we have an ensemble of systems prepared, with all of them having similar values of both R_1 and R_2. However, the "optical" phase difference, ϕ, in different members of this ensemble ranges over many intervals of 2π, with a uniform probability, so that the average of $\cos\theta$ is zero.[2] The average of the inverse of the dimensionless conductance, g,

[2] This will be the case if the average distance between the obstacles is much larger than the wavelength of the electrons. Their ratio is $\sim 10^2$ for $300A$ of a ballistic metallic wire.

$$g \equiv \frac{G}{e^2/\pi\hbar}, \tag{5.38}$$

is thus given by

$$(g^{-1})_{av} = \frac{R_1 + R_2}{(1 - R_1)(1 - R_2)}. \tag{5.39}$$

This result is already quite surprising. Ohm's law of series addition of resistances, $g^{-1} = g_1^{-1} + g_2^{-1} = R_1/(1 - R_1) + R_2/(1 - R_2)$ is not valid in general! It is only valid in the limit of good transmission, or small resistance, $R_i \ll 1$. This has very serious and important consequences that have not been easy to accept initially. In fact, if one combines good transmittances ($R \ll 1, T \sim 1$) in series, the resulting resistance, G^{-1}, first increases linearly with n as it should, but once n is so large that the total transmittance is smaller than unity, then

$$(g^{-1})_{av,n+1} = \frac{R_n + R}{T_n} = (g^{-1})_{av,n} + \frac{R}{T_n}. \tag{5.40}$$

Thus, we add a good transmittance ($R \ll 1$) as the $(n + 1)$th element to the chain of n such elements and the resistance increases by $R/T_n > R$. One may form a "renormalization group" (RG)-type equation for the length-scale (n) dependence of $(g^{-1})_{av,n}$,

$$\frac{d}{dn}(g^{-1})_{av,n} = R[(g^{-1})_{av,n} + 1], \tag{5.41}$$

so that the (dimensionless) resistance, after having increased linearly with n to $O(1)$, will then increase *exponentially* with the length n. This is the phenomenon of 1D localization, as was discussed in chapter 2.

The above is not entirely satisfactory, however, as already remarked by Landauer. The distribution of the resistances in the ensemble is not narrow and thus, as emphasized by Anderson et al. (1980), the results depend on *what quantity is being averaged.* Anderson et al. also pointed out the proper way to average for large n. One needs an object that will behave like an ordinary extensive quantity with both average and mean-square average increasing linearly with n. Such a quantity is $\ln(1 + g^{-1})$. This is so, because $1 + g^{-1} = 1 + R/T = 1/T$ so that $\ln(1 + g^{-1}) = -\ln T$. $-\ln T$ plays the role of the extinction exponent and one would expect it to be additive for two scatterers if the relative phase is averaged on. Indeed, from eq. 5.37 one finds, using (Anderson et al. 1980)

$$\int_0^{2\pi} d\theta \, \ln(a + b\cos\theta) = \pi \ln \tfrac{1}{2}[a + (a^2 - b^2)^{1/2}], \tag{5.42}$$

that $\langle \ln T_{12} \rangle = \ln T_1 + \ln T_2$. Thus, the exact scaling of the 1D resistance with n is given by

$$\langle \ln(1 + g_n^{-1}) \rangle = \rho_1 n, \tag{5.43}$$

with ρ_1 being the resistance, in units of $\pi\hbar/e^2$ of a single obstacle. This indeed increases first linearly and then exponentially with n, differing only quantitatively from the results of eq. 5.41. This establishes the 1D localization, manifested in a measurable quantity—the resistance. The localization of all eigenstates in 1D (Mott and Twose 1961, Borland 1963) is well known and has been rigorously proved. (All this is, of course, at "zero" or appropriately low temperatures, as before. At finite temperatures, one returns to the considerations of section 2.4 and combines resistances for $L \gg L_\phi$ classically.)

Interesting effects may exist in the transport through a given sample on top of the average behavior, as discussed by Azbel (1981, 1983; Azbel and Soven 1983). For a given, finite, system if the energy (and hence the optical path difference between scatterers) is varied, the phase changes, θ, will be modified and as a result T will vary too. In fact, T will show sharp "transmission" resonances (see problem 4). These may show up at low temperatures as sharp oscillations of the resistance as a function of the electron density (which is variable in a MOSFET device; Ando et al. 1982) or the magnetic field (if the system is not strictly 1D).

Now, we are also in a position to give another interpretation of the Landauer T/R result for large transmission coefficients (Imry 1981a). Given a $T \simeq 1$, $R \ll 1$ we combine n such obstacles so that T_n is a number C smaller but on the order of unity. Thus, $|\ln C| = n|\ln T| \simeq nR$, but as long as C is small eq. 5.7 still roughly holds for $g_n \sim T_n = C$. Since Ohm's law roughly holds as long as $g \gtrsim 1$ we can now obtain g_1—the conductance of a single obstacle—as ng_n (since $n \sim 1/R, C \sim 1, T \simeq 1$); that is,

$$g_1 = nC = O(1)\frac{T}{R}, \tag{5.44}$$

which agrees within an order of magnitude with the Landauer result, eq. 5.6.

Parallel Addition of Quantum Resistors, A–B Oscillations of the Conductance

The next resistance addition problem we shall treat is that of resistors in parallel. Here, too, we find that the quantum effects cause the usual classical addition law to be invalid, once the system is coherent ($L_\phi >$ length of the system). An extreme case occurs when one of the parallel conductances vanishes ("open stub"). Such a dead-end may still influence the total transmission of the device. When an A–B flux is introduced in the region between the two conductors, oscillations with period h/e follow. It is simplest to

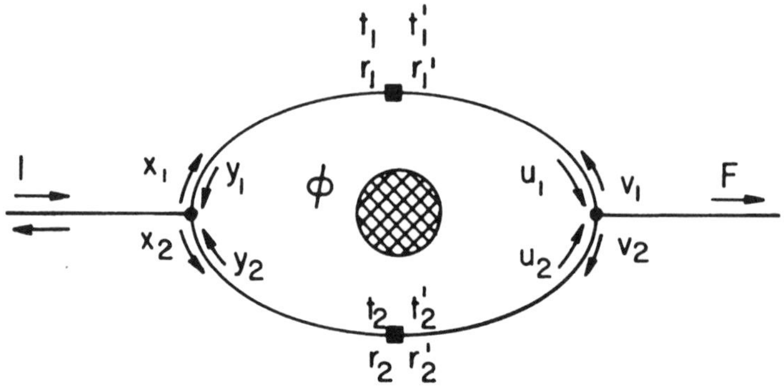

Figure 5.4 Schematic picture of the parallel resistors (ring) system. The arrows denote the various transmitted and reflected amplitudes, defined close to the junctions. The phases accumulated through the channels are absorbed in the scattering coefficients (r_i, r_i', t_i, t_i').

consider this problem using the Landauer formulation with single-channel conductors (Gefen et al. 1984a).

The geometry of the system is described in Fig. 5.4. Each branch of the ring is described schematically as a single scatterer connected to an ideal, mathematically one-dimensional channel. All phases and scattering effects along the channels are absorbed in the parameters describing each scatterer. These parameters are t_i and t_i', the transmission amplitudes from the left and from the right, respectively, and $r_i(r_i')$, the reflection amplitudes on the left (right) of the scatterer $(i = 1, 2)$. Notice that time-reversal and current conservation requirement, which imply $t_i = t_i'$ and

$$-t_i/t_i'^* = r_i/r_i'^* \tag{5.45}$$

(the asterisk denotes complex conjugation), are also satisfied when the phases of each path are absorbed in t_i, etc. Moreover, when an A–B magnetic flux Φ is applied through the center of the ring, the usual (appendix C) gauge transformation for the transmission and reflection amplitudes yields $t_1 \to t_1 e^{-i\theta}$, $t_1' \to t_1 e^{+i\theta}$, $t_2 \to t_2 e^{+i\theta}$, $t_2' \to t_2' e^{-i\theta}$, $r_i \to r_i$, $r_i' \to r_i'$ $(\theta \equiv \pi\Phi/\Phi_0)$, and the transformed t's and r's still satisfy eq. 5.45. Following Shapiro (1983a) each three-terminal junction is described, for example, by a 3×3 unitary scattering matrix S,

$$S = \begin{pmatrix} 0 & -1/\sqrt{2} & -1/\sqrt{2} \\ -1/\sqrt{2} & 1/2 & -1/2 \\ -1/\sqrt{2} & -1/2 & 1/2 \end{pmatrix}, \tag{5.46}$$

where the diagonal elements, S_{ii} $(i = 1, 2, 3)$ denote the reflection amplitude of the ith channel, and the off-diagonal elements $S_{ij}(i \neq j)$ are the transmission amplitudes from channel i to j. In Fig. 5.4, channel 1 of the left-hand junction is chosen to be that of the incoming amplitude (unity) whereas channel 1 of the right-hand junction is that of the outgoing amplitude (F). In this example no reflection occurs in channel 1 and there is symmetry between channels 2 and 3. We do not expect our results to depend qualitatively on the choice of the junction's scattering matrix, except for the trivial effect that these junctions are themselves scatterers and add to the total resistance of the device. For a particular model showing resonances, see Büttiker et al. 1976.

Writing down the linear relationships among the various amplitudes at the junctions and scatterers and using sum and difference variables (e.g., $x_1 \pm x_2$, etc.), we find after some algebra that the total transmission amplitude of the ring is given by

$$F = 2\frac{t_2 t_2(t_1' + t_2') + t_1(r_2 - 1)(1 - r_2') + t_2(r_1 - 1)(1 - r_1')}{(t_1 + t_2)(t_1' + t_2') - (2 - r_1 - r_2)(2 - r_1' + r_2')}. \tag{5.47}$$

This can be rewritten as

$$F = 2\frac{Ae^{i\theta} + Be^{-i\theta}}{De^{+2i\theta} + Ee^{-2i\theta} + C}, \tag{5.48}$$

where $A = t_1^2 t_2 + t_2(r_1 - 1)(1 - r_1'), B = t_1 t_2^2 + t_1(r_2 - 1)(1 - r_2'), D = E = t_1 t_2$, $C = t_1^2 + t_2^2 - (2 - r_1 - r_2)(2 - r_1' - r_2')$. The transmitted intensity which determines the conductance via the Landauer formula is (using $\phi \equiv 2\theta$)

$$T \equiv |F|^2 = 4\frac{\alpha + \beta\cos\phi + \beta'\sin\phi}{\gamma + \delta\cos\phi + \delta'\sin\phi + \epsilon\cos 2\phi + \epsilon'\sin 2\phi'}, \tag{5.49}$$

where $\alpha = |A|^2 + |B|^2$, $\beta = 2\ \mathrm{Re}(AB^*)$, $\beta' = -2\ \mathrm{Im}(AB^*) \equiv 0$, $\gamma = |D|^2 + |E|^2 + |C|^2$, $\delta = 2\ \mathrm{Re}(DC^* + EC^*)$, $\delta' = -2\ \mathrm{Im}(DC^* - EC^*) \equiv 0$, $\epsilon = 2\ \mathrm{Re}(DE^*)$, $\epsilon' = -2\ \mathrm{Im}(DE^*) \equiv 0$. It is straightforwardly checked that β', δ', and ϵ' vanish identically, as they should by Onsager symmetry.

Let us first consider the case where no magnetic flux is present ($\phi = 0$). Even in this case T may exhibit oscillations as a function of the phases of the t's and r's (which influence the coefficients $\alpha, \beta, \delta, \epsilon$ in eq. 5.49). When $t_1 = 0$, an appropriate choice of the phases of r_1 and r_1' may result in $T = 1$ or $T = 0$. That is, by tuning the nonconducting branch we can dramatically affect the transmission through the other channel. This effect is present also when $|t_1|$ and $|t_2|$ are finite. In particular, we can obtain $T = 0$ even when $|t_1| \ll |t_2|$. On the other hand, one can improve the conductance of a scatterer by connecting to it in parallel a very poor, tunable, conductor. Notice that these resonances will disappear once the inelastic diffusion length $L_\phi = \sqrt{D\tau_\phi}$ (τ_ϕ is the inelastic dephasing time) becomes of the order of the size of the ring, L. Thus we may obtain dramatic changes (either increase or decrease) of G with the temperature. Another interesting effect may occur if we expose one of the channels (say channel 1 with $|t_1| \ll |t_2|$) to, for example, electromagnetic fields whose effect is similar to inelastic scattering. This may cause a dramatic change in the transmission of the weakly scattering channel. When $t_1 = t_2 = t$ eq. 5.47 reduces to $F = t$. We expect this result to hold also for an n-branch system. Thus, when the temperature increases, the conductance should increase from

$$G_t = \frac{e^2}{\pi\hbar}|t|^2/(1-|t|^2)$$

to Ohm's law $G = nG_t$.

We now turn on a magnetic flux Φ. In general T is periodic in Φ with a period Φ_0, in agreement with the Byers–Yang theorem (appendix C). A finite first harmonic (periodicity of $\Phi_0/2$) exists in general, as do higher harmonics. These oscillations may be very strong even in the limit of strong scattering ($l_{el} \ll L \ll L_\phi$). Assume, for example, that $|t_1| \sim |t_2| \sim t \ll 1$. In that case, unless very special phase relations hold, $\alpha \sim \beta \sim \delta \sim t^2$, $\epsilon \sim t^4$ and $\gamma \sim 1$. These are then oscillations with a period Φ_0 and an amplitude $\sim |t|^2$ appearing on a constant background of order $|t|^2$ (thus the oscillations are as large as the average, and T may vanish for certain values of ϕ). In addition, there are also (first harmonic) oscillations of a period $\Phi_0/2$ and a smaller amplitude $\sim |t|^4$. If $|t_1| \ll |t_2|$ the relative size of the oscillations becomes smaller (and again, the harmonics with a period $\Phi_0/2$ are even smaller). These oscillations have received ample experimental confirmation (Webb et al. 1985a,b, Chandrasekhar et al. 1985, Datta et al. 1985), as shown in Fig. 5.5, taken from Webb et al. (1985b). The key for observing the h/e oscillation in the experiments was the separation of the field scales for them and for the "slow" fluctuation, to be discussed later.

Prior to the above calculation and its experimental confirmation, a seemingly contradictory effect existed. One of the most interesting predictions of the weak localization theory has been that by Altshuler, Aronov, and Spivak (AAS) (1981a) on periodic oscillations of the (Kubo-type) conductance of rings or cylinders of small diameter (but having many conducting channels) as a function of the Aharonov–Bohm flux, Φ, through their opening. One surprising aspect of this calculation has been that the fundamental period of the oscillations was not $\Phi_0 = h/e$, as the general Byers–Yang theorem demands, but $\Phi_0/2$. The $\Phi_0/2$ period is the "first harmonic" of the Φ_0 oscillation. Thus this periodicity does not contradict the above theorem. The question is only why the fundamental, Φ_0, periodic does not appear in those calculations.

Before answering this question we mention that the prediction of the $\Phi_0/2$ oscillation has received, starting with the beautiful pioneering work by Sharvin and Sharvin (1981), very convincing experimental support (Altshuler et al. 1982b, Ladan and Maurer 1983, Gordon 1984, Gijs et al. 1984). In the more

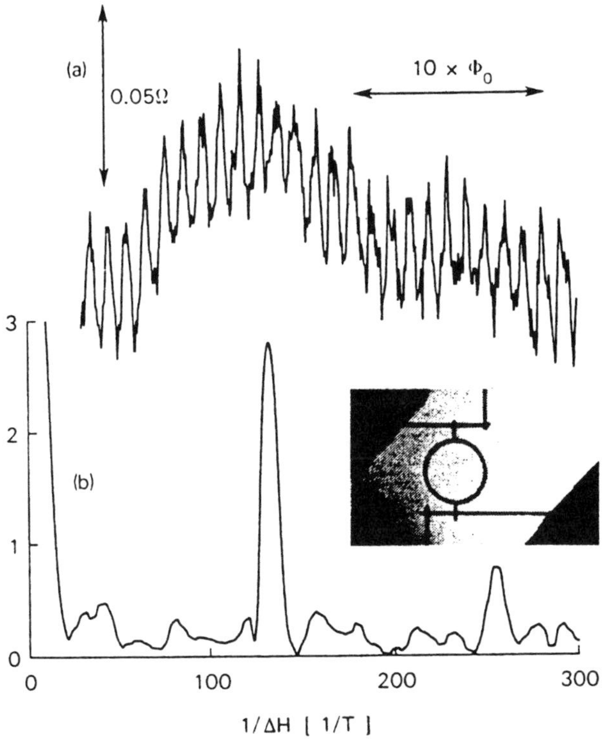

Figure 5.5 (a) The magnetoresistance of the gold ring, shown in the inset and having an inside diameter of ~8000 Å and a width of ~400 Å. The arrow corresponds to 10 flux quanta in the hole of the ring. (b) Fourier power spectrum, in arbitrary units, showing peaks at h/e (and $h/2e$) corresponding to the visible fast oscillations. The low-frequency peak corresponds to the slow modulation due to the flux in the "arms" of the ring which is a conductance fluctuation, to be discussed later. (From Webb et al. 1985b.)

recent experiments on long cylinders an almost quantitative agreement with the full theory (taking into account the non-Aharonov–Bohm magnetic field inside the material) was achieved. The $\Phi_0/2$ oscillation has also been clearly seen in experiments on large arrays of many small "rings" (Pannetier et al. 1984, 1985, Bishop et al. 1985, Licini et al. 1985a, 1985b, Dolan et al. 1986). In all these experiments the fundamental period, Φ_0, has not been seen. Preliminary experiments (Umbach et al. 1984, Webb et al. 1984) on single rings were inconclusive, but did show traces of perhaps both Φ_0 and $\Phi_0/2$ oscillations, with an additional very important aperiodic structure (the slow modulation of Fig. 5.5) that will be discussed later. Convincing Φ_0-periodic oscillations in single small rings were only reported subsequently (Webb et al. 1985a,b; Washburn et al. 1985; Chandrasekhar et al. 1985; Datta et al. 1986 and much later work).

The answer to the dilemma of where the Φ_0-periodic oscillations are in the many experiments mentioned and in the weak localization theories is the following (Gefen 1984, private communication, see also Browne et al. 1984; Büttiker et al. 1985, Murat et al. 1986, Imry and Shiren 1986, Stone and Imry 1986): Both the theory of Altshuler et al. (1981a) and the experiments on cylinders and arrays involve effectively an ensemble averaging over many microscopically distinct systems prepared with the same overall macroscopic conditions. Thus, all rings in the array have similar average impurity concentrations, but the precise configuration of the impurities is obviously different. In the cylinder experiments, the resistance is measured *along* an approximately 1-cm-long cylinder which consists of around 10^4 pieces of length L_ϕ added classically in series. In the perturbative theoretical calculations, one performs ensemble averaging from the very beginning in order to use propagators that depend only on relative distances (apart from boundary effects). The work reviewed above on single rings with contacts suggests that the Fourier coefficient corresponding to the Φ_0-periodic part of the oscillation does not have a definite phase. On the other hand, the AAS $\Phi_0/2$ Fourier coefficient does have a definite phase (e.g., $G(\Phi)$ is minimal[3] at the origin $\Phi = 0$; Altshuler et al. 1982b, Bergmann 1984, Lee and Ramakrishnan 1985). This definite phase is due to the nature of the weak-localization correction as discussed in section 2.6 (see also the analysis of the classical paths in section 4.2). In order for it to be flux-sensitive, a path must encircle the ring at least once. The time-reversed path will encircle the ring in the opposite sense. Their sum will add in phase at $\Phi = 0$ without spin–orbit scattering and will thus be maximal (minimum conductance) at $\Phi = 0$. For $\Phi \neq 0$ these two paths will acquire phases of $\pm\phi = 2\pi\Phi/\Phi_0$ and the sum will oscillate with a period $\Phi/2$, but with a definite decrease from $\phi = 0$ (with no spin–orbit scattering). A beautiful demonstration of the qualitative validity of this picture was given by Altshuler et al. (1982b), where replacing magnesium by lithium (which has a smaller spin–orbit scattering) changed the phase of the $\Phi_0/2$ oscillations in the expected fashion. Thus, the ensemble averaging (*if done on a broad enough ensemble*) eliminates the Φ_0 fundamental component, but the AAS $\Phi_0/2$ one survives. Therefore, experiments on single rings (Webb et al. 1984, 1985a,b, Washburn et al. 1985, Chandrasekhar et al. 1985; Datta et al. 1986) were needed to see the h/e period.

These experiments had in fact been preceded, following an initial insight by Gefen (1984, personal communication), by model calculations (published later) demonstrating the need for single rings to observe the h/e oscillation, by Imry and Shiren (1986) on the Kubo conductivity of closed 1D rings and by Murat, Gefen and Imry (1986) on 1D rings with contacts and, later, by Stone and Imry (1986) for multichannel rings with contacts. Results from Murat et al. are displayed in Fig. 5.6, in which, instead of ensemble averaging, the conductivity was calculated at a series of increasing temperatures T. It turns out that in 1D, once $k_B T \gg \Delta$, different electrons in the "thermal band" of width $k_B T$ around E_F have different phases associated with propagation around the ring. Thus, high enough temperatures, at which the relevant energy scale in 1D is the level

[3] Note that for systems with strong spin–orbit scattering the minimum becomes a maximum.

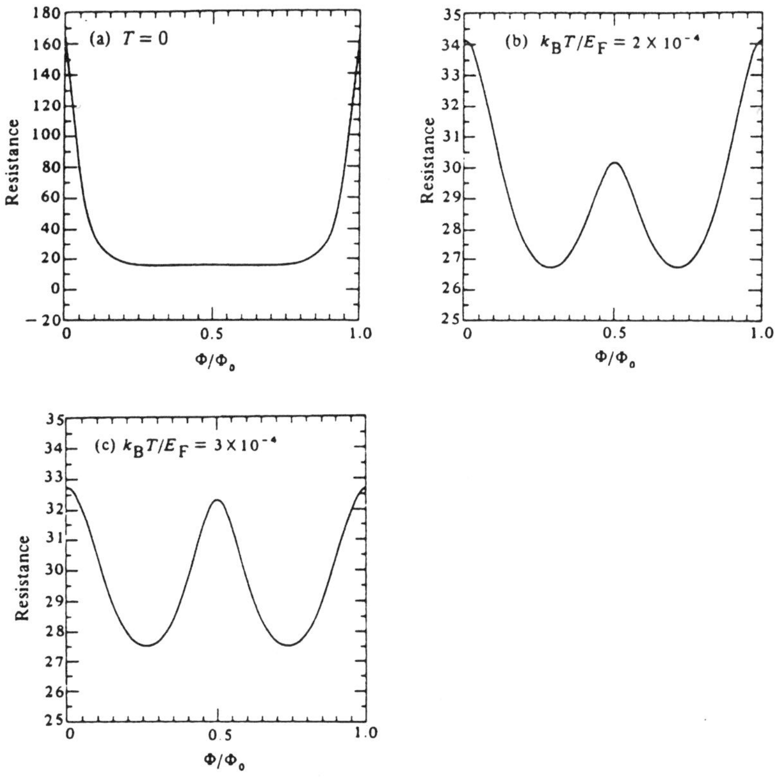

Figure 5.6 Similar to Figs. 4 and 5 of Murat et al. (1986). Resistance as a function of flux for a 1D ring with long arms with scatterers, for three temperatures, exhibiting the self-averaging-out of the Φ_2-periodic component at higher temperatures.

separation Δ, provides self-averaging which is similar to ensemble averaging. This will be discussed further below. However, a glance at Fig. 5.6 is enough to appreciate the better averaging out of the Φ_0-periodic component obtained with increasing T. The idea that ensemble averaging may lead to a $\Phi_0/2$ fundamental periodicity was also discussed by Carini et al. (1984) and Browne and Nagel (1985). However, since they considered a certain static quantity (participation ratio) averaged over a *whole band*, they also found that the Φ_0-periodic component decreased like $1/L$ (L being the length of the system) even for a *single* ring. Averaging over the whole band is like thermal averaging with $k_B T$ comparable with the *electron band width, which is usually very large.* This is not applicable to the low-temperature conductivity.

The results shown in Fig. 5.6 demonstrate the gradual tendency for "effective ensemble averaging" with increasing temperature. At low temperatures we see the basic periodicity of Φ_0, but at higher temperatures the basic period reverts to $\Phi_0/2$. The crossover is obtained, in 1D, when $k_B T$ becomes comparable to the level separation. This should be expected, since consecutive levels in 1D are defined by having $O(2\pi)$ more phase variation along the whole system. It was also confirmed that the effectively ensemble-averaged result (c) is insensitive to changes of phases of the scatterers which strongly influence the Φ_0 oscillation.

Such "energy-averaging" is effective also at higher dimensions. Generally, there will exist some characteristic energy scale ΔE, so that energies differing by ΔE will have significant differences in the interference along the system. In 1D, $\Delta E \sim \Delta$ since there is no diffusive metallic regime. Once $g \lesssim 1$ and $L \gtrsim l$, we have localization (chapter 2). An important question is what is the "energy correlation range," ΔE, in more general cases. ΔE is in fact generally of the same nature as the Thouless parameter E_c which, as discussed in chapter 2, measures the sensitivity of the energy levels to boundary conditions (i.e., a phase difference across the system, or an Aharonov–Bohm type flux through the ring). One would thus heuristically identity the "energy correlation range" ΔE with E_c. This follows from the discussion following eq. 4.16, according to which the phase of a diffusive path around the system increases by $O(\pi)$ for an energy change of E_c. This argument was first given by Stone and Imry (1986), who also demonstrated the validity of $\Delta E \sim E_c$ numerically for the multi-channel case. This is also consistent with the results of Lee and Stone (1985) in the weakly localized regime for the related fluctuation problem, to be discussed in the next section. Thus, the condition that the thermal "Fermi" smearing will *not* wash out interference effects, such as the Φ_0-periodic component in a ring, is

$$k_B T \lesssim E_c. \tag{5.50}$$

Using the Thouless relation $E_c = \hbar D/L^2$, eq. 5.50 is found to be equivalent to

$$L \leq \sqrt{D\hbar/k_B T} \equiv L_T, \tag{5.51}$$

that is, the sample is shorter than the thermal length L_T defined by eq. 5.51. We note that τ_ϕ is typically larger by one to two orders of magnitude than $\hbar/k_B T$ in many real systems (especially in metals, related to the validity of the Fermi-liquid theory; these two times appear to approach similar orders of magnitude in some very dirty systems). Comparing the condition 5.51 with the one requiring that L be smaller than the inelastic diffusion length, we find that our condition in eq. 5.51 is more restrictive, but not hopelessly severe. This is especially so because, for $k_B T \gg E_c$, averaging may be expected to reduce the relative oscillation by $\sim(E_c/k_B T)^{1/2}$. For example, for a 500 Å × 1500 Å gold wire with a length of 5000 Å and a resistance of 20 Ω, $\Delta \sim 1$ mK and $E_c \sim 0.05$ K. Hence the condition in eq. 5.51 is experimentally feasible and the effect should be reduced by only a factor of 3 at ~ 0.5 K by energy averaging. This is the range of parameters relevant for several experiments.

Until now we have discussed only the effect of the Aharonov–Bohm type flux through the opening of the ring. One might also enquire about the effect of the magnetic field in the material. In fact, the initial experiments on rings, where a substantial fraction of the magnetic flux penetrated their arms, as well as similar experiments on singly-connected fine lines, showed aperiodic fluctuations (Umbach et al. 1984) in the resistance of these systems. The random-appearing structure was reproducible for a given system as long as the latter was not effectively annealed. This is the same as the "slow" structure of Fig. 5.5, which is also an aperiodic conductance fluctuation. The magnetic field scale of this structure corresponds to a flux of the order of a flux quantum through the wire. This is reminiscent of the oscillations found by Dingle (1952) in the equilibrium properties of free electrons in the geometry of, say, a disk, as function of the flux through it. One might imagine (Blonder 1984) that the aperiodic nature of the resistance change in the real system might be due to the random specific stacking of impurities and defects in a given system. It will then follow that each given stacking should produce as its own "fingerprint" a specific $R(H)$ curve.

The multichannel conduction formula provides an ideal tool to quantitatively check the above idea. A disordered tight-binding Anderson model can be used to represent the (noninteracting) system. It is possible to calculate the S matrix for a given model numerically either by multiplying the transfer matrices (Pichard and Sarma 1981a,b; Azbel 1983) and variations thereof, or by the Green's function method of Thouless and Kirkpatrick (1981), generalized to 2D by Fisher and Lee (1981). With the latter method it is reasonable to reach 2D models of $\sim 40 \times 400$ sites, for example. The convergence properties of this method appear to be appropriate for the cases of interest here. These calculations have been performed by Stone (1985). The effect of the magnetic field was taken into account by modifying the phases of the matrix elements of the Hamiltonian so that the sum of all phases around each loop is given by the flux inside it ("Aharonov–Bohm effect in each loop"). The elements of the transfer matrix are changed accordingly. Typical results (Stone 1985) where the computer experiments and real experiment have the field scales determined by the condition of a flux quantum through the system are depicted in Fig. 5.7. The large difference in the vertical scale, with the fluctuations in the computer experiments larger by more than two orders of magnitude than in the real experiments, is due mainly to the different channel numbers (about 40 in the former, 2×10^4 in the latter), leading to different resistances. This and many other similar results obtained by Stone (1985) provide convincing evidence that the physics of the aperiodic reproducible oscillations is indeed the modification of the electron interference by the magnetic flux through the system.

The qualitative picture is as follows: Each t_{ij}, for example, can be obtained as a sum over all paths through the sample of the transmission amplitude from i to j via the given path, which is clearly an interference phenomenon. A magnetic field having a flux through the system on the order of Φ_0 changes the relative phases of the most distant paths by the order of 2π. This determines the field scale on which the resistance may fluctuate. These ideas are discussed

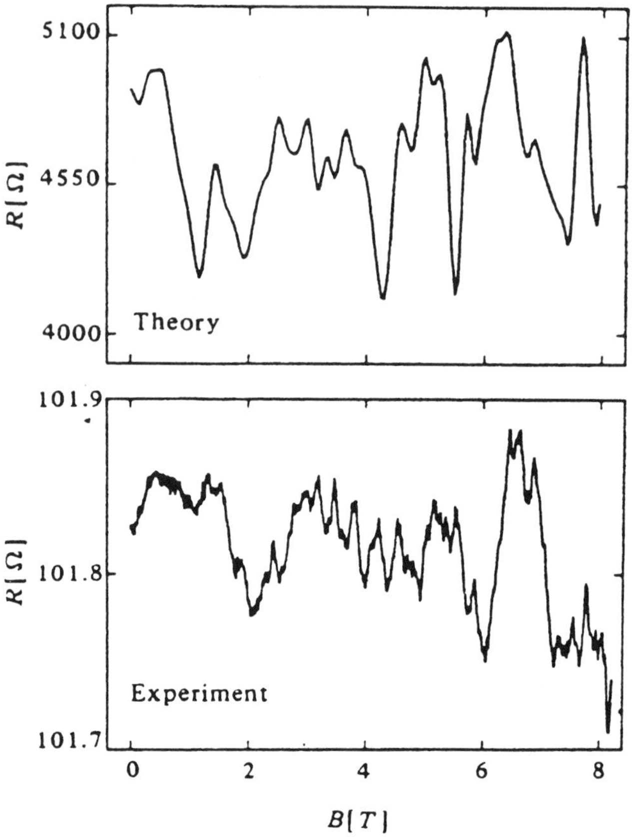

Figure 5.7 Resistance as a function of flux for a small wire. Theory and experiment normalized to have similar horizontal scales. Vertical scales are discussed in the text. (From Stone 1985.)

more fully by Stone (1985) and Stone and Imry (1986). The concept of the energy correlation range, E_c, alluded to above, emerges very clearly from these numerical results. More general ideas (Altshuler 1985; Lee and Stone 1985; Altshuler and Khmelnitskii 1985; Imry 1986; Lee et al. 1986) on the magnitude of these conductance fluctuations will be discussed in the next subsection.

We conclude this section by summarizing the differences between the h/e and AAS $h/2e$ oscillations in rings. The former, being sample-specific, is sensitive to all sorts of ensemble averaging, including energy averaging. Since the latter is not energy sensitive, and is already the result of averaging, it is observed without the "contamination" of the aperiodic conductance fluctuations. On the other hand, a result of the experiment of Webb et al. (1985b) has been that the Φ_0-periodic oscillations did not appear to fade with increasing magnetic field. In fact, they persisted to $\sim 10^3$ oscillations with no noticeable weakening. This was quite surprising, given expectations that the non-Aharonov–Bohm flux inside the arms would make Φ ill-defined and eventually smear the structure, *as in fact happens* with the AAS-type $\Phi_0/2$-periodic oscillations (Altshuler et al. 1981a, 1982c, Sharvin and Sharvin 1981). It is possible, however, at least in the case where the scales of the periodic (Aharonov–Bohm) and aperiodic (due to the field inside the material) structures are well separated, to give a heuristic argument (Stone and Imry 1986) for the persistence of the periodic structure in the Φ_0-periodic component.

The magnetic field applied to the real ring will create both an Aharonov–Bohm flux Φ through the ring's opening and a (classically relevant) flux Φ_c through the ring's arms. The ratio of these two fluxes is just a geometrical factor

$$\frac{\Phi}{\Phi_c} = A. \tag{5.52}$$

For H perpendicular to the plane of the ring, A will be the aspect ratio—the ratio of the area of the hole to that of one of the arms. Now recall that, for example, by the results of Stone (1985), the scale of changes of Φ_c which markedly alters the conductance of an arm of the ring is $\Delta\Phi_c \sim \Phi_0$. Thus, for $A \gg 1$ it follows that when H is changed so as to span a range Φ_0 of Φ, Φ_c will change only by Φ_0/A. This will cause just a very small change of the background contribution to the conductance due to Φ_c, which will result in a slow variation of $G(H)$ on top of which the faster oscillation due to Φ will occur. Further analysis (Stone and Imry 1986) along the lines of the Landauer formula with the introduction of the effects of Φ_c indeed reveals that the Φ_0-periodic oscillation survives the existence of Φ_c even when $\Phi_c \gg \Phi_0$, although Φ_c causes slow (determined by $\Delta\Phi_c \sim \Phi_0$) amplitude and phase modulations of the oscillation. This is also consistent with numerical simulations, and with the diagrammatic calculations (Altshuler 1985, Lee and Stone 1985) as well as with the experiment (Webb et al. 1985b).

The Φ_0-periodic oscillation is (Büttiker et al. 1985, Stone and Imry 1986; see next section) *the result of an almost canceling addition of many random-phased terms*. This is the essential reason why this oscillation survives the large magnetic fields in the ring's arms. The AAS contribution is, as discussed above, of a fundamentally different nature. It is obtained from the addition of coherent terms. Once these are made incoherent by a large magnetic field (or by, say, a random magnetic field due to static magnetic impurities), this contribution is greatly reduced. Interestingly, there should also exist a typically smaller (Stone and Imry 1986, Gefen et al. 1984b), $\Phi_0/2$-periodic contribution which is the first harmonic of the Φ_0-periodic one and is also due to incoherent terms and *not* to the special coherent-backscattering AAS-type ones. This contribution, due to two unrelated paths encircling the ring in opposite directions, should also survive large magnetic fields. The immunity of the Φ_0-periodic type oscillation to (uniform or random) fields is well summarized by the statement that "you cannot kill a dead horse."

It is interesting to note how these processes have different sensitivities to the various parameters and thus each is observable under different experimental conditions. A final remark is in order about the strengths of these phenomena. At higher temperature, when $L_\phi \ll L$, the h/e and AAS oscillations should die out, respectively, like e^{-L/L_ϕ} and e^{-2L/L_ϕ} due to the double path length of the latter contribution. It turns out that the aperiodic fluctuations die out more slowly with increasing temperature.

On the Universality of the Conductance Fluctuations

The conductance of different samples belonging to the same "impurity ensemble" (i.e., having the same average disorder but differing in the detailed defect arrangements) varies from sample to sample. Let us stay in the diffusive metallic regime ($l \ll L \ll \xi$, or, equivalently, $1 \ll g \ll N_\perp$), discuss the quasi-1D situation (otherwise only some numerical factors change), and use the two-terminal Landauer formula for the dimensionless conductance, g,

$$g = \text{tr}\, tt^\dagger = \sum_{ij} T_{ij}. \tag{5.53}$$

For multiterminal situations this discussion applies to the conductance between a given pair of contacts. The conductance fluctuation of each sample is

$$\Delta g \equiv g - \langle g \rangle, \tag{5.54}$$

where $\langle\ \rangle$ denotes an ensemble-average. What about the mean square fluctuations $\langle \Delta g^2 \rangle$? Since it is by itself an ensemble-averaged quantity, it can be calculated diagrammatically. This was done by Altshuler (1985) and by Lee and Stone (1985). The unexpected, extremely intriguing result was that

$$\langle \Delta g^2 \rangle = C, \tag{5.55}$$

where C is a universal constant depending only on the effective dimensionality and the general symmetries (see below), *not* on the conductance g itself and certainly not on other microscopic details of the system or the defects!

It was found that $c \simeq 0.862$ for quasi-1D systems with time-reversal symmetry and with no spin–orbit coupling. C has somewhat different but universal values for 2D and 3D and depends on the magnetic field and the spin–orbit scattering (which break the symmetry of the system from orthogonal to unitary (Dupuis and Montambaux 1991) or symplectic, respectively). The result is valid when the sample is in the coherent quantum limit

$$L \ll L_\phi, L_T. \tag{5.56}$$

But otherwise the result can be simply estimated by, for example, dividing the sample into smaller "coherent" parts and adding them classically. This remarkable result has been corroborated by numerical computations and by experiment. The fact that a property of all dirty systems with the same general shape is the same at low temperatures is even more remarkable than the universality in the critical behavior of extremely pure and carefully handled systems. It is possible (Altshuler and Khmelnitskii 1985) to consider, more generally, the correlation function of the conductance taken as a function of E_F and the magnetic field, H,

$$F(\Delta E_F, \Delta H) = \langle g(E_F, H) g(E_F + \Delta E_F, H + \Delta H)\rangle - \langle g\rangle^2, \tag{5.57}$$

which also has universal behavior. The correlation range in E is E_c and that for H corresponds to a flux quantum in the area of the system, as may be expected. For a strong enough H the time-reversal symmetry is broken and C is multiplied by 1/2. Universal changes occur also with spin–orbit scattering.

In the Landauer-type formulation, this universality is intimately connected with the universal correlations in the spectrum of the transmission matrix. This will be discussed briefly in appendix I. Here we give a very simplified and nonrigorous version of the argument.

Consider first (Büttiker et al. 1985) one typical $T_{ij} = |t_{ij}|^2$. t_{ij} is given by a sum of an exponentially large number $\mathcal{N}$, of terms due to the various paths donated by m, connecting input channel j with output channel i. Thus, apart from normalization,

$$T_{ij} \propto \sum_{1}^{\mathcal{N}} 1 + \sum_{m \neq n} e^{i(\phi_m - \phi_n)}, \tag{5.58}$$

where the first term is the "diagonal" classical one and the second is the quantum interference one, whose ensemble average is zero. However, the typical value of the second term, which is a sum of $\mathcal{N}^2$ random contributions ($\mathcal{N} \ggg 1$) is $O(\mathcal{N})$, the same as the classical term! This argument and the rather surprising result that the relative fluctuation in T_{ij} is of order unity were known and appreciated by Rayleigh in the theory of scattering of light from random media.

The question is now how many independent T_{ij}'s are there? One might argue that there should be $N_\perp$ independent "conduction channels" and thus $N_\perp^2$ independent T_{ij}'s. Therefore, $\sqrt{\langle \Delta g^2}/g \sim 1/N_\perp$. This is, however, not the case. For very long systems, $L \gg l$, most of the eigenvalues are exponentially small and drop out of the game. For a conductance $g \sim N_\perp l/L$ the number of "effective conducting channels," having transmission coefficients close to unity (Imry 1986), is

$$N_{eff} \sim \frac{\xi}{L} \sim g \sim N_{\perp} \frac{l}{L}. \tag{5.59}$$

This is obtained by writing the eigenvalues as $\exp(-L\mu_n)$ and taking the inverse localization lengths, μ_n, to be, roughly, equally spaced, as explained in appendix I. The physical localization length, ξ, is such that its inverse is the smallest of the μ_n's. One expects that $\langle \Delta g^2 \rangle$ is only due to those eigenvalues which are relevant, that is, to the fluctuation in the number of eigenvalues in an interval which has N_{eff} eigenvalues on average. Replacing $N_{\perp}$ by N_{eff}, one obtains

$$\langle \Delta g^2 \rangle = O(1), \tag{5.60}$$

which is the required result.

A more convincing presentation of this argument, having to do with eigenvalue repulsion ("spectral rigidity"; Dyson 1962), is briefly discussed in appendix I. One writes g as a "linear statistic" of the eigenvalues of a certain random matrix. The assumption that the spectrum of this matrix obeys random matrix theory (RMT) rules yields, in fact, a constant universal value for $\langle \Delta g^2 \rangle$. This is close, though not exactly equal, to the right one. There has been much work on the precise spectral properties of the relevant matrix (see, for example, Muttalib et al. 1978, Mello 1988, Mello and Pichard 1989, Mello 1990, Pichard 1991, Macêdo and Chalker 1992, 1994, Stone et al. 1991, Slevin et al. 1993, Jalabert et al. 1993). Recently Beenakker and Rejaei (1993) found the needed distribution, which is not exactly RMT and yields the precise value of C for the quasi-1D geometry. At any rate, this point of view, even though approximate, does give one some insight as to the origins of the universality, why it is sensitive to some symmetry breakings, and how it can be qualitatively carried over into, for example, the strongly localized regime. For a discussion of the conduction paths through the disordered system, see Oakshott and MacKinnon (1994).

Problems

1. (a) Use the result of eq. D.1 at small q to obtain the matrix elements of $\hat{x}$ and $\hat{p}$ between exact eigenstates.
   (b) Show that these matrix elements lead with the Kubo formula to the usual Drude σ.
   (c) Show that these same matrix elements can be used to obtain the "sensitivity to boundary conditions" and the Thouless relation.
   (d) Show that the above matrix elements are also obtained (Thouless) from the postulate that the eigenstates are random phase mixtures of $\boldsymbol{k}$ states in a shell of "thickness" $\hbar/\tau$ around the Fermi surface.

2. For a small metallic particle, use the Kubo formula and the results of problem 1 above to get the $\sigma(\omega)$ at low frequencies, assuming level width $\gtrsim$ level spacing.
3. Sketch $\sigma(\omega)$ for a small particle in the localized regime, with the size of the particle, $L \gg \xi$ (loc. length).
4. (a) Discuss the effect of resonant tunnelling on the conductance of a tunnel barrier using the Landauer approach and a simple 1D model (see Baym 1969, p. 104, for a general reference).

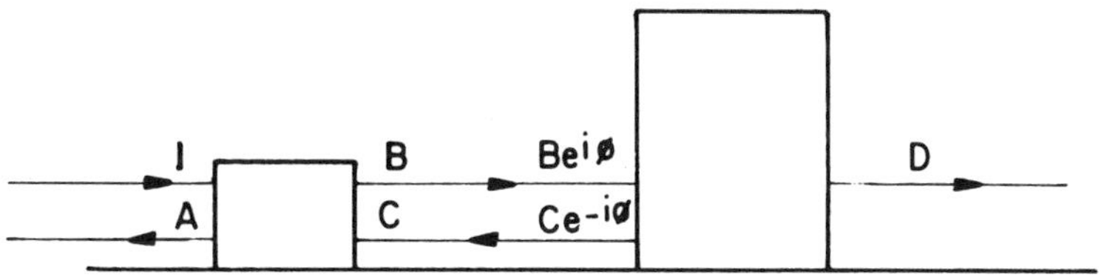

The well inside the barrier has a single quasi bound state at E_0, having lifetimes due to tunneling to the left and right leads of τ_l and τ_r, respectively. Calculate first the transmission coefficient $T(E)$ through this barrier as a function of E and explain physically the results. Then, obtain the conductance, G, for given μ_1 and $\mu_2 = \mu_1 - eV$. When E_0 can be changed with a gate, sketch G as a function of that gate voltage. Discuss how, when $E_0 > \mu_1 > \mu_2$ initially, the resonance will contribute to the nonlinear conductance $G(V)$.

   (b) Using the above, discuss qualitatively the effect of localized states on the conductance in a 1D disordered chain at low temperatures (Lifschitz and Kirpichenkov 1979, Azbel 1981).
5. The "Coulomb blockade": Suppose that $\mu_2 < E_0 < \mu_1$, but that the energy to put a second, opposite-spin electron in the state of the well is $E_0 + U$. U is called "the Hubbard repulsion" and is often parametrized as $U = e^2/2C$ where C is an effective capacitance. Show that when $E_0 + U > \mu_1 > \mu_2 > E_0$, the conductance is exponentially small. What is the activation energy? What value(s) of V will cause resonant conductance? Generalizations of this simple phenomenon have caused a lot of recent interest. For references see Likharev and Zorin (1985), Ben Jacob and Gefen (1985), Averin and Likharev (1991) and Grabert and Devoret (1992).

6

The Quantum Hall Effect

1. INTRODUCTION

While the quantum Hall effect (QHE) is almost a macroscopic phenomenon (except that it occurs for 2D electrons), its physics has much in common with many aspects of mesoscopic physics. The quantized Hall current is very similar to a persistent current over large length-scales, and many other considerations are related as well. Thus, we devote this chapter to the QHE. Small-size effects on the QHE will be an interesting research direction.

Let us consider magnetotransport in a 2D electronic system. The elementary Drude result for the resistivity tensor $\tilde{\rho}$, giving the fields

$$\begin{pmatrix} E_x \\ E_y \end{pmatrix}$$

in terms of the current densities as

$$\begin{pmatrix} \rho_{xx} & \rho_{xy} \\ \rho_{yx} & \rho_{yy} \end{pmatrix} \begin{pmatrix} j_x \\ j_y \end{pmatrix}$$

is

$$\hat{\rho} = \begin{pmatrix} \sigma_0^{-1} & B/nec \\ -B/nec & \sigma_0^{-1} \end{pmatrix}, \tag{6.1}$$

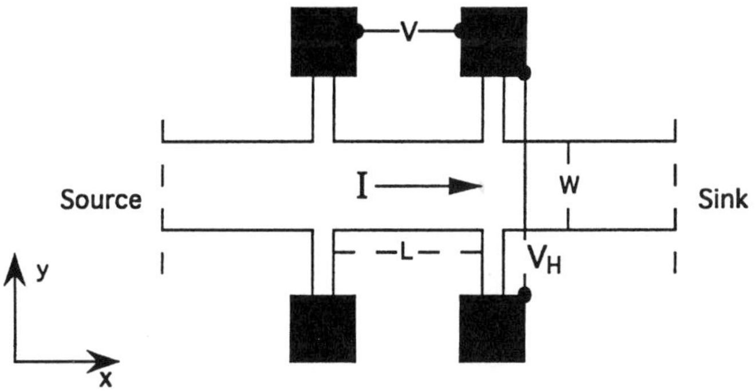

Figure 6.1 A schematic magnetoresistance and Hall effect measurement in a typical Hall bar. A current I is supplied from "source" to "drain." V_H is measured *across* the current, $V_L = V$ *along* it.

σ_0 being the Drude conductivity $\sigma_0 = ne^2\tau/m$. In the usual experiment we have a "Hall bar" with a supplied current I_x, and $I_y = 0$. These conditions imply (where V_H and V_L are defined in Fig. 6.1)

$$E_y = -\frac{B}{nec} j_x, \qquad V_H = -\frac{B}{nec} I_x, \tag{6.2}$$

$$E_x = j_x/\sigma_0, \qquad V_L = \frac{L}{W}\frac{I_x}{\sigma_0}, \tag{6.3}$$

for a sample with width W and length L (see Fig. 6.1). B is perpendicular to the plane of the figure along the positive z direction. Assuming homogeneity, $I_x = j_x W$, the Hall voltage is $V_H = WE_y$ and, the longitudinal voltage is $V_L = LE_x$. The Hall resistance and the resistance per square are equal in 2D to the resistivities: $R_{yx,\square} = \rho_{yx} = -B/nec$, and $R_{xx,\square} = \rho_0$. (The independence of R_{xx} on B reflects the well known absence of of magnetoresistance in the simple Drude theory.) The measured quantities, whether the sample is homogeneous or not, are the resistances (not the resistivities). The former (and the conductances rather than conductivities) are the fundamental quantities for the QHE, unlike the usual case. The conductivities are given by inverting eq. 6.1 (it is useful to note that $\omega_c\tau/\sigma_0 = B/nec$ (ω_c is defined in eq. 6.7 below):

$$\sigma_{xx} = \sigma_{yy} = \frac{\sigma_0}{1 + (\omega_c\tau)^2},$$

$$\sigma_{xy} = -\sigma_{yx} = \frac{-\sigma_0\omega_c\tau}{1 + (\omega_c\tau)^2}. \tag{6.4}$$

The symmetries of the nondiagonal components of $\hat{\sigma}$ and $\hat{\rho}$ presume that the samples are rotationally invariant. For mesoscopic samples this necessitates ensemble averaging. The more general Onsager relationships follow from time-reversal symmetry and imply:

$$\sigma_{ij}(H) = \sigma_{ji}(-H), \tag{6.5}$$

and a similar relationship for $\hat{\rho}$.

The result B/nec for ρ_{xy} can be understood from the elementary condition that the Lorentz force, $F_y = ev_x B/c$ on the moving electrons, remembering that

$$j_x = -nev_x, \tag{6.6}$$

is balanced by the Hall field E_y. The latter is generated by the surface charge layers accumulated on the top and bottom edges of the sample since $j_y = 0$. An equivalent derivation of the above for *samples with translational invariance* is accomplished by noting that in a frame moving with the drift velocity $v_x = cE_y/B_z$, the electric field vanishes (we take linear response, $E_y \ll B_z$, the situation is "electric field-like"). Thus, the current in the laboratory frame is the one required by the Drude theory. We see, therefore, that to obtain deviations from the above, one must have *breaking of translational symmetry*. This is provided by defects, impurities, and so on. We disregard the periodic potential—it should be well approximated by the effective-mass approximation for the small electron densities in semiconducting systems which are of interest in the QHE connection. We will go on with that assumption but it should be kept in mind that there are subtleties associated with the competing periodicities due to the lattice and the magnetic field. We also remember that the above Drude theory is certainly a major oversimplification, although often (but not always) the high- (rather than low-) field Hall effect is described rather well by eq. 6.2, yielding a reasonable carrier density and sign (Ashcroft and Mermin 1976).

At high magnetic fields and low temperatures the effects of the quantization start to be important. For electrons with no scattering we know that the allowed energy levels become the discrete Landau levels (in 2D, with B perpendicular to the 2D plane the motion is fully quantized):

$$E_n = (n + \tfrac{1}{2})\hbar\omega_c, \qquad \omega_c = \frac{eB}{mc}, \tag{6.7}$$

where the mass m is understood to be the effective mass of the relevant carriers and ω_c is the cyclotron frequency. Each Landau level E_n has a huge (extensive) degeneracy

$$p = \frac{BA}{\Phi_0} = \frac{A}{2\pi l_H^2} \qquad (\Phi_0 = hc/e), \tag{6.8}$$

given by the number of flux quanta of B in the total area $(A = LW)$ of the sample. The number of states is, of course, conserved in the sense that the number of 2D, $B = 0$, states in an interval $\hbar\omega_c$ equals p.

Even at $T \to 0$, one must consider here the effects of elastic scattering by impurities and defects, characterized by an effective elastic scattering time τ. Translational symmetry is broken, hence the above Galilean transformation argument does *not* work. It is now impossible to solve the problem exactly as in eqs. 6.7 and 6.8. A very reasonable expectation is that the Landau-level degeneracies split and at least for

$$\omega_c \tau \gg 1, \tag{6.9}$$

which will be assumed from now on, the Landau levels broaden into narrow bands of approximate width $\hbar/\tau$ and with very small overlaps. Thus, the Landau levels become relevant and one expects the usual de Haas–van Alphen and Schubnikov–de Haas effects to prevail. These are due to the consecutive filling of higher Landau levels with increasing (2D) density n or decreasing B, obtained when

$$nA = jp, \qquad \text{or} \qquad \frac{1}{B} = \frac{j}{\Phi_0 n}, \tag{6.10}$$

with j an integer, yielding the well-known periodicity as a function of $1/B$ for a given n (see, e.g., Ashcroft and Mermin 1976). In two dimensions the situation is the simplest and the Schubnikov–de Haas oscillations are expected to be very strong due to the lack of smearing by the motion parallel to B. In fact, the observation of this oscillation was the convincing proof for having a truly 2D system (Fowler et al. 1966, see Ando et al. 1982, Kawaguchi and Kawji 1982). It should of course be kept in mind that the spin of the electrons will lead to the additional spin splitting of the levels. The mass appearing in the gyromagnetic ratio is in general not the same as the one in the cyclotron frequency. The former could naively be expected to be the bare electronic mass, but different, sometimes "anomalous" values for the g-factor are possible and often occur. The exchange interaction will increase the spin splitting.

The results, shown in Fig. 6.2a, that were obtained in the experiments of von Klitzing et al. (1980) were striking and unexpected from the above general discussion. Si MOSFET devices were used, where the density n could be controlled electrostatically by a gate and a rather high magnetic field was applied so that small values of the integer filling j could be achieved. The later (Fig. 6.2b) results on better-quality GaAs heterostructures show the effect even more strikingly. At low temperatures the Schubnikov oscillations became so deep that σ_{xx} was very small, practically vanishing at its minima. At the same values of n (or B) where an integer number, j, of Landau levels were full, the deviation from the $R_{xy} = B/nec$ "straight line" was as follows: R_{xy} and G_{xy} developed plateaus, sticking for finite ranges of B or n to the values they have at the completely full integer number of Landau levels, j. The values on the plateaus were constant to a relative accuracy of about 10^{-7} and even in the original

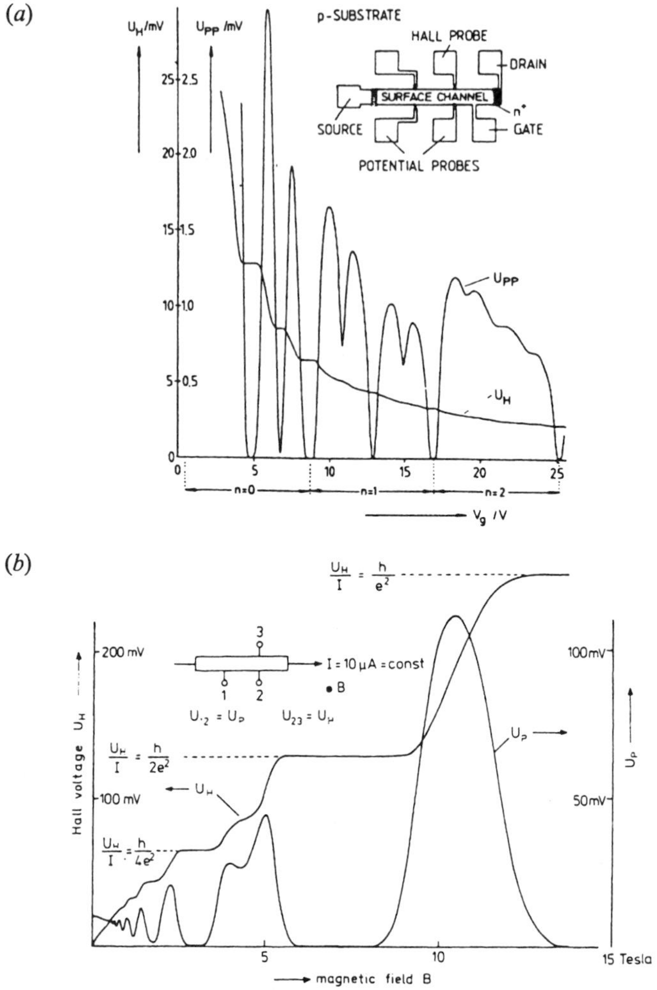

Figure 6.2 (a) The original QHE results of von Klitzing, Dorda and Pepper (1981) for a Si MOSFET as function of the gate voltage controlling the electron density. (b) Later results (from von Klitzing 1982) in gaAs as a function of B, exhibiting broad steps of ρ_{xy} with almost vanishing ρ_x at 1.6 K.

publication the absolute value was found to be equal to ($nA = jp$, as in eq. 6.10):

$$G_{xy} = \frac{-nec}{B} = j\,\frac{-e^2}{h} \tag{6.11}$$

to within 10^{-5}. Currently, relative flatness of "steps" to within 10^{-8} and absolute values of accuracy of 10^{-7} are achieved. This quantum of conductance is the best determination of the fine-structure constant $\alpha = e^2/\hbar c$. This result has a real metrological significance.

We remark that for $G_{xx} = 0$ the inverse of the matrix

$$\begin{pmatrix} 0 & G_{xy} \\ -G_{xy} & 0 \end{pmatrix}$$

is

$$\begin{pmatrix} 0 & -1/G_{xy} \\ 1/G_{xy} & 0 \end{pmatrix}$$

so it should not come as a surprise that both $G_{xx} = 0$ and $R_{xx} = 0$, while G_{xy} and R_{yx} are the inverse of each other.

It has been clear that this phenomenon is associated with the dissipationless nature of the transport and that the occurrence of the step must be associated with the "pinning" of the Fermi level between two Landau levels in the range where states are localized—somehow not changing the transport when n or B is varied. Our first task is to understand these qualitative ideas more systematically.

2. GENERAL ARGUMENTS

It is easiest to do the theory for the idealized geometry of Fig. 6.3a, where the 2D electron gas is wrapped on a cylinder of circumference L_y with a radial Hall magnetic field B. The Hall voltage, $V_H = V_y$, is supplied by a time-varying A–B flux, $\Phi = \bar{\Phi} + cV_y t$, and the Hall current is in the x direction. This is a convenient modification of the geometry originally considered by Laughlin (1981). A potential $V_0(x)$ is assumed, at first, to vary only in the x direction (see Fig. 6.3b). $V_0(x)$ can be thought to include the averaged effects of impurities, possible potential barriers, and so on. It has to vary slowly on the scale of

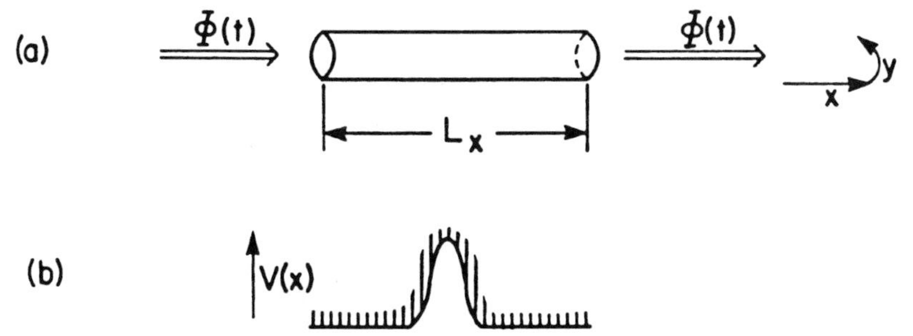

Figure 6.3 (a) A theorist's sample: the Hall voltage is inductively applied with a time-dependent A–B flux Φ. (b) A potential $V(x)$ shown is the case of a local barrier. The equally spaced thin lines depict the positions of the Landau-level centers, x_j for a given Φ. What happens at the edges of the cylinder is not considered yet.

the magnetic length l_H. It is simpler to think about the case where V_0, the amplitude of V, satisfies $V_0 \lesssim \hbar\omega_c$, but the treatment is valid in particular regimes also for larger V_0. Slow variations of L_y as a function of x are also straightforward to include. Some modest variation of V_0 with y can also be handled in the quasiclassical approximation (next section).

In the Landau gauge $A = (0, Hx, 0)$, the usual separation of variables $\psi(x,y) = e^{ik_j y}u(x)$ works, $u(x)$ satisfies a 1D Schrödinger equation with a potential

$$V(x) = V_0(x) + \tfrac{1}{2}m\omega_c^2(x - x_j)^2 \tag{6.12}$$

with $\omega_c = eH/mc$, $x_j = -l_H^2 k_j$, and $l_H^2 = \hbar c/eH$. Consider first the Hall current I_x. Due to the modification of the boundary conditions in the y direction induced by the gauge transformation used to remove the A–B flux from the problem (appendix C), the allowed values of k_j are given by $k_j = (j + \Phi/\Phi_0)2\pi/L_y$. Thus, when Φ is varied linearly with time, all x's (see Fig. 6.3b) move uniformly to the left at a velocity $\bar{v}_x = cV_y/HL_y$ leading to an average Hall current e^2V_y/h for each full Landau level with spin degeneracy removed. This establishes that for free electrons with *any* potential $V(x)$ the Hall conductance per Landau level with degeneracy will be

$$G_{xy} = \frac{-e^2}{h}. \tag{6.13}$$

The general gauge argument due to Laughlin (1981) is based on generalizing the above. We present here a simplified version due to Imry (1983). For $V_y \to 0$ the flux is changed adiabatically and linearly in time. After every cycle, when Φ is changed by Φ_0, the system must, by the Byers–Yang theorem, come back to itself. (In the above example, the set of orbit centers x_j moved by one unit and comes back to the same situation.) It is possible, however, that electrons have moved along the cylinder during this process (in fact, in the above example, one electron per single full Landau level was carried along the system from its right- to its left-hand side). Coming back to the *same* state implies that only an *integer* number, j, of electrons can be thus transferred per cycle. This establishes the quantum Hall effect (QHE) in full generality:

$$G_{xy} = -j\frac{e^2}{h}, \tag{6.14}$$

where in the above example j was the number of full Landau levels including degeneracies (spin and degenerate valleys of the band structure in k-space). This demonstrates the generality of the quantum Hall result, for systems for which Φ can be varied adiabatically keeping them in the ground state (at $T = 0$ or in the equilibrium state at low temperatures. A conducting system has a continuum of extended states at E_F which are sensitive to flux; this is why there will always be some dissipation for finite $V = -(1/c)\dot{\Phi}$ and the above argument does not work. This is our second way to see the importance of having localized states at E_F for establishing the QHE.

Note that there is a built-in time periodicity in this problem, when Φ increases linearly in time the period is given by h/eV and the frequency by

$$\omega = eV/\hbar \tag{6.15}$$

in analogy with the a.c. Josephson effect (chapter 7). The d.c. Hall current along the cylinder should have a small a.c. component with the above frequency (Imry 1983a). It remains to be seen whether this can be obtained in a more realistic weak-link type geometry (Imry 1988).

Another possible modification to the above can occur if the system (for a reason which has to do with complications which we have not discussed yet) has several degenerate ground states and if upon adding Φ_0 to Φ it goes from one ground state to another. It can then take an integer number of flux quanta, $m > 1$, to come back to the same state. This is a way in which the Byers–Yang theorem can be extended to allow, with the appropriate filling, for a quantum Hall conductance of e^2/mh per single Landau level, as in the fractional quantum Hall effect (FQHE, section 4; see Thouless 1990, Thouless and Gefen 1991, Gefen and Thouless 1993).

An important modification of the gauge argument was given by Halperin (1982). It leads to the realization of the possible relevance of edges ("sample surfaces" in 2D) for the QHE problem and to a further understanding of the irrelevance of the details of the disorder. Halperin considered the Corbino-disk geometry (Fig. 6.4) which is topologically equivalent[1] to the above cylindrical

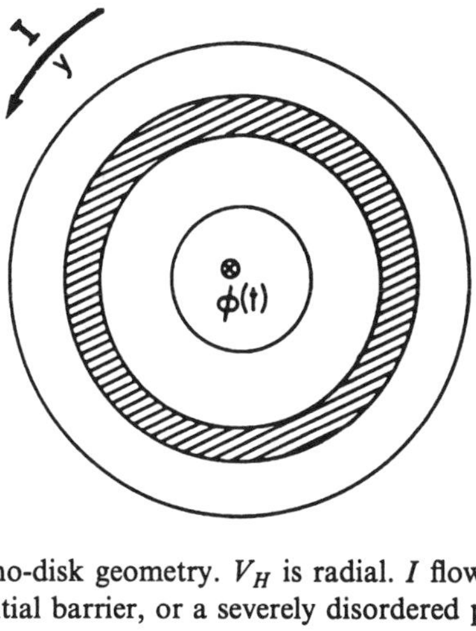

Figure 6.4 A Corbino-disk geometry. V_H is radial. I flows azimuthally. The shaded region depicts a potential barrier, or a severely disordered part of the sample.

[1] But note that in Halperin's treatment the current flows azimuthally and V_H is radial. This is rotated with respect to our earlier argument, see below.

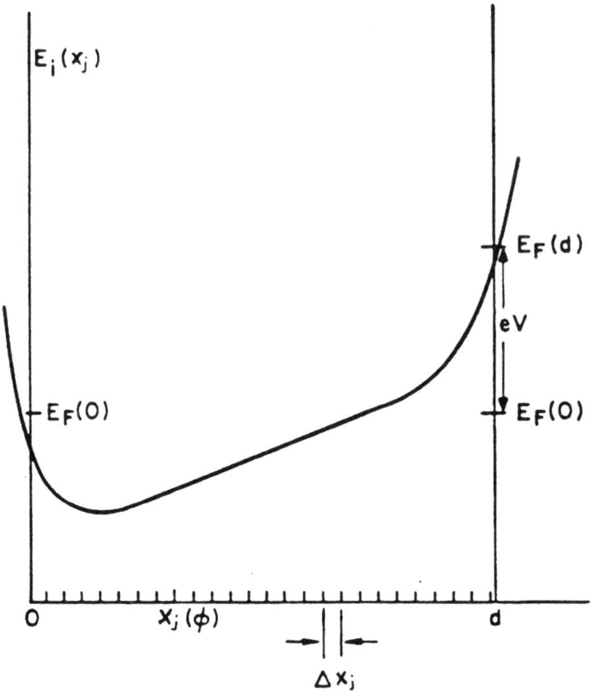

Figure 6.5 A schematic dependence of the energy of a given Landau level with edges but no disorder on the center coordinate x_j. The leads are filled so that $E_F(d) - E_F(0) = eV$. The markings on the x axis signify the x_j's. Their distances are all equal to $\Delta x_0 = 2\pi(l_H^2/Ly)$.

one. He introduced the potential walls at the inside and outside edges of the ring. We neglected those in the model of Fig. 6.3a, since we assumed that the edges there were somehow connected to current leads. The voltage V is now introduced radially (what was before the x and current direction) as in the original Laughlin argument. The energies of the states as functions of the positions of the orbit centers x_j are now as depicted in Fig. 6.5. Now the A–B flux in the hole of the ring is only needed in order to demonstrate that the current in each state is given for a large ring by the difference in energy of consecutive states along x, divided by Φ_0. The same follows, of course, from the expression $v_y = (1/\hbar)\,\partial E/\partial k_y$ for the group velocity. The total current in the occupied levels is thus, for the Landau level under consideration:

$$I_{tot} = -\frac{c}{\Phi_0}\sum(E_{j+1} - E_j) = -\frac{ceV}{\Phi_0} = -\frac{e^2}{h}V, \tag{6.16}$$

in agreement with eq. 6.14. A equivalent way of presenting the above is the following: When Φ is slowly changed by Φ_0, the quantized extended states on the edges move continuously each into its consecutive one. The net result *must* be that one electron per full Landau level is transferred between the two edges of the sample. How it happens through the disordered regime in the bulk is rather nontrivial and will be discussed in the next section. However, even without understanding that in detail, we see that the total energy of the system will change by eV per active Landau level, when Φ changes by Φ_0. Using eq. 6.16 for the current establishes that a current of $(e^2/h)V$ per occupied Landau level flows in the azimuthal direction. This quantum Hall current is thus analogous to the persistent current discussed in chapter 6. It will be an *equilibrium* current in the presence of the voltage V as $T \to 0$ since, because $\sigma_{xx} = 0$, there is no dissipation even in the presence of V. Besides emphasizing the role of the edges (for linear transport), this shows that having localized states in the bulk is immaterial for the current. If the shaded area in the figure is a barrier or a well (along x), energy variations due to it will cancel. Clearly, this should happen for arbitrary potentials in the bulk, including those that can localize states. We see very vividly how details of the "inside" potential do not matter at least as long as the walls are steep and can be approximated by sharply increasing (decreasing) functions of the radial coordinate, x on the outer (inner) edge. Clearly, this does not mean that "the whole" current flows on the edges. A lot of current can flow, in various directions, in the bulk, but the final *net* current is given by eq. 6.16. An important byproduct of the above argument is that for most values of n or B for which the Fermi level is within most of the bulk states or the edge states, the system is on a given plateau. The transitions between plateaus occur when the Fermi level goes through the very narrow band of delocalized bulk states. Thus, this explains the broadness of the Hall plateaus for large systems and $T \to 0$. It is instructive to note that at equilibrium the Maxwell relation (equality of the mixed derivatives of the free energy with respect to μ and Φ) implies

$$\frac{\partial I}{\partial \mu} = c\frac{\partial N}{\partial \Phi}.$$

The current on each edge is e/h per unit added chemical potential on that edge, because of the property that a flux quantum corresponds to a state per Landau level (MacDonald and Girvin 1988, MacDonald 1995).

3. LOCALIZATION IN STRONG MAGNETIC FIELDS AND THE QHE

Let us remember that at $B = 0$ a 2D system with disorder should have its states localized at all energies according to the discussion of chapter 2. On the other hand, the gauge argument of the previous section requires extended states to have flux sensitivity and provide the Hall current. At the same time (Aoki and Ando 1981), ranges of energy with only localized states are needed to pin E_F there and have finite plateaus. Thus, the introduction of the magnetic field must, at least when it is strong enough to be in the QHE regime, delocalize *some* states. Actually, such extended states may be provided quite generally for a finite sample having potential walls and edge states. These states play a principal role in the Halperin discussion (see the previous section). Their role in carrying the current is particularly clarified when the experimental arrangement, such as that of Fig. 6.1, is analyzed according to the Landauer formulation, where the channels moving to the right/left are fed by the l.h.s./r.h.s. reservoirs and the quantized Hall current corresponds to unit transmission in edge states of full Landau levels. The voltage probes were considered by Büttiker (1990). However, it is of fundamental interest to see how the magnetic field delocalizes enough states for having the QHE in the bulk of the system as well. This discussion should also strengthen our understanding of the irrelevance of the localized states in carrying the current and determining the values of the conductivities. These states are, of course, crucial, as already mentioned, for pinning the Fermi energy for finite ranges of B or n (see also Kiss et al. 1990).

The effects of disorder are reviewed by Prange (1990). We are going to discuss here mainly the high magnetic field quasiclassical "guiding center" picture (Iordanskii 1982, Kazarinov and Luryi 1982, Trugman 1983, Joynt and Prange 1984). Its results are in agreement with numerical studies by Ando (1983, 1984).

Before presenting the discussion, it is essential to clarify orders of magnitude. $\omega_c\tau \gg 1$ is the condition for the classical cyclotron motion to be relevant. Remembering that the classical cyclotron radius, l_c, for an electron at the Fermi energy is given by v_F/ω_c, the condition $\omega_c\tau \gg 1$ implies $l_c \ll l$, again a purely classical condition. In terms of the quantum mechanical length $l_H = \sqrt{\hbar c/eB}$, $l_c \sim k_F l_H^2 \sim l_H\sqrt{E_F/\hbar\omega_c} \sim l_H\sqrt{j} \gtrsim l_H$, where j is the number of full Landau levels. Thus what would be the quantum strong field condition $l_H \ll l$, usually implies a *weaker* field than the classical one. The range $l \ll l_H$ is the weak localization regime. The intermediate range $l_H \ll l \ll l_c$ is an interesting one, to be briefly discussed later. We shall first concentrate on the regime $\omega_c\tau \gg 1$, $l_c \ll l$.

Classically, in 2D with strong B perpendicular to the plane and a given not-so-strong electric field E in the plane, the fast cyclotron revolution of the electron develops a drift perpendicular to E, that is, the center of the cyclotron motion drifts along a constant potential line at a velocity v_d so that the Lorentz force balances the force due to E:

$$\frac{v_d}{c} = \frac{E}{B}. \tag{6.17}$$

(In a more comprehensive description, the electron performs, for small E, the fast cyclotron revolution, where the center of the circle drifts slowly along a constant potential contour. For larger E, the electron "skips," performing motion along "arcs" and reflecting from the sloping potential; the motion along the sample edges is a good example.) Let us now generalize this to a smooth but otherwise arbitrary potential $V(\boldsymbol{x})$ that does not change very much (i.e., less than $\hbar\omega_c$ in the quantum case) on the scale of the cyclotron motion. Again, there will be a separation of scales: fast cyclotron motion with a characteristic time $1/\omega_c$ and slow drift of the center along an equipotential line with a velocity given at each point by eq. 6.17 with the local E. Consider first a closed equipotential line (e.g., around a potential hill or valley, or along the edge in a finite sample, or along the inner or outer edge of a disordered Corbino disk). The time of drift around this contour will be given by $\oint(dl/v_d)$, dl being an element of length around the contour. The frequency of the slow motion will thus be given, by using eq. 6.17, by

$$\omega_s = \frac{2\pi}{\oint \frac{dl}{v_d}} = \frac{(2\pi c/B)}{\int \frac{dl\, d_\perp x}{dV}} = \frac{2c}{B}\frac{\Delta V}{\Delta A}, \tag{6.18}$$

$dx_\perp$ being an element of length in the direction of the gradient and $A(V)$ the area subtended by the equipotential contour $V(x) = V$. The fast motion is quantized with energy separation $\hbar\omega_c$. The slow periodic motion will be quantized for each Landau level into locally equidistant levels separated by $\hbar\omega_s$. In this approximation the energy of each state equals the sum of the cyclotron energy $\hbar\omega_c(j+\frac{1}{2})$ and the potential energy, eV. Equation 6.18 implies that the area between two such quantized orbits will be given by

$$B\,\Delta A = \frac{hc}{e} = \Phi_0; \qquad \Delta A = 2\pi l_H^2. \tag{6.19}$$

Again, we find that the flux of B in the area per state in a given Landau level corresponds to a flux quantum, as for free electrons. (We remark that the above picture is reminiscent of the semiclassical theory due to Onsager for quantizing the motion of a Bloch electron in k-space. Since the orbit in r-space is obtained from the one in k-space upon rotation by 90° and scaling by l_H^2, the quantized area between equal-energy contours due to consecutive states in k-space is $2\pi/l_H^2$. This is why, for example, the Schubnikov oscillation has to do with the equal energy areas in k-space.)

Now consider the states due to a given Landau level in a smooth random potential with amplitude smaller than $\hbar\omega_c$ (so that we can surely neglect inter-Landau-level mixing). The potential energy of the appropriate quantized equipotential contour has to be aded to $(j+\frac{1}{2})\hbar\omega_c$. Starting with the higher energies in the "jth Landau band," they correspond to equi-V contours running around the hills in the potential. Similarly, the low energy levels in this Landau band are due to orbits running around valleys or "lakes." Both of these states are localized (the electron "runs" around finite closed contours). It turns out that in 2D there is one and only one energy at which the equipotential curves span the whole (very large) system. This can be understood intuitively by looking at the "terrain" of $V(x, y)$ and imagining filling it with water up to a given height V_0. For small V_0 we have isolated lakes whose areas increase with V_0. For large V_0 we have isolated islands in a continuous sea. In both cases the "coast lines" (i.e., the $V(x, y) = V_0$ contours), do *not* extend across the whole system. There is only one particular energy, E_c, at which *both* continents and sea extend across the whole very large system. To prove this intuitively obvious property, we may note that by symmetry, if the coastlines for given E extend across the system in the x direction they must also do so in the y direction and hence they *must* intersect (this is where the 2D nature comes in). However, equipotential curves of different energies must not intersect, by definition. Thus, this infinite extension of the equipotential lines happens at *an isolated energy*. A crossing point of the contours at this energy is obviously a saddle point of the potential. Using percolation theory language, this situation is described by saying that the lakes percolate along the whole system at low E, the islands at high E, and that there is only a single E_c at which both percolate. For an infinite system we are guaranteed to have a delocalized state at that energy, E_c (actually, there will be a finite DOS at E_c but the delocalization occurs mathematically at E_c only).

It is clear that the characteristic size ξ_p, which may be defined by the r.m.s. of the area enclosed by the equipotential contours for localized states, must blow up as $E \to E_c$, that is,

$$\xi_p \sim |E - E_c|^{-\nu_p}. \tag{6.20}$$

ν_p is thought to be of the order of but larger than unity. Of course the contour is rather jagged and its length is therefore much greater than $O(\xi_p)$. There has been a lot of work using percolation concepts in applying this picture to the QHE, which we shall not review here.

We are now in a position to understand how the existence of the special percolating bulk extended state in a given Landau level is essential for having a quantum Hall current in that level. We return to the Halperin picture defining the current via the change of the total energy due to an A–B flux quantum. We can now easily understand how, when Φ is changing by Φ_0, a single electron is transferred between the two edges of the sample to increase the energy by eV. We remember that extended states on the edges did move, each one into the next, in the same spatial direction, upon that change of flux. What we have to understand is exactly how the net transfer of one electron in the bulk is accomplished. Consider for simplicity one saddle point of the potential in the bulk in a finite system (it may be argued that the exact percolation will always happen through the last, bottleneck saddle point). This saddle point (see Fig. 6.6) is situated at some definite energy E_c. For a general Φ, E_c will *not* be one of the quantized levels defined by eq. 6.19. It will have two nearest levels of that kind, 1 and 2 in Fig. 6.6, whose orbits run around the ring and do not intersect, except at E_c. When Φ is changed by Φ_0, one of these two states *must* go into the other. This will happen because at some particular value of Φ in the given

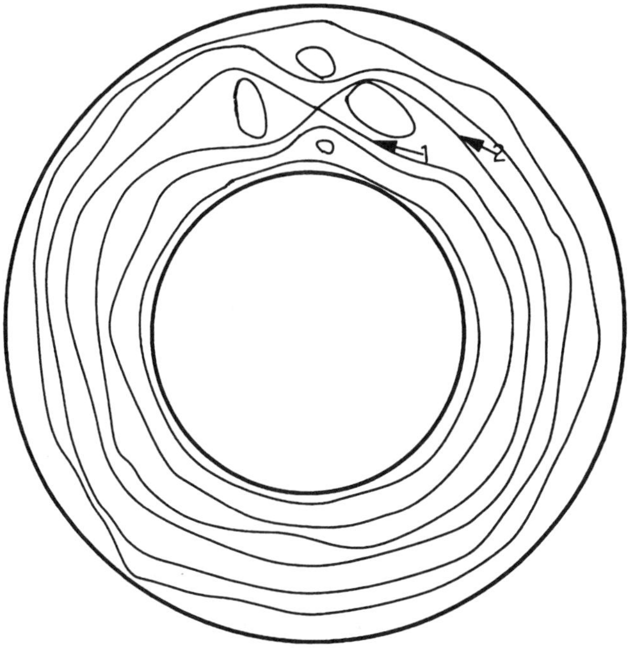

Figure 6.6 Schematic contours for a ring with edges and one saddle point. The quantized extended states along the ring for a general value of Φ are shown by full lines. The state at E_c, having the saddle point energy is shown by contours 1 and 2. It is the intermediay through which states of the type 1 go into those of type 2 at some point in every flux interval of Φ_0.

interval E_c will, by continuity, be precisely the allowed quantum state (as in Fig. 6.6). Thus the two states go into each other via the special state at E_c.

The above argument establishes, for $\omega_c\tau \gg 1$, that each Landau level which has one extended energy below E_F will contribute $(e^2/h)V$ to the Hall current. Alternatively, we can consider the dual way of presenting the gauge argument, applying the Hall voltage inductively as in the argument leading to eq. 6.13 and measuring the current in the radial direction. Here the above picture establishes how the radial current is carried through the bulk via the saddle-point extended state.

Repeating the same discussion for the long cylinder model of Fig. 6.3 can provide a clear case where the existence of edges is not a necessary condition for having a QHE. If the circumference of the cylinder (voltage direction) is much shorter than the length, the above x–y symmetry is broken and percolation will occur in the radial direction more easily than along the cylinder. Still the mechanism of moving electrons from the radially extended flux-sensitive states on the left of the saddle point to those states on the right of the saddle point will be the same as above. Interesting questions remain regarding the influence of the deviations from the semiclassical approximation, when

l_c/l is not small enough. Such deviations should be most significant for the above saddle point process and they should have physical significance for questions such as the breakdown and the frequency dependence of the QHE. Tunnelling between such states (Jain and Kivelson 1988) was discussed by Chalker and Coddington (1988) and Milnikov and Sokolov (1988); see also Macêdo and Chalker (1994).

We conclude this section by briefly reviewing an intriguing generalization by Khmelnitskii (1984a) of the QHE theory to the next, weaker, magnetic field regime $l_H \ll l \ll l_c$, in which $\omega_c\tau \ll 1$ and the Landau levels lose their meaning. He assumes a generalization (Khmelnitskii 1983, Levin et al. 1983, Pruisken 1984, 1985) of the single-parameter RG equation given by eq. 2.29 and Fig. 2.2 for the range where the magnetic field is large enough for g_{xy} to be important, that is, $g_{xy} \gtrsim 1$. Such an RG flow was also obtained by Pruisken (1984, 1985) and is summarized in Fig 6.7, which received direct support from the experiments of Wei et al. (1986). These RG equations were solved (see problem 1 at the end of this chapter) using initial (small scale) values of, for example, the Drude conductivities (eq. 6.4). For $\omega_c\tau \gg 1$ this leads to the usual QHE behavior. However, for $\omega_c\tau \ll 1$, it is immediately seen that for $\sigma_0 \sim k_F l \gg 1$ there is a whole regime where, although $\omega_c\tau \ll 1$, the initial values for σ_{xy} are in the "domains of attraction" of QHE fixed points with j values smaller than $E_F\tau$ (the same values of j as for conventional QHE steps at higher fields). Thus, it follows that in this regime there are further quantum Hall plateaus, *unrelated to Landau levels*. Associated with them are novel extended-state energies that are "levitated" with respect to where the Landau

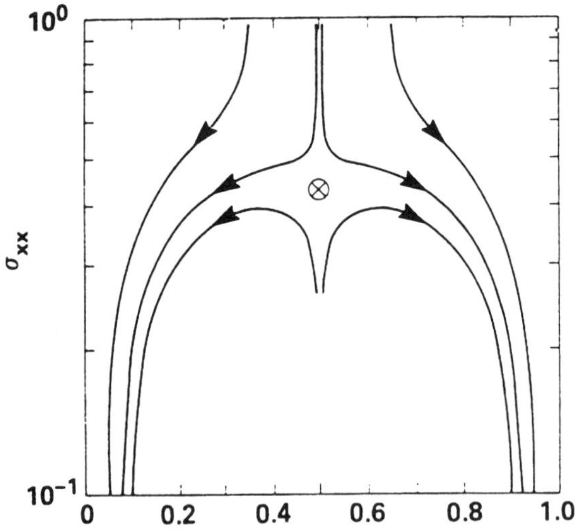

Figure 6.7 Theoretical RG flow diagram for the two-parameter scaling of g_{xx} and g_{xy} for the integer QHE. (From Pruisken 1985.)

levels would have been for weaker scattering. The validity of this surprising picture by Khmelnitskii rests only on the above-mentioned RG flows. There are recent experiments (Jiang et al. 1993, Wang et al. 1994, Hughes et al. 1994) which seem to be consistent with such QHE steps and extended states. The picture was generalized to the fractional QHE (FQHE) regime by Kivelson et al. (1992).

4. BRIEF REMARKS ON THE FRACTIONAL QUANTUM HALL EFFECT (FQHE)

Once the general arguments establishing that σ_{xy} in the QHE must equal an integer multiplying e^2/h were crystallized and became very convincing, experiment played another trick on the theory. Tsui, Störmer and Gossard (1982) discovered, making magnetotransport measurements on high-quality heterostructure GaAs samples, that there was an apparent QHE step at a filling of one-third of the lowest Landau level, with $g_{xy} = \frac{1}{3}e^2/h$. This step was also accompanied, as in the IHQE, by a dip in g_{xx}. Their results are shown in Fig. 6.8. Later experiments found many more fractional Hall plateaus.

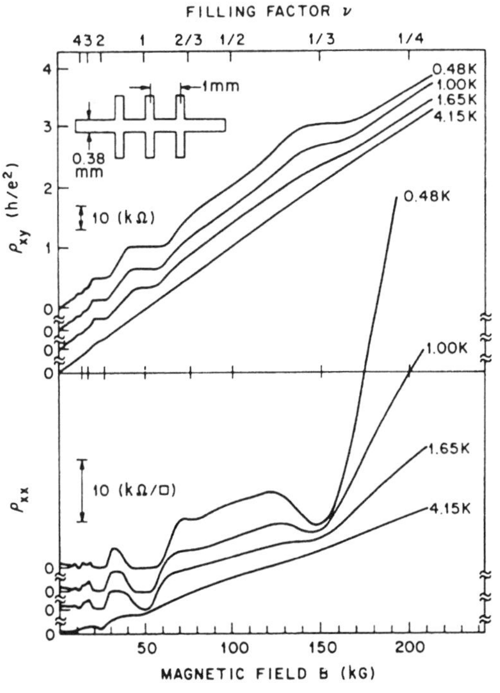

Figure 6.8 The original FQHE at 1/3 filling of the lowest Landau level from Tsui et al. (1982). Note the formation of a dip in ρ_{xx} and a plateau in ρ_{xx}, with decreasing temperature.

This clearly necessitates a revision in our thinking about the QHE and it cannot be explained with a model of noninteracting electrons. The electron–electron Coulomb interaction should be reckoned with. It is expected to make the (3D or 2D) electron gas form a "Wigner crystal" at low densities and temperatures. In this crystalline phase, the electrons systematically avoid each other in order to minimize Coulomb repulsions. It is expected that this electronic crystal should be pinned by defects in the underlying material and therefore become immobile (for small fields). Thus both σ_{xx} and σ_{xy} vanish in the linear regime.[2] It is suggested that a less drastic condensation of the gas into a *correlated* liquid, where electrons "stay away" from each other, could be the cause of the fractional effect. In analogy with other quantum liquids, Laughlin (1983) constructed a variational wavefunction for such a system which seems to capture most of its very interesting physics. The validity of this picture has been checked against various numerical computations and it constitutes an excellent approximation. This correlated ground state at a filling of 1/3 of the lowest Landau level has some binding energy and its lowest excited states have an energy gap. This gap immediately explains the dip in σ_{xx}, and the tendency of σ_{xy} to have a plateau can be understood as well. The elementary excitations above the correlated ground state have a fractional charge and a nontrivial statistics, which have caused major interest.

Below, we summarize the main features of the Laughlin ground state and then briefly discuss the elementary excitations above it. Following that, we mention some interesting recent developments having to do with the special role of the 1/2 filling.

The Hamiltonian is that of a 2D electron gas in the x–y plane with (ordinary, 3D, $1/|\mathbf{r}_1 - \mathbf{r}_2|\epsilon$) Coulomb interactions and a large magnetic field B in the z direction, described in the symmetric gauge by the vector potential

$$A = \frac{B}{2}(y, -x). \tag{6.21}$$

For a filling of the lowest Landau level close to 1/3, it appears very reasonable to stay in the basis of the (highly degenerate) single-particle wavefunctions of the lowest Landau level, which are up to normalization

$$\psi_{0m}(x, y) = z^m \exp\left(-\frac{1}{4l_{\mathrm{H}}^2}|z|^2\right), \tag{6.22}$$

with $m \geq 0$ and using the convenient notation $z = x + iy$. It is easy to show that $\langle m|L_z|m\rangle = \hbar m$ and that $\langle m|r^2|m\rangle = 2(m+1)l_H^2$. Thus the difference in the flux encircled by consecutive states is again ϕ_0. The degeneracy of the level is

[2] However, this issue is far from being settled. There is an opposite point of view according to which the highly correlated crystal can slide unhindered over the defects. One would expect that that might happen in a metastable state, but all this is still not fully understood.

$$m_{max} + 1 = p \simeq \frac{R^2}{2l_H^2} = \frac{A}{2\pi l_H^2}$$

as in (6.8), R being the radius of the sample.

For noninteracting electrons, we would just distribute the given N electrons in the $p \simeq 3N$ available basis states, and antisymmetrize the result. The straightforward way to introduce the correlation (Laughlin 1983) is to multiply the noninteracting wavefunction by a product of functions of the type $f(z_i - z_j)$, where f is expected to vanish when z_i and z_j approach each other. These f's are called "Jastrow factors" and were introduced very successfully by Jastrow in calculations of liquid helium to take care of the strong short-distance repulsion between each pair of atoms. In principle, the whole function f can be regarded as a set of variational parameters: $f(z)$ is varied to minimize the expectation value of the Hamiltonian in the assumed ground state. It will turn out, as found by Laughlin, that f is almost fully determined by symmetries and not much variational freedom is left. In the large magnetic field limit, the kinetic energy, as we saw when discussing the motion with electric fields, is mostly in the fast cyclotron revolution. Thus, the potential energy plays the major role in the nontrivial aspects of the problem. We also remark that due to the f-factors, the overlaps between the electrons are quite small. Just making the function f antisymmetric will take care of the Fermi nature of the electrons, and there is no need to have the usual Slater determinant.[3]

Thus, the Laughlin wavefunction is taken to be

$$\psi(z_1, \ldots, z_N) = \prod_{j<k} f(z_j - z_k) \exp\left(-\frac{1}{4l_H^2} \sum_{l=1}^{N} |z_l|^2 \right), \tag{6.23}$$

Where the last factor (up to center-of-mass motion) can be absorbed in the Jastrow part. f must be taken to be odd, $f(z) = -f(-z)$, to assure antisymmetry of ψ. Since the system is invariant to rotations, the total angular momentum, L_z, is a good quantum number. This implies that $\prod_{j<k} f(z_j - z_k)$ must be a homogeneous polynomial in $z_1, \ldots, z_N$. To get a total angular momentum M, the polynomial should be of degree M (to verify this statement, operate with $\hat{L}_z = (\hbar/i)\sum_j (\partial/\partial\phi_j)$ on such a wavefunction, here ϕ_j is the phase of z_j). This and the oddness of f imply that

$$f(z) = z^m, \qquad m \text{ odd}. \tag{6.24}$$

Since the number of pairs z_j, z_k is $N(N-1)/2$,

[3] Another reason for dropping the noninteracting function is that in our case it has, due to eq. 6.22, the form of a Vandermonde determinant. This introduces a factor $\prod_{j<k}(z_j - z_k)$, which can be absorbed in the definition of $f(z_i - z_k)$.

$$M = \frac{N(N-1)}{2} m.$$

Thus, the only variational freedom left, given a filling N/p, for which we take $N/p = 1/3$, is in choosing m. M being a good quantum number means that each ψ has only a single m. Laughlin devised an elegant picture to make the determination of m in a way which is intuitively clear. He represented $|\psi|^2$, now having a single variational parameter, m, as a classical probability density

$$|\psi_m(z_1, \ldots, z_N)|^2 = e^{-\beta\phi(z_1, \ldots, z_N)}. \tag{6.25}$$

β is the inverse of a fictitious temperature whose value should be immaterial. Laughlin took $\beta = 1/m$, just to make the picture simple.

$$\phi(z_1, \ldots, z_N) = -2m^2 \sum_{j<k} \ln |z_j - z_k| + \frac{m}{2l_H^2} |z_l|^2. \tag{6.26}$$

We remark that a similar, physically transparent, representation had been used by Dyson (1962) in the theory of spectral correlation of random matrices, alluded to in chapter 4. Equation 6.26 is the potential energy of a set of changes, m, in 2D, repelling each other with the usual 2D logarithmic Coulomb interaction and attracted to the origin by a single-particle potential $(1/2l_H^2)|z_l|^2$. The latter can be regarded as due to a homogeneous positive charge density $\rho_+ = 1/2\pi l_H^2$. The lowest energy of this classical plasma will, of course, occur when the negative charge density $m/3.2\pi l_H^2$ exactly balances the positive charge density and produces charge neutrality. This implies

$$m = 3 \qquad \text{for} \qquad N/p = \tfrac{1}{3}. \tag{6.27}$$

When the total energy of the fluid is plotted against N/p, it is found to have a negative cusp at $N/p = 1/m = 1/3$. Thus the special state $m = 3$ has an extra stability (due to the "charge neutrality" of the plasma analogue) at the right filling. The existence of this extra stability means that there is an effective gap at the 1/3 filling in the spectrum and thus $\partial\mu/\partial N \to \infty$, and the system is incompressible. This is a generalization of the discontinuous behavior of $\mu(N)$ at complete fillings for the pure noninteracting 2D gas. The important connection between this incompressibility and the integer or fractional QHE is discussed by MacDonald (1995). It is this extra stabilization which makes ρ_{xx} vanish when $T \to 0$. It also helps to lock the Hall conductance, given by $\frac{1}{3}e^2/h$ at $N/p = 1/3$, on a step when the elementary excitations are pinned by a small disorder. There is by now a large body of computations verifying that eq. 6.23 is very close to the correct ground state. Had we taken $N/p = 1/5, 1/7, \ldots$, the best values of m for the variational wavefunction eq. 6.23 would have been $5, 7, \ldots$. It must be taken into account, however, that for small enough filling, the Wigner crystal will replace the correlated fluid as being the true ground state of the system.

Fillings complementing the odd denominators, say 2/3—which was in fact the next fraction to be discovered—can be treated in the same way using holes in the full lowest Landau level. The whole hierarchy (Haldane 1983) of odd-denominator rational filling factors, such as 2/5, 3/7, and so on, presents a complex question that we shall briefly mention after discussing the elementary excitations of the 1/3 state.[4]

Before doing that, it is advantageous to return to, for example, the disk geometry of Fig. 6.4. Without interactions, the *single particle* states move into one another when the SA–B flux is changed adiabatically by Φ_0. On the average, for a 1/3 filling, *one* electron is transferred across the system after three quanta are added. Thus, the Hall current will be given by $\frac{1}{3}(e^2/h)V$, as in the Drude picture for the same filling. The problem is to understand how the interaction stabilizes this situation for a finite range of N/p.

Laughlin devised an ingenious way of obtaining the quasiparticles of his theory. He imagined piercing the system with an infinitely thin flux line and adiabatically increasing its strength by Φ_0. At all the intermediate situations the sole effect of this flux, as the A–B one (appendix C) is to dictate a phase change of $2\pi\Phi/\Phi_0$ of the many-body wavefunction around it. By the time Φ reaches Φ_0, we return to the original Hamiltonian and boundary conditions, and it can be argued that we have now created an excited state. However, as in the A–B case, a single-electron wavefunction z^k evolves with this change of flux through the origin into z^{k+1} or z^{k-1}, depending on the sign of the flux (except the $m = 0$ state which goes under flux removal to $z^* \exp(-|z|^2/4l_H^2)$ which is a state in the next Landau level). Thus, if we consider the charge inside some circle around the origin, and since our many-body wavefunction is made of the single-electron states, changing the flux by Φ_0 amounts to transferring in or out of the circle the average charge per state, which is $1/m$. The angular momentum of this single-particle wavefunction has changed by unity. The "motion" of the single particle states is of the same nature as that discussed in the A–B context, and it means not only that the Hall conductance will be $\frac{1}{3}e^2/h$, but also that the effective charge of the quasiparticle thus created is 1/3 of an electron charge, since the total change of charge by creating the quasiparticle in a large cylinder surrounding the fluxon is $e/3$ for $m = 3$. Laughlin has given approximate forms for the wavefunctions of the quasiparticle (Haldane 1983, Halperin 1984) based on the discussion above.

A similar charge counting has been used by Su and Schrieffer (1981) in interacting 1D electronic systems, also resulting in possible fractional charges of 1/3. Much thought has been given in the 1D problem to the question of how these fractionally charged quasiparticles can be observed. It turns out that that is very nontrivial, since most experiments use ordinary electrons which "dress

[4] The FQHE states of the quasiparticles of the $1/m$ state, on top of the "parent" state for electrons, lead to the next stage of the hierarchy.

up" only while entering the system and—when they are finally observed outside of the system—the correct bookkeeping yields a charge e. To observe the charge $e/3$, an experiment must be performed inside the FQHE system. It has been suggested that an A–B oscillation due to quasiparticles moving on an effective ring inside the 2D gas might yield periodicity with a flux quantum of $3hc/e$. There are experimental indications that this may be the case (Simons et al. 1989, 1991). The nontrivial theoretical picture has been discussed by Thouless (1990), Thouless and Gefen (1991) and Gefen and Thouless (1993).

Another strange property of the Laughlin quasiparticles is their statistics (Halperin 1984). It turns out to be neither Fermionic nor Bosonic but a "fractional statistics" in between (hence the name "anyons"). This appears to be due to the fact that the quasiparticles may be considered as "riding" on flux quanta and thus, when two quasiparticles are exchanged, there is an extra, A-B type, phase due to these fluxes (Halperin 1984, Arovas et al. 1985, Haldane and Rezayi 1988). The association of fluxons and particles is a very helpful theoretical device. However, whether it can be regarded as a physical reality is a separate question.

We next review an interesting idea due to Jain (1989) for the hierarchy of FQHE states. The picture developed from this idea has recently received impressive experimental support, both on the special role of $N/p = \nu = \frac{1}{2}$, and on the steps around it (see Fig. 6.9, from Du et al. 1993a), and it appears to be at least close to the truth (Willet et al. 1993, Goldman et al. 1994). The theory pertaining to the picture has been developed by Halperin, Lee and Read (1993). The picture is based on two observations on the experimental data: that there is something peculiar happening at a filling of 1/2 and that the most prominent FQHE plateaus are at fillings of $m/(2m \pm 1)$, m being an integer. At a filling of 1/2 there is just a not-too-prominent σ_{xx} minimum (but no step in σ_{xy}) and the transport properties are rather similar to those of electrons at zero magnetic field. The special role of $m/(2m \pm 1)$ is not easy to understand from the "hierarchical" scheme (Haldane 1983), although in the end this and the composite Fermion picture described below seem to be related.

For a naive version of this picture one assumes that around a 1/2 filling, two flux quanta in opposite direction to that of B are attached to each electron (which therefore behaves as a Fermion, henceforth termed a "composite Fermion"). Consider what happens at some rational filling ν. We have ν electrons per (external) flux quantum. The flux "riding" on the composites is 2ν, so the remaining flux Φ_{eff} per composite is

$$\Phi_{eff} = \frac{1 - 2\nu}{\nu}. \tag{6.28}$$

For a 1/2 filling $\Phi_{eff} = 0$. For a $1/m$ filling $\Phi_{eff} = m - 2$, corresponding for $m = 3$ to an integral QHE for the composite Fermions and for $m = 5, 7, \ldots$ to simple odd-denominator fillings of $1/3, 1/5, \ldots$. (However, see below for a

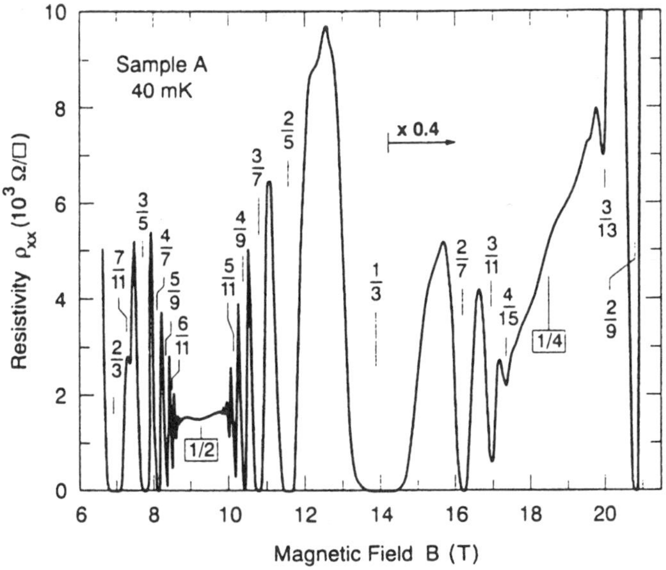

Figure 6.9 Overview of the diagonal resistivity ρ_{xx} in the vicinity of $\nu = 1/2$ and $\nu = 1/4$ at $T = 40$ mK in sample A (from Du et al. 1993). Landau level filling fractions are indicated. For fields higher than 14 T the data are divided by a factor of 2.5. Note the special role of $\nu = 1/2$. The behavior around it is of Schubnikov-type oscillations which develop as they usually do around $H = 0$. Note also the special role of $\nu = m/(2m \pm 1)$. To quote from this paper: "In fact, these features due to the FQHE at $\nu = m/(2m \pm 1)$ replicate the features due to the IHQE at $\nu = p$ and $B = 0$."

better way[5] to obtain these higher fractions.) The complementary filling of $\frac{2}{3}$ corresponds to $\Phi_{eff} = -\frac{1}{2}$, again an IQHE for the composites. What about the prominent fillings of $m/(2m \pm 1)$? They correspond to Φ_{eff} of $\pm 1/m$, that is, to higher IQHE plateaus. When m increases, these two dual series of plateaus at $m/(2m \pm 1)$, "converge" to the 1/2 filling case, which is like the $H \to 0$ limit of the ordinary high-j QHE or Schubnikov series (whose attainment is, of course, limited by the impurities and by the temperature). This explains the main features of Fig. 6.9, including the Schubnikov–de Haas type of behavior around 1/2 filling.

The generalization of this picture to the hierarchy of fractions views them as due to an IQHE of different types of composite fermions made out of an electron and p flux pairs. For them we have, generalizing (6.28),

$$\Phi_{eff} = \pm\frac{1}{j} = \frac{1}{\nu} - 2p \Rightarrow \nu = \frac{j}{2pj \pm 1}$$

[5] It turns out that it is advantageous to view, for example, the 1/5 FQHE as the IQHE of composite fermions based on electrons and four flux quanta. This is easily seen to be consistent with the appropriate version of (6.28).

where j is the index of the IQHE plateau of the composites. The $-$ sign is for integer plateaus for an effective field which is in an opposite direction to the original B.

This picture not only provides a nice way to look at most existing data on the IQHE and FQHE using one paradigm but it is also very appealing, being simple and physically suggestive. Fundamental developments of this picture were given by Halperin et al. (1993) in terms of the appropriate gauge picture and the Chern–Simmons theory of the Hall effect. The above trick to eliminate some of the flux should be valid only as a mean field description. Thus the effect of fluctuations around this picture must be considered. It should be a good approximation if the resulting state is "incompressible." It can be argued (Read 1994) that the phases similar to those resulting above from associating flux with the particles are physically due to vortices in the system. Arguments for the apparent physical validity of the composite Fermions were given by Kang et al. (1993), Willett et al. 1993a,b and Störmer et al. (1994). The Fermi-liquid picture around a 1/2 filling appears to be valid from numerical comparisons for finite systems (Rezayi and Read 1994).

Problem

1. Assume that Fig. 6.7 is valid. (a) Explain why in the $L \to \infty$, or zero-temperature, limit almost all samples will exhibit a QHE. (b) Show that only under very special conditions will a sample have $\sigma_{xx} = \frac{1}{2}$ and σ_{xy} quantized at half-integral values. (c) Sketch the behavior of a sample differing minutely from the behavior of (b). (d) Repeat the argument of Khmelnitskii (1984a): Find the domains of attraction of the QHE fixed points in terms of $\sigma_{xy}^{(0)}$ as given in eq. 6.4. Find the appropriate QHE plateaus for $\omega_c\tau \gtrsim 1$. Show that for $\omega_c\tau \ll 1$ the new plateaus are obtained when E_F crosses the "levitated extended states" at $E_j \cong (j+\frac{1}{2})(\hbar/\omega_c\tau^2)$. Here these states are unrelated to the Landau levels.

7

Mesoscopics with Superconductivity

1. INTRODUCTION

Much of what we have considered in mesoscopics has had to do with phase-coherent effects of normal electrons. Each electron retains phase coherence over the length L_ϕ and different electrons in the energy band of width k_BT around the Fermi energy display a similar interference pattern over the length L_T. The superconducting state is characterized by having a macroscopic wave-function, retaining its coherence ideally over *arbitrarily large* lengths. It is of great interest to enquire how these two (normal and super) coherence effects couple with each other. This is not an obvious issue, since in the normal conductor the active charge carrier is a single electron or a hole, while in the superconductor the coherence is of the wavefunction of the condensed Cooper pairs. It is known, however, that superconductivity can penetrate into a normal metal over a length-scale of L_T (which has been referred to in this context as "the normal metal coherence length, ξ_N"[1]). This is the well-known proximity effect (e.g., Deutscher and de Gennes 1969).[2] Two superconductors can quite generally be correlated with each other and exchange a supercurrent through a normal metal barrier which can be L_T (or L_ϕ, see section 5) long. This is called

[1] When the "normal" metal is itself a superconductor with a lower critical temperature, T_{cn}, T is replaced by $T - T_{cn}$ in the definition of ξ_N.

[2] There are strong indications that longer penetration is possible in mesoscopics (Petrashov et al. 1993a).

"a superconducting weak link" (e.g., Likharev 1979). It is akin to (but not exactly the same as) the usual Josephson effect, which occurs via tunneling through a thin insulating barrier.

The supercurrent through a weak link is a *periodic function of the phase difference* between the two superconductors. Its maximum magnitude is called "the critical current of the weak link." The above phase difference is sensitive to electric fields between the two superconductors and to magnetic fluxes there and in adjacent loops. Various types of weak links have numerous real-life applications, due to their phase-related sensitivity to weak electromagnetic fields. For example, such weak links are important in extrasensitive detectors for weak static magnetic fields and/or high-frequency radiation. The built-in nonlinearity of such weak links due to the periodic dependence on the phase differences introduces a whole set of interesting phenomena and applications which are outside the scope of this book.

The combination of superconducting and normal components brings in a host of novel and interesting phenomena. As hinted in chapter 4, the supercurrent through the normal weak link is quite similar to the persistent current along a ring made from this weak link (Altshuler et al. 1983, Altshuler and Spivak 1987), we emphasize that this correspondence, related to the Andreev process discussed later (section 5), is valid only for a "long" system, whose Thouless energy is smaller than the energy gap of the superconductor). Since the critical current of the weak link is proportional to its conductance, the critical current in a ballistic point contact can show quantized steps related to those of the conductance (Beenakker and van Houten 1991a,b,c). It was experimentally discovered that adding superconductors as "Andreev mirrors" (Petrashov et al. 1993b, 1995) to a normal loop showing A–B resistance oscillations may increase the amplitude of the latter by more than two orders of magnitude. Interesting new effects exist also for super–normal combinations with a single interface (Beenakker 1992a; van Wees et al. 1992).

The case where the normal "metal" is a semiconductor is of special interest due, for example, to the realizability of ballistic effects, and to the possibilities of going through the metal–insulator transition and of inducing superconductivity in the semiconductors. Using electrostatic gates or optical excitation to control the properties of the semiconductor will open up a host of interesting new situations. The technology of producing good superconductor–semiconductor contacts is developing[3] and hopefully such contacts will soon be made under enough control for systematic measurements to be possible (van Wees 1993, van Wees et al. 1994, den Hartog et al. 1995). This promises to be one of the most exciting new directions of research.

We shall, for the most part, describe the superconductor using the Ginzburg–Landau (G–L) theory which was originally developed phenomenologically and later derived from the microscopic theory. We shall briefly review and use the G–L theory, without going into the microscopic picture. The

[3] One useful combination of materials is InAs with Nb.

former is powerful enough to describe many experimentally relevant situations. Good presentations may be found in textbooks such as de Gennes (1966) and Tinkham (1975). The rest of this section is devoted to an elementary review of the G–L theory. Thin rings and wires, with an emphasis on fluctuation effects, will be considered in section 2, Section 3 will discuss weak links and a brief introduction to vortices will be given in section 4. The Andreev process along with some of its applications will be described in section 5, where some interesting experimental results and open questions in this area will also be mentioned. Much of our discussion will be elementary and hopefully pedagogic, emphasizing issues that are relevant to mesoscopics.

The superconducting state is described by a complex "order parameter" $\psi(r)$ which physically signifies the "macroscopic wavefunction" envisioned earlier by London. The free energy is described as a functional, F, of the "field" $\psi(\boldsymbol{r})$, where $\boldsymbol{r}$ ranges through the whole sample. The equilibrium state is given by the minimum of F, and the probability for a configuration having a $\psi(\boldsymbol{r})$ is proportional to $e^{-\beta F[\psi]}$.

$F[\psi]$ for the superconductor is given in the G–L theory by

$$\int d^3x \left\{ -\psi^* \left[\left(\hbar \frac{\partial}{\partial x} - \frac{2ieA}{c} \right)^2 /2m - a \right] \psi + \tfrac{1}{2} b |\psi|^4 \right\} = F[\psi], \tag{7.1}$$

where a and b are constants, A is the vector potential and $2e$ is the pair charge, as known following the BCS theory. In the homogeneous case, $\psi = \text{const}$ and $A = 0$, F/Vol is given by $f = a|\psi|^2 + \frac{1}{2} b |\psi|^4$, having a minimum at $\psi = \psi_0$. In the normal state, $\psi_0 = 0$ so $a > 0$. In the superconducting state, $|\psi|$ is finite and $a < 0$. Hence, the coefficient a must change sign at the critical temperature T_c below which superconductivity sets in.

A valid description near T_c and a qualitatively reasonable one at most temperatures of interest would be

$$a = \bar{a}\, \frac{T - T_c}{T_c}, \qquad \bar{a} = \text{const} > 0, \qquad b = \text{const} > 0. \tag{7.2}$$

Below T_c, the order parameter and the gain of free-energy density of the superconducting state compared with the normal one, are given by

$$|\psi_0|^2 = -\frac{a}{b}; \qquad f_n - f_s = \frac{1}{2} \frac{a^2}{b}. \tag{7.3}$$

It is known from the microscopic theory that

$$f_n - f_s = \tfrac{1}{2} n(0) \Delta_s^2, \tag{7.4}$$

where Δ_s is the superconducting energy gap, satisfying $2\Delta_s \simeq 3.5 k_B T_c$. In a convenient normalization,

$$\frac{\hbar^2}{2m|a|} = \xi^2(T), \tag{7.5}$$

$\xi(T)$ is the superconducting coherence length, given at low temperatures by $\xi_0 \sim \hbar v_F/\Delta_s \sim 10^3$–$10^4$ Å for clean superconductors. For a dirty superconductor, $l \ll \xi_0$, $\xi(T \to 0) \sim \sqrt{\xi_0 l}$. In both cases, near T_c,

$$\xi(T) \sim \xi(0)\left(\frac{T_c}{|T - T_c|}\right)^{\frac{1}{2}}. \tag{7.6}$$

As with any wavefunction of charged particles, spatial variations in the phase of ψ, and/or having a nonzero vector potential A, result in a current density

$$j = \frac{e\hbar}{im}(\psi^* \nabla \psi - \psi \nabla \psi^*) - \frac{4e^2}{mc}\psi^* \psi A, \tag{7.7}$$

which is a gauge-invariant. On the open sample surface, the boundary condition on ψ is

$$j_n = 0. \tag{7.8}$$

The equation for the extrema of $F[\psi]$, called the G–L equation, reads

$$\frac{1}{2m}\left(i\hbar\nabla + \frac{2e}{c}A\right)^2 \psi = a\psi + b|\psi|^2\psi. \tag{7.9}$$

Together with eq. 7.7 and eq. 7.8 it gives a full description of most of the situations of interest when time dependence is not considered and the spatial variations of the parameters are slow on the scale of $\xi(T)$ (this is the condition for keeping only the lowest-order term in the expansion in the gradients of ψ as in eq. 7.1). In cases where the solution of eq. 7.9 is not unique, one has first to find the minima of $F[\psi]$, and among them the deepest one is the equilibrium state. Higher minima are *metastable* states and the transitions among them are understood in simple cases (Little 1967, Langer and Ambegaokar 1967, Halperin and McCumber 1970) as due to fluctuations involving deformations of $\psi(x)$ going through saddle points of $F[\psi]$.

Generally speaking, due to the small parameter $[k_F\xi(0)]^{-1} \ll 1$, the fluctuations around the G–L minimum are rather small, in fact typically negligible in bulk superconductors. In 0D grains and 1D wires the superconducting transition is broadened by thermal fluctuations in a way which is rather well understood (Shmidt 1966, Mühlschlegel et al. 1972, Gunther and Gruenberg 1972, Scalapino et al. 1972). In 2D the fluctuations cause qualitatively new effects: the nature of the long-range order is changed and the transition occurs via the fundamentally interesting mechanism of vortex unbinding (Berezinskii 1971, Kosterlitz and Thouless 1973, Halperin and Nelson 1979). 2D arrays of Josephson-coupled superconducting "dots" are particularly interesting systems in which particlelike vortex motion was beautifully demonstrated (Mooij et al. 1990, Lenssen et al. 1992, Mooij and Schön 1992, Elion et al. 1992, van der Zant et al. 1991a,b, 1992a,b, Tighe et al. 1991). We shall briefly return to discussing some of the fascinating issues related to vortices in section 4.

The Meisser effect—the ability of a large enough piece of a superconductor to expel a not-too-large magnetic field—is immediately obtained from the G–L description. In the simplest case, a uniform magnetic field generates supercurrents on the surface of the specimen, shielding the magnetic field from its interior. The current and field decay away from the surface with a characteristic length called the London penetration depth,

$$\lambda_L^{-2}(T) = \frac{16\pi e^2}{mc^2}\psi_0^2, \tag{7.10}$$

as long as the field is smaller than some critical magnitude. This is valid when the G–L parameter κ, defined by

$$\kappa = \frac{\lambda_L(T)}{\xi(T)} \tag{7.11}$$

is small compared with unity. When it is large, one has type II superconductivity, where due to the negative normal–super interface energy in this case, quanta of flux (with the associated current circulations around them) can penetrate the superconductor relatively easily. Note that at low temperatures, when $\psi_0^2 \sim n$, λ_L is on the order of the Thomas–Fermi screening length multiplied by c/v_F (or divided by the fine structure constant α).

2. SUPERCONDUCTING RINGS AND THIN WIRES

Let us now consider our usual thin wire of cross-section A (to be distinguished from the vector potential $\boldsymbol{A}$) and length L along x, bent into a ring, so that ψ will satisfy periodic boundary conditions along x. We start with no magnetic fields, in the superconducting state, $T < T_c$. Solutions of the G–L equation 7.9 with $|\psi| = \psi_0$, as given by eq. 7.3, are immediately found by inspection. These solutions, which are the local minima of F, are given by

$$\psi_n(x) = |\psi_0| \exp(ik_n x), \qquad k_n = 2\pi n/L. \tag{7.12}$$

They have phase changes of $2\pi n$ along the ring and free energies

$$F_n = F_0 + \frac{4\pi^2 n^2 A |\psi_0|^2}{2Lm} \tag{7.13}$$

so that the $n = 0$ state is the stable one and the $n \neq 0$ states are metastable. For a *given* A–B flux Φ, the associated phase change around the ring is $2\pi\Phi/(hc/2e) = 4\pi\Phi/\Phi_0$. Because of the occurrence of $2e$ in the G–L equation, the effect of the gauge transformation as in appendix C involves the superconducting flux quantum, which we shall now denote as Φ_s, leading to the phase change ϕ:

$$\phi = 2\pi\frac{\Phi}{\Phi_s}, \qquad \Phi_s = \frac{hc}{2e} = \frac{\Phi_0}{2}. \tag{7.14}$$

The stationary solutions are still being given by eq. 7.12, but calculating the G–L free energy for the nth solution yields that n is replaced by $n - \theta$, where

$$\theta \equiv \frac{\phi}{2\pi} = \frac{\Phi}{\Phi_s}. \tag{7.15}$$

Now, many solutions exist for a given θ. One expects that the most stable one (deepest minimum of F) would be the one with $n = m$ where m is the integer closest to θ. This is due to the occurrence of $(n - \theta)^2$ in the energy. In chapter 4, the persistent currents in the normal case were so small that their effect on θ (or Φ) was negligible, so that $\theta = \theta_{ext}$ (applied flux). Here the situation can be very different and the self-generated fluxes may be of crucial importance. Let us take as an example the case where the thickness of the wire in the direction perpendicular to the flux is much larger than λ_L. We know that inside the wire, farther than λ_L from the boundaries, both the current, $\boldsymbol{j}$, and the magnetic field vanish. Let us integrate the current in eq. 7.19, that is, evaluate $\int \boldsymbol{j} \cdot dl$, around a line encircling the ring and going through its interior (in the sense of being much farther from the surface than λ_L). Since $j = 0$ along the integration contour, we find that

$$\Phi = \oint \boldsymbol{A} \cdot dl = \frac{hc}{2ie} \oint i\nabla\theta \cdot dl = n\frac{hc}{2e}, \tag{7.16}$$

$2\pi\theta$ being the phase of ψ, which must change by n due to ψ being single valued (this is the "physical" ψ, before the gauge transformation of appendix C has been made; i.e., $\boldsymbol{A}$ has not been eliminated yet). We find that the flux is quantized in units of Φ_s, as is well known to be the case in superconductors.

More generally, when the thickness of the wall of the ring, d, is not much larger than λ_L, the quantized object is the flux plus a quantity proportional to $\oint \boldsymbol{j} \cdot dl$. This is called "the fluxoid," and its quantization reduces to that of the flux, for $d \gg \lambda_L$.

Since one could put any amount of external flux θ_{ext} in the ring, the flux quantization *implies* that the self-generated fluxes are very important and cannot be neglected as in the normal case. This means, of course, that the persistent currents here are larger by orders of magnitude than those in the normal case. The magnitude of λ_L gives us a measure of that strength.

The deep reason for the above is that here ψ is *not* a single-particle eigenstate. It is a "macroscopic wavefunction," reflecting the fact that in the superconductor the whole electron gas forms a new ground state and a finite fraction of this electron fluid participates in the current. To understand the order of magnitude of the persistent current in the superconducting ring we note that from eqs. 7.1–5 (cf. eq. 7.13), for the "clean" case ($\xi_0 \ll l$),

$$\frac{\partial^2 E_n}{\partial \theta^2} \sim (f_n - f_s)\left(\frac{A}{L}\right)\xi_0^2 \sim n(0)\Delta_s^2 \cdot \mathrm{Vol}\left(\frac{\xi_0}{L}\right)^2 \sim N_\perp^2 \Delta. \tag{7.17}$$

This corresponds to the London idea that only the A^2 term in the Hamiltonian gives the flux sensitivity ("rigid wavefunction"). We emphasize that Δ_s is the superconducting gap and Δ the level spacing in the normal case. For the dirty case, the result is multiplied by $O(l/\xi_0)$ and the last expression is replaced by $N_\perp \Delta_s l/L$. Equation 7.17 has to be compared with the result for normal electrons where (see eqs. 4.13 and 4.14) the same quantity is given in the diffusive case by $E_c \sim N_\perp \Delta l/L$. The magnitude of the currents for clean superconductors, for example, is larger by $N_\perp L/l$, which is five or six orders of magnitude for the typical metallic rings used in the persistent current experiments. For the dirty case, this ratio is $\Delta_s/\Delta \sim 10^4$. The above ratio would be much less overwhelming for a good-quality and small-$N_\perp$ semiconducting ring.

It is helpful to remember that the induced flux generated by a current I in the ring,

$$\Phi_i = \mathcal{L} I, \tag{7.18}$$

is determined by the self-induction coefficient, $\mathcal{L}$. For the ring of interest to us here, $\mathcal{L} \sim L/c$, except for logarithmic factors. For the largest persistent current possible in single-channel normal rings, $I \sim I_0 \sim ev_F/L$ (chapter 4) the induced dimensionless flux would be on the order of (Altshuler 1991, private communication)

$$\Phi_i/\Phi_0 \sim \alpha \frac{v_F}{c} \sim \alpha^2 \sim 10^{-4}. \tag{7.19}$$

Here, α denotes the fine-structure constant. Therefore Φ_i is totally negligible (see also Loss and Martin 1993; a different opinion is expressed by Azbel 1993). It is easy to see that $\Phi_i \ll \Phi_0$ also for many-channel situations, where $\Phi_i \propto \sqrt{N_\perp}$ in the super-pure case (see discussion following eq. 4.11), and it is independent of $N_\perp$ in the ordinary ballistic and diffusive ones. On the other hand, for thick enough *superconducting* rings, $\Phi_i/\Phi_0 \gtrsim 1$, so that the real flux Φ is affected substantially by Φ_i. The relevant thickness scale is λ_L, which we found to be of the order of the microscopic length divided by α. It turns out that the small amplitude of the normal persistent currents does not allow metastable current-carrying states in typical *normal* systems. It would be of great interest if an unusual case could be found where such "bootstrap"-type magnetic moments could exist in a nonsuperconducting system. Systems of interacting rings might show related effects at low temperatures (Szopa and Zipper 1995).

We recall the idea due to Kohn (1964) (and the related ideas by Edwards and Thouless (1972), for the single-particle levels) that the flux sensitivity (r.h.s. of eq. 7.17, cf. eqs. 4.1 and 4.14) of the ground-state energy or free energy of a system determines whether it is an insulator or a conductor. We see now that the characterization of the superconductor vs. a normal metal is via the scaling of $\partial^2 E/\partial\theta^2$ with the dimensions of the system. This appears to be a fundamentally interesting point of view for characterizing these three states of matter (Scalapino et al. 1991, Scalapino 1993).

A useful general picture helping to visualize the effect of the self-induced flux was given by Bloch (1970), starting from the total free energy of the system $F(\Phi)$, Φ being the total flux. Given the external flux Φ_{ext}, one has to minimize the appropriate free energy F_t including that of the system, F, and the electromagnetic part, $(1/2\mathcal{L})(\Phi - \Phi_{ext})^2$, with respect to Φ,

$$F_t = F + \frac{1}{2\mathcal{L}}(\Phi - \Phi_{ext})^2. \tag{7.20}$$

Minimizing this with respect to Φ yields

$$\Phi = \Phi_{ext} + \mathcal{L}I \tag{7.21}$$

with $I = -\partial F/\partial\Phi$ as before. This simply means, as it should, that the total flux is equal to the external flux plus the flux generated by the circulating currents.[4] It is convenient to present this pictorially as suggested by Bloch (1968, 1971) by displaying on the same figure the function $I(\theta)$ and the linear relationship eq. 7.21, which will be presented as $I = q(\theta - \theta_{ext})$. From the general Byers–Yang theorem, the equilibrium $I(\theta)$ must be an odd periodic function of θ with a period 2, but due to the superconducting pairing it has a period 1. The branch $I_n(\theta)$ for each n (cf. eq. 7.12) is just a straight line. The function $I(\theta)$ consists, for each value of θ, of the branch $I_n(\theta)$ with the smallest I. The straight line $I = q(\theta - \theta_{ext})$ may have many intersections with the various branches (dashed lines in Fig. 7.1) and that with the lowest current (θ which is "almost" quantized with a value closest to θ_{ext}) is the most stable. When the wall of the ring (or cylinder) becomes thicker, the intersections approach the integral values of θ and the quantization of flux becomes more accurate.

Gunther and Imry (1969) have considered the long, thin-walled cylinder and have given some expressions for various regimes of d/λ_L which show how flux quantization becomes more precise with increasing d/λ_L. At finite temperatures a few n-states will become populated and $I(\theta)$ will round off at the "transitions" among different values of n.

[4] In the last few equations we have taken $c = 1$.

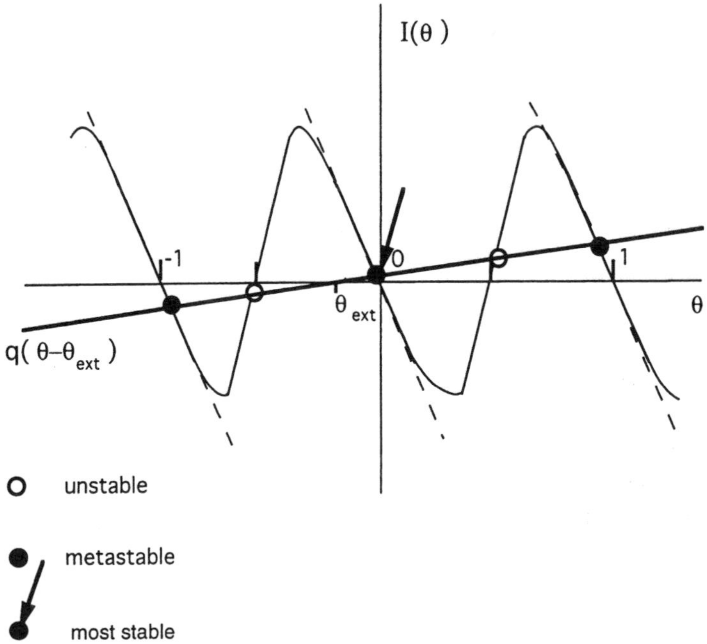

Figure 7.1 Schematics of $I(\theta)$ for a superconducting ring. The dashed lines show the currents $I_n(\theta)$ for the solutions (7.12). The full line is the equilibrium $I(\theta)$ neglecting the induced flux at low temperatures. The full straight line is $I = q(\theta - \theta_{ext})$ (same as eq. 7.21). Its intersections (full dots) with $I_n(\theta)$ are locally stable solutions, the one with the smallest current is the most stable (identified by an arrow). The other set of intersections presents unstable states. Flux quantization follows when the dashed lines are much steeper than the full straight line.

The underlying assumption behind the G–L description and the whole theory of the superconducting state is that the "macroscopic wavefunction" is associated with long-range order in the correlation function of the field $\psi(r)$. In usual, bulk systems the effect of the fluctuations of ψ can be considered as a small deviation (except when the temperature is incredibly close to T_c) from the average situation (the minimum of $F[\psi]$). These deviations are very small and can for the most part be treated within a gaussian approximation. In low-dimensional systems the situation is very different and the fluctuations can in fact wash out the long-range order of $\psi(r)$. We are now going to discuss this issue for the thin-wire case and demonstrate that the physical consequences are serious but not necessarily overwhelming (Gunther and Imry 1969, Imry 1969a,b,c).

The quantity of interest to us is the fluctuation of the order parameter $\psi(x)$ and its correlation function (which should be long-ranged),

$$\langle \psi(x)\psi^*(x')\rangle \qquad \text{for large } |x - x'|. \tag{7.22}$$

It was found by Rice (1965) that this equilibrium correlation function loses its long-range order at one and two dimensions, mainly because of phase fluctuations. Amplitude fluctuations are not crucial due to their additional finite free energy "price." The fundamental importance of the phase fluctuations, related to their ability to generate dissipation, will become clearer in the next section. Writing $\psi(x) = |\psi| e^{i\phi(x)}$, one considers, taking the simplest case, where $\langle \phi \rangle = \text{const}$ and can be taken to be zero,

$$\langle e^{i[\phi(x)-\phi(x')]} \rangle = e^{-\langle \Delta\phi^2 \rangle /2}, \tag{7.23}$$

where $\Delta\phi \equiv \phi(x) - \phi(x')$ and the equality is valid for the case of gaussian fluctuations, which is relevant for the treatment below. We are concerned with the behavior of $\langle \Delta\phi^2 \rangle$ at large separations, $|x - x'|$.

Thermodynamic fluctuation theory (see Landau and Lifschitz 1959) tells us that the probability of a fluctuation of some extensive variable X, of a given small subsystem, is proportional to

$$P(X) = \text{const} \times \exp(-\beta F(X)), \tag{7.24}$$

where $\beta = (k_B T)^{-1}$ and $F(X)$ is the free energy of the given subsystem in which the above variable is uniform (i.e., its density, X per unit volume, is constant throughout the subsystem, which is taken to be homogeneous) and has the value X. Local fluctuations may be thought of as having already been averaged over, in obtaining $F(X)$. Our whole system is the ring plus a heat bath and we shall first regard the whole ring as the subsystem. For $|x - x'| = O(L)$, $\Delta\phi = \phi(x) - \phi(x')$ is an extensive variable; in order to calculate its square fluctuation, we have to weight each value of $\Delta\phi$ by its probability, eq. 7.24. The states to be sampled are those in which $d\phi/dx$ is a constant. These are exactly the states given in eq. 7.12. By eq. 7.24, the probability to be in the state n is proportional to

$$P(n) \propto \exp(-\beta 2\pi^2 n^2 \hbar^2 A |\psi|^2 / mL), \tag{7.25}$$

so that

$$\langle n \rangle = 0$$

and

$$\langle n^2 \rangle = \frac{mLk_B T}{4\pi^2 \hbar^2 A |\psi|^2}, \qquad \langle k^2 \rangle = \frac{mk_B T}{A|\psi|^2 L\hbar^2}; \tag{7.26}$$

also

$$\langle \Delta\phi^2 \rangle = \frac{mk_B T}{\hbar^2 A |\psi|^2} \frac{|x - x'|^2}{L}. \tag{7.27}$$

Suppose that we now choose as our subsystem the region of the wire between x and x' ($|x - x'| \ll L$). Again, to consider a thermodynamic fluctuation one takes $d\phi/dx = \text{const}$ in the present subsystem. Thus, one can sample all these states without any reference to transitions between the various n subensembles. The value of $\langle\Delta\phi^2\rangle$ turns out to be

$$\langle\Delta\phi^2\rangle = \frac{mk_BT}{\hbar^2 A|\psi|^2}\,|x - x'|, \tag{7.28}$$

which is larger by a factor of $L/|x - x'|$ than eq. 7.27 (this is Rice's (1965) result).

To understand the seeming discrepancy between the two last equations, one has to note that in obtaining eq. 7.27 one samples states which have constant phase gradients only in the region between x and x', while in eq. 7.28 a more restricted ensemble, of states having constant phase gradients over the whole ring, was used. Clearly, these two results represent two different types of averages. We shall show below that each of them is valid within a different time scale, and we shall discuss their physical meanings. Let us first examine some common features of both, valid when $|x - x'| = O(L)$ but $|x - x'| \ll L$ (these two conditions mean that a set of systems with increasing L is considered, with $|x - x'|/L$ being a small but finite fraction when L increases).

In the limit where L is large, $\langle\Delta\phi^2\rangle$ "diverges" and, from eq. 7.23, there is no long-range order.[5] It should also be noted that in two dimensions the situation is more interesting: the correlation function decays like a negative power of $|x - x'|$. It may thus be of infinite range in the sense, say, that its integral over all space diverges at low enough temperatures.

When the magnetic flux through the ring is nonzero, eq. 7.13 is modified as above, to $E_n = 4\pi^2\hbar^2 A|\psi|^2(n - \theta)^2/2mL$, where θ is the flux in units of $\hbar c/2e$. The lowest free energy state is the one in which n is closest to θ. For θ_{ext} nonintegral, the induced current may fluctuate, but it will not decay. The fluctuation analysis is valid for the deviation of $\Delta\phi$ from its average.

Note that the fluctuations of intensive variables, such as $k = d\phi/dx$ or the current I, and the relative fluctuation of extensive variables, like $\langle(\Delta(\Delta\phi))^2\rangle^{1/2}/\langle\Delta\phi\rangle$, tend to zero at the large L limit. Thus, eqs. 7.27 and 7.28 do not, in fact, represent any "divergence," and are the usual normal thermodynamic fluctuations.

We will now discuss the issue of the *time scales* of the fluctuations. We shall start by considering the time domains characteristic of the ensembles leading to eqs. 7.27 and 7.28. We shall denote by τ_H a typical hopping time between two

[5] However, $\hbar^2 A|\psi|^2/2mk_BT$ can be quite large (Imry 1969); a typical value, with $A = 10^{-8}$ cm^2, $T \approx T_c/2$, is $\approx$1 cm. Thus, ordering over the finite size of a real sample may well be possible with no effects due to the fluctuations. We shall still consider below the more subtle case where $2mk_BTL/\hbar^2 A|\psi|^2 \gg 1$, and there is no long-range order over the sample.

neighboring n states and by τ_r the relaxation time of the whole ring within a given subensemble n. τ_r is the time that it takes to establish equilibrium in the ring for a given n. We have in mind the case where $\tau_H \gg \tau_r$. The states which lead to the fluctuation in eq. 7.28 are such that $d\phi/dx$ is constant between x and x' and arbitrary elsewhere. Such states can exist only for times much shorter than τ_r (note that $|x - x'| \ll L$). Thus if we coarse-grain over a time interval of order τ_r, the above fluctuation will be ironed out to zero. Then if we wait a much longer time, larger than τ_H, the various n states will be sampled and only the (smaller) fluctuation corresponding to eq. 7.27 will develop.

The main question is: What is the physical relevance of the above two types of fluctuations? As long as $\tau_H \gg \tau_r$, metastable "persistent" currents may not decay for the times in which eq. 7.28 is significant. This strongly suggests that for this case the fluctuations given by eq. 7.28 represent a very short-time effect with which one could deal, say, by coarse-graining, before one looks at overall features of the ring. This corresponds to internal fluctuations of the subensemble of the ring with a given persistent current. The global fluctuations in which n is changed modify the quantum state of the system and this can be shown to lead to a finite steady-state resistance.

The mechanism leading to the hopping between n states has been of fundamental interest. The relevant case occurs when $\tau_H \gg \tau_r$. Following Little (1967), one realizes that in determining the above mechanism, $|\psi|$ fluctuations have to come in as well.[6] The time, τ_H, of transitions across saddle points of F was calculated by Langer and Ambegaokar (1967, see also Langer 1971). The free energy barrier ΔF is given by $(8\sqrt{2}/3)(f_n - f_s)A\xi$, that is, the free energy gain of the superconducting state in a length ξ of the wire. (Note that eq. 7.17 is smaller than this by the usual "Bloch wall" factor ξ/L, see section 4.1). Halperin and McCumber (1970) gave the correct description of the prefactor (the rate multiplying $\exp(-\beta\Delta F)$). This appears to agree with experiments (Webb and Warburton 1968, Newbower et al. 1972).

Using the above description, we make the following points:

1. The equilibrium supercurrent exists and never decays once $\tau_H \gtrsim \tau_r$, for nonintegral θ_{ext}. This was the first example of a persistent current in a system with no long-range order and possibly with impurities (Gunther and Imry 1969).
2. Superfluid effects such as metastable supercurrents on time scales comparable to or shorter than τ_H, which is a characteristic lifetime for the decay of these currents, are entirely possible.
3. As long as τ_H is much larger than a "time of attempt," τ_0—the inverse of the frequency at which the system attempts to cross the free-energy

[6] Little has shown that phase fluctuations (between different subensembles) can occur only via intermediate states with an amplitude fluctuation. This was further treated by Langer and Ambegaokar (1967), who actually found a plausible evaluation of the free-energy barrier for the above transitions. For our purposes here it is enough to assume that there *exists* a mechanism which carries the system between the various n subensembles, without treating this mechanism specifically.

barrier between neighboring n states—the system will spend most of the time near the free energy minima, and a negligible time "in transit." Since t_0 can reasonably be expected to fall within a few orders of magnitude of τ_r, this agrees with the condition $\tau_H \gg \tau_r$. This picture will, of course, break down when $\tau_H \lesssim \tau_0$.

4. For $\tau_H \gg \tau_0$, characteristic properties of the system, which are common to all the n states, will *not* be destroyed by the averaging over n. Flux quantization and the Meissner effect, which are examples of this, will be considered below.
5. From the calculation of the "phase slip" (see eq. 7.31, and Langer and Ambegaokar 1967) resistivity due to the above fluctuations, ρ_s, one may obtain, denoting by τ the relaxation time of normal electrons and by ρ_n their resistivity, that

$$\rho_s/\rho_n = S\tau/\tau_H, \tag{7.29}$$

where S is a numerical factor $S = \hbar^2 An/2Lmk_BT$ ($S \sim 10^4$ in typical cases: $L = 1$ cm, $A = 10^{-8}$ cm^2, $T \approx 2$ K). Thus, assuming that it is possible to detect experimentally resistivities of the order $\epsilon\rho_n$ (ϵ is, say, 10^{-6}), the time scale of fluctuations leading to a detectable resistance is

$$\tau_H \sim \epsilon^{-1} S\tau \approx 10^{-1}\ \text{s}. \tag{7.30}$$

Thus, the behavior of the system can be characterized by three regimes, in terms of τ_H.

(a) $\tau_H \lesssim \tau_0$ ($\approx 10^{-11}$ s): the system is presumably normal.
(b) $\tau_0 \ll \tau_H \leq \epsilon^{-1} S\tau \approx 10^{-1}$ s: resistance is detectable but other superconductive properties such as persistent currents are possible (see below).
(c) $\tau_H \gtrsim \epsilon^{-1} S\tau$: resistance exists in principle but is undetectable; metastable currents have lifetimes of order τ_H.

Note that there are many orders of magnitude separating the time scale for the beginning of regime (c) and that in which the metastable persistent currents have undetectably long lifetimes.

We will now consider the equilibrium phenomena: persistent currents and flux expulsion As long as the system stays in the subensemble belonging to a given n, and no attempts are made to measure times with a resolution better than τ_r, the fluxoid is constant and is equal to n. Once we are out of regime (a) above, the system spends most of the time at the various n states, and the fluxoid is quantized most of the time. For a nonintegral external flux, there will be a finite equilibrium persistent current, given by the thermal average of the currents in the various n-states. This equilibrium persistent current is of the superconducting type, that is, its period is $h/2e$ and its magnitude scales as in eq. 7.17. This is in contrast to regime (a) in which the persistent current should be of the normal-metal type, as in chapter 4. We also note that the Josephson effect is possible, in spite of the phase fluctuations, using the Bloch (1968) interpretation discussed at the beginning of this section and in the next section. Its magnitude will depend on whether $\tau_J \gg \tau_H \gg \tau_0$, or $\tau_H \gg \tau_J \gg \tau_0$. Here τ_J is the inverse of the Josephson frequency ω_J, defined in eq. 7.32.

For wall thickness $\gg \lambda_L$, the total flux, rather than just the fluxoid, is quantized, as discussed above. Moreover, the Meissner effect exists, in the sense that magnetic fields are expelled from the inside of the cylinder's wall (this is independent of n).

The specific heat around the superconducting transition is of interest. We note that when ξ is much larger than the range of interaction, the G–L theory can say nothing about the behavior of the correlation function at distances comparable to the interaction range (Fisher and Langer 1967, Fisher 1967). Therefore, the same applies to the energy of the system, which is rigorously given by the values of the correlation function in the range of interactions, and to its temperature derivative. This observation is crucial here, because it shows that the specific heat is not so dramatically affected by the long-range fluctuations. The phase transition is smeared, but the peak around the critical temperature without fluctuations generally remains (Gunther and Gruenberg 1972, Scalapino et al. 1972).

3. WEAKLY COUPLED SUPERCONDUCTORS, THE JOSEPHSON EFFECT AND *SNS* JUNCTIONS

The Bloch Picture

Consider a gedanken experiment in which the superconducting ring of the last section (which can also be taken to be rather massive, not necessarily in the mesoscopic regime), has a small part (the shaded section in Fig. 7.2) which is gradually being eroded into worse and worse conducting material. This erosion can be achieved in principle by, for example, damaging this section of the ring with some ionizing radiation which will gradually reduce its conductivity and eventually evaporate it altogether. Obviously, the amplitude of the oscillations of the persistent current $I(\theta)$ curve of the ring (as well as the related free energy, $F(\theta)$) will decrease monotonically with the increasing damage, but will become vanishingly small only when the "gap" created in the superconductor is thick enough. Let us stop the erosion process at some intermediate stage where $I(\theta)$ still has a finite amplitude, which is small enough to guarantee a *single* intersection with the straight line given by eq. 7.21 (see Fig 7.2b)). Thus, there are no metastable states to worry about and it can be expected that if θ_{ext} is varied slowly enough the system will follow the equilibrium state (the intersection of $I(\theta)$ with the straight line). At this stage we are not surprised to find that a finite persistent current can flow in equilibrium *if* θ is not an integer. The magnitude of this current and its dependence on geometry, if the "gap" in the ring is not too wide, can still be taken to be smaller than, but in the

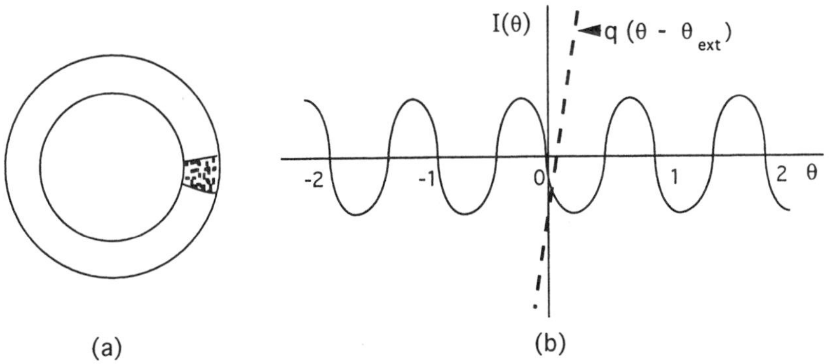

Figure 7.2 (a) A ring with an artificial weak link (shaded area). (b) $I(\theta)$ curve and $I = q(\theta - \theta_{ext})$ line for a superconducting ring having an appropriate weak link (schematic).

same range as in a superconductor (see the discussion following eq. 7.17). Moreover, it can then be demonstrated (see the next subsection) that the ring may be opened and a finite supercurrent, with no detectable dissipation (except for special extreme cases) can flow between the two superconductors through the weak section. This supercurrent is called "the d.c. Josephson effect." An even more intriguing situation occurs when an e.m.f. is applied in the ring by letting θ_{ext} increase slowly and linearly with time. The total θ (intersection of $I(\theta)$ with the straight line in Fig. 7.2) will follow and increase in time with the same average rate. The e.m.f. is given by

$$V = -\frac{1}{c}\frac{d\Phi}{dt} = -\frac{h}{2e}\frac{d\theta}{dt} = -\frac{\hbar}{2e}\frac{d\phi}{dt} \tag{7.31}$$

since $\theta = \Phi/\Phi_s = \phi/2\pi$. There is no d.c. current, but the (time-dependent) current will oscillate with a period given by the time for θ to increase by unity. The frequency of this oscillation is the "a.c. Josephson frequency"

$$\omega_J = \frac{2eV}{\hbar}. \tag{7.32}$$

Thus we have a "d.c. to a.c. converter" with a voltage–frequency ratio given *exactly* by a combination of universal natural constants! Note, that any system which has a (A–B) flux-sensitive persistent current, must have an a.c. Josephson-type effect as above (Büttiker et al. 1983a, Imry 1983b).

Actually, the above is *not* how Josephson (1962, 1965) originally arrived at his a.c. effect. Bloch devised this gauge argument later (1968) to account for the extreme precision of the effect and its total insensitivity, for example, to solid state effects. The Laughlin gauge argument (1981) for the QHE (chapter 6) is obviously of a similar nature. Yang (1989) has emphasized the historical development of the flux → vector potential → quantum phase idea, from Faraday through Maxwell, the Dirac monopole and the A-B effect, to the a.c. Josephson effect and the QHE.

The way to visualize the physics of what happens using only the above very general arguments (a more specific picture will be given in the next section) is the following. The phase ϕ of eq. 7.31 represents the total phase change around the ring. Let us assume that the phase of the order parameter is more or less constant in the superconducting part and that it changes mainly across the "weak link" (damaged part of the ring). This is a very reasonable assumption since, for example in the G–L picture (eq. 7.1), the energy price for changing the order parameter is smallest where superconductivity is "weakest" (and the coefficient a is less negative or even positive). Thus, eq. 7.31 is the relationship between the rate of phase change and the voltage across the weak part of the superconductor. This relationship is exceedingly general and follows from general principles (even the appearance of the charge $2e$, which follows from the BCS theory, can be argued to arise on a more general level (Byers and Yang 1961, Bloch 1968)). Thus, one may apply it in more general situations.

For example, let us cut the ring and connect a current source to the wire comprising it. A small current is now driven by the source through this wire. In fact, it is now clear that the interesting physics is occurring just in the weak link, which can be between any two superconducting "banks." This more usual singly-connected system should display the Josephson effects, as will be discussed directly in the next subsection.

Should the relative phase across the weak link change with time—for example, via the saddle-point activation processes discussed in the last subsection—a voltage would be generated across this "phase-slip center." This is due to the phase not being rigid but changing, due to these fluctuations, in a direction that would reduce the current on the average. A phase change of 2π occurring at the average rate of $1/\tau_H$ will contribute, using eq. 7.31, a voltage of

$$V_{slip} = \frac{h}{2e\tau_H}, \tag{7.33}$$

that is, the rate of increase of ϕ due to V_{slip} will exactly balance the decrease due to the fluctuations. This is how the estimate of eq. 7.29 was arrived at.

The condition for the applicability of the Bloch argument with equilibrium currents is that the voltage V be so small, that $\tau_J \gg \tau_H \gg \tau_0$. If $\tau_H \gg \tau_J \gg \tau_0$, it appears still possible to have some a.c. Josephson effect, but with metastable currents. Here τ_J is the inverse of ω_J. We reiterate that

$$\tau_H \propto e^{-\beta \Delta F}, \tag{7.34}$$

where ΔF is the Langer–Ambegaokar-type barrier for the weak-link wire and is determined by the Josephson coupling energy (see next section) for a tunneling junction. If the wire has a narrow section, ΔF will be smallest there and the resulting "weak" section will become the phase-slip center. Once $\Delta F \gg k_B T$, V_{slip} approaches zero exponentially and may easily become immeasurably small.

The Josephson Junction and Other Weak Links

A most economical derivation of the Josephson effect may be found in the Feynman lectures on physics (Feynman et al. 1965). Assume that the two macroscopic wavefunctions ψ_1 and ψ_2 on the two sides of the junction ($\psi_i = \sqrt{n_i} e^{i\phi_i}$, $i = 1, 2$; n_i being the pair densities) are coupled with a coupling constant K and a voltage V is applied between them. With an appropriate choice of the zero of the energy, the coupled time-dependent Schrödinger equations (for particles with charge $2e$) read

$$\begin{aligned} i\hbar\dot{\psi}_1 &= -eV\psi_1 + K\psi_2, \\ i\hbar\dot{\psi}_2 &= eV\psi_2 + K\psi_1, \end{aligned} \tag{7.35}$$

The equations for $\dot{n}_1 = \dot{\psi}_1^* \psi_1 + \psi_1^* \dot{\psi}_1$ and $\dot{n}_2$ are

$$\hbar \frac{\partial n_1}{\partial t} = -\hbar \frac{\partial n_2}{\partial t} = 2K\sqrt{n_1 n_2} \sin \phi \tag{7.36}$$

where $\phi = \phi_2 - \phi_1$. Taking for simplicity $n_1 = n_2 = n$, we find for the phase difference:

$$\frac{d\phi}{dt} = \frac{2e}{\hbar} V. \tag{7.37}$$

Since the current I_{12} from 1 to 2 is given by $2e\dot{n}_1 = -2e\dot{n}_2$, we rewrite eq. 7.36 as

$$I_{12} = J_c \sin \phi = J_c \sin \omega_J t. \tag{7.38}$$

Here J_c is the Josephson current amplitude, $J_c = (4e/\hbar)Kn$, and ω_J is given by eq. 7.32. Equations 7.32, 7.37, and 7.38 are the fundamental equations of the Josephson junction. Equation 7.38 constitutes the a.c. Josephson effect and its static case, $\phi = \text{const}$, is the d.c. Josephson effect. The accuracy of the voltage–frequency universal relationship 7.32 follows from the Bloch argument presented in section 2 and at the beginning of this section (where the voltage is applied inductively by varying the flux in a ring). In the most general situation the current is just an odd periodic function of ϕ, not necessarily a pure sine. It can be expected that J_c and K increase with the coupling between the superconductors 1 and 2. In fact, from the microscopic theory, one obtains for a tunnel barrier, as $T \to 0$,

$$J_c = (\pi\Delta_s/2e)G_n, \tag{7.39}$$

where G_n is the normal state conductance of the barrier, given by eq. 2.19. When $T \to T_c$, J_c vanishes proportionally to Δ_s^2 (Ambegaokar and Baratoff 1963). Equations 7.36 and 7.37 can be interpreted as the equations of motion for the number of transferred pairs and the relative phase across the junction, which are conjugate variables.

An important remark about the Josephson current amplitude, J_c, is that it is of the same order, $|t|^2$, as the single-electron transmission probability (to which G_n is proportional, see eq. 2.19). This may seem surprising, in view of the fact that *two* electrons are being transferred. In the microscopic derivation the process is second order in t since it goes through an intermediate state where one electron is transferred, having an energy denominator Δ_s. At the final state, which is degenerate with the initial one, the second electron is transferred as well and is paired with the first into the "condensate." The extra factor Δ_s^2 needed for eq. 7.39 is obtained from the so-called "coherence factors" (admixture of the spin-up electron at $k\uparrow$ and the spin-down hole at $-k\downarrow$) in the BCS–Bogoliubov ground state (see, e.g., de Gennes 1966).

The above picture applies specifically to the Josephson effect across a tunneling barrier. In fact, the effect occurs in a large variety of "bridges" and "weak links" of various types between the two superconductors, since the relevant issue is, obviously, just having some pair coupling, K, as in eq. 7.35. This ranges through several situations where the "weak link" is due to a normal metal (or semiconductor) separating the two superconductors. Its conductance can be made small enough to cause the coupling (eq. 7.39) to be sufficiently weak in several ways. These include geometrical constrictions (various types of point contacts) or simply a low enough conductivity of the weak link, which in the semiconducting case might even be controlled with a gate. Questions such as how K depends on the parameters of a specific type of weak link, and what is the effective capacitance of the weak link, are interesting detailed issues for each different type of junction (for a review, see Likharev 1979). The general properties, embodied in eqs. 7.37 and 7.38, remain *qualitatively* the same. This also includes the sensitivity to relatively small magnetic fluxes (see below). Our discussion in chapters 2 and 4 would lead us to believe that the weak link, *as long as it is coherent*, is simply a "black box" with certain scattering characteristics (see, e.g., Beenakker 1992a) for impinging electrons, which determine its normal conductance. it might be expected, then, that the Josephson current amplitude, for a junction made of a normal conductor or insulator, will be given more generally by an expression such as eq. 7.39 at low temperatures. This expectation appears to be qualitatively correct by and large, although specific mechanisms for the coupling should still be considered. This includes the "proximity effect"—inducing weak superconductivity in the normal bridge (e.g., Deutscher and de Gennes 1969) and the Andreev (1964) process, to be discussed in the next section. The condition for coherence of the normal electrons, within the thermal band, in the weak link is that its length L should satisfy $L \leq \xi_N$, where ξ_N, the "normal metal coherence length," is similar to the length L_T defined previously. Since in all known cases, $\xi_N \leq L_\phi$, this guarantees coherence of the partial waves of each electron.

We conclude this subsection with a short discussion of the effect of magnetic fluxes. The underlying relationship is the one explained in appendix C between the flux in a loop and the phase change around it, except that the charge of the particle under discussion is $2e$, hence eq. 7.15 applies. The other ingredient in this picture is the rigidity of the phase in the superconducting part, so that the phase change occurs in the junctions only. As an instructive example, imagine the ring of Fig. 7.2, but with two equivalent junctions with phase changes across them of ϕ_1 and ϕ_2 in the up–down direction. Supppose that one now tries to pass a superconducting current from the top to the bottom of the ring. The Josephson current will be given by

$$I = I_J(\sin\phi_1 + \sin\phi_2). \tag{7.40}$$

The flux Φ in the ring will cause, by eq. 7.15,

$$\phi_2 - \phi_1 = 2\pi\frac{\Phi}{\Phi_s} \equiv \phi. \tag{7.41}$$

Thus, the supercurrent I has a maximum amplitude given by

$$I_{max} = 2I_J \cos\frac{\phi}{2}. \tag{7.42}$$

The two critical currents of the junctions add constructively for $\phi = 0$, for example, but completely annihilate each other for $\phi = \pi$. This is in complete analogy with a two-slit interference situation or to the A–B conductance oscillation discussed in chapter 5. This double junction can be made into a device (DC SQUID) that is sensitive to very small fractions of Φ_s in the loop (and is in principle the method used to detect the persistent currents mentioned in chapter 4). This phenomenon is obviously very general. In a planar junction parallel to the x–y plane (Josephson current in the z direction), the phase difference increases along, say, the x direction, due to a magnetic field having a component in the y direction.

4. BRIEF REMARKS ON VORTICES

It turns out that a quantum of magnetic flux with the associated phase change of 2π and a current loop (obtained using (eq. 7.38) the phase dependence of the current) around it, behaves physically like a single entity. This has been referred to as a "fluxon," "soliton," or "quantized vortex." Vortices enclosing a flux quantum are well known to exist in long Josephson junctions (see, e.g., Scalapino 1969) and as Abrikosov vortices in type II superconductors (e.g.,

Tinkham 1975). Their dynamics in the latter case dominates the important issues of how much supercurrent can be carried by the superconductor and what is the resistance (see below) that may be associated with the vortex motion (cf. eq. 7.33). More generally, vortices (with a continuous vorticity) exist and are of tremendous importance in classical flow problems and quantized vortices are important in superfluid helium (Donnely 1991). 2D arrays (for a simple picture and experimental consequences, see Imry and Strongin 1981 and Hebard and Paalanen 1985) of Josephson junctions can now be made with good homogeneity, displaying nontrivial aspects of vortex motion (Mooij et al. 1990, Fazio et al. 1991a,b, Fazio and Schön 1991, Tighe et al. 1991, Elion et al. 1992, Lenssen et al. 1994, Mooij and Schön 1992, van der Zant et al. 1991a,b, 1992a,b, van Otterlo et al. 1993).

When a vortex parallel to z moves along the x direction, the phase difference across it in the y-direction changes by 2π. The Josephson relationship implies an induced voltage in the perpendicular, y, direction. A current in the y direction causes a force on the vortex in the x direction. The vortex may also experience frictional forces hindering its motion. All this suggests treating the vortex as a composite particle, where the current and voltage roles are exchanged with respect to the usual ones, and with a 90-degree rotation. This vortex–particle analogy can be fully demonstrated for the simple case of a long Josephson junction (Bergman et al. 1983). This duality should also hold for superconducting films and arrays, and there have been experimental demonstrations of the motion of vortices as particles (Mooij et al. 1990, Mooij and Schön 1992) in the latter.

Another interesting aspect is the quantization of the vortex's motion. When the Coulomb capacitative energy is introduced for the long junction, the vortex acquires a mass (for a short junction, the classical equation of motion for ϕ acquires a kinetic energy term; for consequences see Fulton and Dolan 1987). The ensuing quantization can be treated rather extensively (Jackiw 1977). Quantization exists also in more general situations such as the array (Eckern and Schmid 1989) and the type II superconductor (Blatter et al. 1994). In the latter case there still exist questions about the mass of the vortex. Once quantization is understood, one is led to expect a host of quantum mesoscopic phenomena (coherent band motion in a periodic system, localization due to disorder, A–B-related oscillations (van Wees 1990a,b), and so on) which are dual to the phenomena we have treated for electrons. A vortex which encircles a charge experiences an A–B phase shift that can be controlled capacitatively through the above charge. This is detectable (van Wees 1990a,b) via transport measurements in a Josephson array. Although not much is known about dephasing and energy-averaging for vortices, this very difficult experiment was successfully performed by Elion et al. (1993). It demonstrated that vortex interference is possible and thus opened the way to many further interesting possibilities. Stern (1994), building on the analogy between vortices and electrons in a strong magnetic field, showed that QHE phases for the vortices are achievable. This is due to the discrete nature of the array. The same analogy earlier motivated Ao and Thouless (1993) to demonstrate the universality of the Magnus force exerted by the moving superfluid on the vortex in a 2D superconductor. This is a topological property and it is independent of the underlying charge of the electrons. For a theoretical review of many interesting phenomena related to the capacitative coupling of the phase dynamics to the charges, see Schön and Zaikin (1990). Much can be learned by transferring information on various physical phenomena from electrons to fluxons and vice versa (Mooij et al. 1990, Fazio and Schön 1991). As an example we mention the Kosterlitz–Thouless–Berezinskii transition which is equivalent to collective opposite-charge pair separation. The latter has recently been observed in thin films (Tighe et al. 1993, Delsing et al. 1994, Liu and Price 1994, Kanda et al. 1994, Kanda and Kobayashi 1995, Katsumoto 1995, Yamada and Kobayashi 1995).

Interestingly, momentum is transferred between the moving and circulating superfluid in the vortex and its normal core (in the type II case) via the Andreev process. This appears to be the relevant physical mechanism for understanding the Magnus-type force *on the normal core*, which may explain, for example, why the normal core tends to follow the overall superfluid velocity (see also Hoffmann and Kümmel 1993, Wingreen et al. 1994). The combination of vortex and Andreev physics might thus be useful for a better understanding of magnetotransport in type II superconductors.

5. THE ANDREEV REFLECTION, MORE ON *SN* AND *SNS* JUNCTIONS

Let us consider the normal–super (N–S) boundary. An electron with energy ϵ_k just above the Fermi energy in n, is hitting the interface with the superconductor. There are two related questions: (1) How is the normal current in N converted to a superconducting current, carried by pairs, in S? (2) How is the information about the phase χ of the superconducting order parameter transferred to the normal electrons in N? The quasiparticle energies E_k in S, measured from the common Fermi level, are known to be given by

$$E_k = \sqrt{\Delta_s^2 + \xi_k^2}, \qquad \xi_k \equiv \epsilon_k - \epsilon_F. \tag{7.43}$$

ξ_k is the excitation energy for the normal metal. For negative ξ_k, we interpret $|\xi_k|$ as the positive energy of the hole excitation. Let us consider for simplicity low temperatures and currents, so that $\xi_k \ll \Delta_s$. The interface is static and thus we can have only *elastic* scattering processes. There is no way for an electron in N with a momentum $\boldsymbol{k}$ to become a quasiparticle in S. Only two possible reflection processes may exist in the low-ξ_k limit: (a) An ordinary specular reflection to an electron with $-k_\perp$ and the same $k_\parallel$ ($-k_\perp$ and $k_\parallel$ are the components of $\boldsymbol{k}$ perpendicular and parallel to the interface). The above is valid assuming a planar interface. For a rough interface, there will be a "diffuse" reflection, but $|k|$ will be conserved. (b) It was found by Andreev (1964, 1966) that the other energy-conserving process is creating a hole at $-\boldsymbol{k}$ (including the reversal of both $k_{\parallel}$ and $k_{\perp}$, for $\xi_k \to 0$. For finite ξ_k it can be seen that to conserve energy, $k_{\perp}$ will be slightly changed).[7] This is called an Andreev reflection. It is described, using pure electron language, as the two electrons k and $-k$ (with opposite spins) going into S and eventually joining the condensate.

Thus, this process, which necessitates a Fermi sea to have the $-k$ electron, accomplishes the transfer of current between N and S. Charge is obviously conserved. For an ideal interface, and when N and S have the same normal electronic structure, the normal reflection vanishes. More generally, it can be argued that typically[8] the probability for the Andreev reflection is almost unity and that for normal reflection is close to zero. This is so because the length scale of the effective "potential" created by the space-dependent Δ_s, via eq. 7.43 is $\xi_0 \gg k_F^{-1}$. Thus it is improbable for the normal reflection to have enough momentum-transfer to reverse $k_{\perp}$. A regular atomic-scale potential barrier or roughening at the interface will obviously switch on the normal reflection. The effect of an interface barrier was systematically treated, along with the additional quasiparticle transmission processes for $\xi_k > \Delta_s$, by Blonder et al. (1982). Here we shall not reproduce the detailed calculations which use an appropriate modification, due to Bogoliubov and later to de Gennes, of the microscopic theory. Below, we present a physical discussion of some of the interesting phenomena at low temperatures, $k_B T \ll \Delta_s$. In the following, we shall need only the above-mentioned facts and the very important phase shift (Andreev 1964, 1966, Kulik 1969) associated with the Andreev process. It is found that the reflected hole at the Fermi energy experiences a phase shift equal to $-\chi + \pi/2$, where χ is the phase of the order parameter of the superconductor. Similarly, when a hole is Andreev-reflected into an electron, a phase-shift $\chi + \pi/2$ is obtained by the latter.

It has been realized already by Andreev that this reflection leads to interesting states of a new kind bound in the normal region in a S–N–S sandwich, for energies below Δ_s. The classical trajectories representing these states are depicted in Fig. 7.3 for $\epsilon \ll \Delta_s$ (ϵ is measured from ϵ_F). An electron hits the N–S boundary, and is reflected as a time-reversed hole with a phase $-\chi_2$; the latter is reflected again at the other boundary as an electron with a phase χ_1; and so on. The difference between the ballistic and diffusive regimes is simply in how the electron trajectory behaves between the reflections. The semiclassical motion is periodic in both of these cases. It is possible to use, for example, the semiclassical quantization conditions in order to obtain the quantized states. This is simplest in the ballistic regime, where quantization

[7] The spin of the electron at $-\boldsymbol{k}$ is opposite to that of the one at $\boldsymbol{k}$. We suppress the spin indices. More generally, the reflected hole is in a time-reversed state with respect to that of the initial electron.

[8] Except perhaps at glancing incidence small enough $k_{\perp}$.

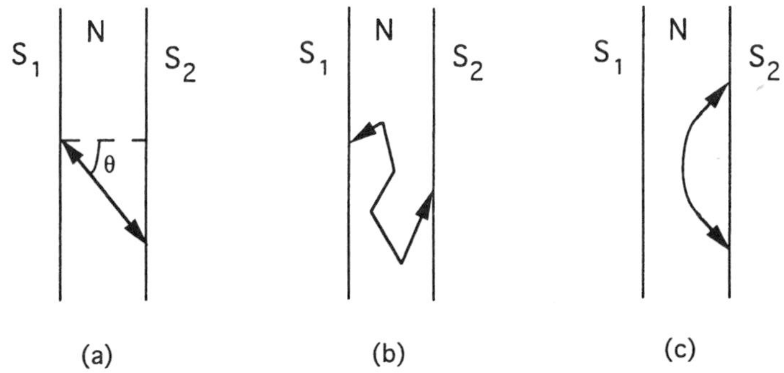

Figure 7.3 An Andreev-reflected bound state (at an energy $\ll \Delta_s$) in the N-region in the ballistic regime (a) and the diffusive one (b). The arrows present the directions of motion of the electron and the hole. (c) The same trajectories but with a large enough magnetic field that will localize the motion to reflections from a single N–S boundary.

(Andreev 1966, Kulik 1969, Zaikin 1982, see also Abrikosov 1988) depends in most situations on the angle θ (see Fig. 7.3a). These normal local states lead to an observable specific heat in the intermediate state of the superconductor (where regions of N and S coexist in a bulk sample) in agreement with experiment. This, as well as other effects, confirms the validity and relevance of the Andreev process. It is interesting that a magnetic field, when large enough, can bend the trajectories so that they do not hit the second superconductor (Fig. 7.3c).

The most interesting property of the Andreev reflection from our point of view is that (for $\xi_k \ll \Delta_s$) the electron obtains in each period an extra phase of $\chi_1 - \chi_2 + \pi$. Thus, the levels are of the same nature as when the N-region is bent into a ring (from now on, we shall consider for simplicity the case where the thickness, L, of the N-region between the two superconductors, is larger than its transverse dimensions) and an A–B flux of

$$\Phi = \Phi_0(\chi_1 - \chi_2)/2\pi + 1/2 \tag{7.44}$$

is applied to it. Obviously, the electronic levels in the ring will depend on $\chi_1 - \chi_2$ and will be periodic in that phase difference with a period 2π, as in the A–B ring. The persistent current, given by eq. 4.5, where F depends on Φ due to the above-mentioned sensitivity of the energy levels to $\chi_1 - \chi_2$, will be dissipationless, like a Josephson supercurrent, flowing from S_1 to S_2 via the N-region (Kulik 1969, 1970a,b). Since we would like the relevant energy range for the normal electrons to be much smaller than Δ_s, we have to take $E_c \ll \Delta_s$ for this correspondence to be qualitatively valid. For N and S having similar Fermi velocities, the above condition is equivalent to $L \gg \sqrt{\xi_0 l}$, which will be referred to as the long normal link case. $\xi_0 = \hbar v_F/\Delta_s$ is the coherence length of the pure S_1 or S_2 and l is the elastic mean free path of the weak link. The length $\sqrt{\xi_0 l}$ is also the ξ_N of the normal link (taken as dirty, $l \ll L$) at a temperature comparable to T_c of the superconductors.[9] We see here a definite physical mechanism by which this supercurrent is carried. In the Andreev reflection at S_2, two electrons go into S_2. In the Andreev reflection at S_1, two electrons go from S_1 to N. The net effect is that in each cycle of the periodic motion a pair is transferred from S_1 to S_2. We believe that this picture gives a vivid, clear demonstration of the relationship between normal persistent currents and the Josephson-type effect. The phase of the hole in the latter case leads however to quantitative differences between the two problems. In the presence of a voltage, the multiple Andreev reflection can lead to a subharmonic gap structure (Bratus et al. 1995; see also Frydman and Ovadyahu 1995).

Before discussing further results on the *SNS* junction, we start with the *NS* situation, including phase coherence effects (Furusaki et al. 1991, Furusaki and Tsukada 1991, Lambert 1991, 1993, 1994, Nakano and Takayanagi 1991, Takane and Ebisawa 1991, 1992, Beenakker 1992a,b, Furusaki 1992, Takagi 1992, Hui and Lambert 1993; Lambert et al. 1993, Lambert and Robinson 1993, Marmorkos et al. 1993, Zaikin 1994). For the microscopic theory, see Fukuyama and Yoshioka 1992, Yoshioka and Fukuyama 1990, 1992). We consider a series addition (the multichannel generalization of eq. 5.36) of a disordered normal section having a scatterring matrix $S_0(\epsilon)$, an ideal normal lead and a planar *NS* interface between the latter and the superconductor S. Thus, there is a spatial separation between the normal and Andreev scattering; S is taken to be in the clean limit. For N channels, the scattering problem for electrons and holes separately is $2N \times 2N$. In the representation where the first/second set of N entries stand for electrons/hole channels, the $2N \times 2N$ reflection part of the scattering matrix of the ideal planar *NS* interface is well approximated by

$$S_A(\epsilon) = \begin{pmatrix} 0 & e^{i\chi} \\ e^{-i\chi} & 0 \end{pmatrix} e^{-i \operatorname{arc\,cosh}(\epsilon/\Delta_s)}. \tag{7.45}$$

Here ϵ is measured from the Fermi energy and χ is the phase of the order parameter in S. The simplest model for the space dependence of the gap $\Delta_s(r)$ was used (Likharev 1979), where $\Delta_s(r)$ changes discontinuously from 0 to Δ_s at the *NS* interface. The total $4N \times 4N$ S-matrix of the normal part (including electrons and holes which do not couple to each other) is given by

$$S_N(\epsilon) = \begin{pmatrix} S_0(\epsilon) & 0 \\ 0 & S_0(-\epsilon)^* \end{pmatrix}. \tag{7.46}$$

Earlier, this problem had been considered by Takane and Ebisawa (1991, 1992a,b). Blonder et al. (1982) obtained the result below for independent

[9] It should be emphasized, however, that we have in mind the case of low temperatures, $k_B T \ll \Delta_s$, where $\xi_N = L_T \gg \sqrt{\xi_0 l}$ and ξ_N is taken at the actual (low) temperature.

channels; see also Shelankov (1984), Zaĭtsev (1980, 1984). By combining S_N and S_A and using the appropriate generalization of the Landauer formula, and for $B = 0$, Beenakker (1992a) obtained for the total conductance of the NS junction:

$$G_{NS} = \frac{2e^2}{\pi\hbar} \sum_{n=1}^{N} \frac{T_n^2}{(2 - T_n)^2}. \tag{7.47}$$

Here the T_n's are the transmission eigenvalues *of the disordered normal part.* The physical meaning of eq. 7.47 can be most easily appreciated in the single-channel case: An electron goes through the disordered normal part with an amplitude t_e and then it is reflected as a hole with an amplitude $-ie^{i\chi}$ (cf. eq. 7.45). The latter has an amplitude t_h to go back through the disordered part, in which case a pair is transmitted with a probability T^2. If the hole were normally reflected (the amplitude for that is $r_h = r_e^*$) from the disordered normal part, it could Andreev-reflect again, with an amplitude $-ie^{-i\chi}$, to upset the above transfer of two electrons, provided also that the electron thus obtained goes back through the disordered part. However, this "secondary" electron can also be reflected by the disordered normal part, with an amplitude r_e, and then be retransmitted by the Andreev process into the superconductor (with an amplitude $-ie^{i\chi}$), and so on. Summing all the amplitudes for the series of these processes, yields

$$e^{i\chi}(-it_e t_h + it_e t_h r_e r_h + \cdots) = e^{i\chi} \frac{-it_e t_h}{1 + r_e r_h} = \frac{-iTe^{i\chi}}{1+R} = \frac{-iTe^{i\chi}}{2-T},$$

for the total transmission amplitude of a pair due to an incident electron. The corresponding probability is $T^2/(2-T)^2$, for the given channel. Equation 7.47 is to be compared with the conductance of the disordered normal part itself (eq. 5.16),

$$G_N = \frac{e^2}{\pi\hbar} \sum_{n=1}^{N} T_n.$$

It can be seen that $G_{NS} \leq 2G_N$. In cases where the transmission eigenvalues, T_n, are mostly zero or unity, it follows from eq. 7.47 that

$$G_{NS} \cong 2G_N. \tag{7.48}$$

This becomes a precise equality for a ballistic point contact, on the conductance plateaus, where the T_n's are exactly zero or unity.

When the normal part is disordered, the ensemble averages for a length L satisfy

$$\langle G_{NS}(L)\rangle = 2G_N(2L)\rangle \cong \langle G_N(L). \tag{7.49}$$

The first equality follows from the properties of the T_n's, as in appendix I (see

Lee et al. 1987). The difference between $\langle G_{NS}(L)\rangle$ and $\langle G_N(L)\rangle$ is mostly in the weak-localization contribution. This contribution is actually enhanced in the *NS* case by a factor of almost 2 compared with the normal case (Beenakker 1992a, 1994, Marmorkos et al. 1993, Macêdo and Chalker 1994, Takane and Otani 1994). This leads to a zero-bias dip in the differential conductance (Lenssen et al. 1994), which should occur in this form only when normal reflections at the *NS* interface are weak enough.

An extremely interesting case occurs when the normal part has a transmission resonance as in a quantum dot (Beenakker 1992a) or in the strong localization regime (see problem 4 at the end of chapter 5; Lifschitz and Kirpichenkov 1979, Azbel 1983). G_{NS} then has a peak, as function of energy, at the resonance energy, in the same way as G_N. However, the resonance of G_{NS} is twice as high and it has a non-Lorentzian shape. This is an example of how the quantum (wave) properties may help the electron pass over large barriers.

Beenakker (1992a) had also considered the effect of an additional barrier at the *NS* interface (Blonder et al. 1982, Volkov 1994) when *N* is in the diffusive regime. He generalized the very interesting prediction of reflectionless tunneling (see van Wees et al. (1992), for the case where the potentials in the normal part are smooth to more realistic potentials, with a deeper theoretical treatment). The combination of the disordered normal region and the *NS* barrier makes the barrier ineffective in decreasing $\langle G_{NS}\rangle$. It is as if the electron normally-reflected from the barrier is reflected back in the disordered normal part, and attempts to Andreev-reflect at the *NS* interface with the barrier again and again, until it succeeds. This may be the explanation for the experimental results of Kleinsasser et al. (1989), Kastalsky et al. (1990, 1991) and a lot of later work (Beenakker et al. 1994; for a review and further results see Beenakker 1995). Physically, such an effect should also exist with a resonant transmission through a localized state in an Anderson insulator, where the interface barrier increases just one of the two barriers existing anyway in the model of problem 4 at the end of chapter 5. The latter effect might have been observed by Frydman and Ovadyahu (1996). Marmokos et al. (1993) confirmed numerically the theoretical predictions in the diffusive regime, including the reflectionless tunneling. The latter was found to hold when the small transmission coefficient of the barrier is still larger than the ratio l/L of the normal part. Marmorkos et al. also studied the conductance fluctuations and confirmed their increase in the *NS* case (Takane and Ebisawa 1991, 1992, Beenakker 1993), as well as the conjecture (Beenakker 1992a) of their insensitivity to a magnetic field.

We shall now discuss some pertinent results (analogous to those in chapter 4) for supercurrents in *SNS* junctions. We start by taking the normal section to be in the diffusive regime. For noninteracting electrons, Altshuler and Spivak (1987) found a component of the conductance of the normal part which is sensitive to the phase difference $\chi_1 - \chi_2$ and is 2π-periodic in it. They had also pointed out that in addition to the average supercurrent whose amplitude[10] is $I_c(\phi) \sim (e\Delta_s/\hbar)g_n$ for low temperatures, $\sqrt{\xi_0 l} \ll \xi_N$ (i.e., $k_B T \ll \Delta_s$), there will be mesoscopic (sample-to-sample) fluctuations in the supercurrent, which they found to be given by

$$\langle \Delta I_c^2 \rangle \sim (eE_c/\hbar)^2. \tag{7.50}$$

This result, which is in agreement with the sample-specific persistent current in chapter 4, is valid for $L \ll \xi_N$ (i.e., $k_B T \ll E_c$), but in addition L should be larger than the transverse dimensions of the normal weak link. Interestingly, the magnitude of the gap Δ_s does not appear. It was found by Beenakker (1991; motivated by results of Beenakker and van Houten (1991a) in the ballistic case) that the result in eq. 7.50 holds only in the limit of a "long sample" (in the sense that $L \gg \sqrt{\xi_0 l}$, which means $E_c \ll \Delta_s$; as above, L is the length of the normal region and ξ_0 the superconducting coherence length in the pure case, still at low temperatures, $L \ll \xi_N$). For a short sample, $L \ll \sqrt{\xi_0 l}$, or $E_c \gg \Delta_s$ (Beenakker 1991), the mesoscopic fluctuations in the critical current are of the order of

$$\langle \Delta I_c^2 \rangle \sim \Delta_s^2 \langle \Delta G_n^2 \rangle \sim \left(\frac{e\Delta_s}{\hbar}\right)^2, \tag{7.51}$$

in agreement with the universality of conductance fluctuations, and being the appropriate analogue to those in our case. As discussed, the condition $L \ll \sqrt{\xi_0 l}$ is equivalent to $E_c \gg \Delta_s$. The physical difference between the cases in which eqs. 7.50 and 7.51 are valid is in the order of magnitude of the window of energies in which most of the current flows, which is E_c and Δ_s, respectively. In this connection, see also Takane and Otani (1994) and Takane (1994).

Since $\langle I_c \rangle \sim (\Delta_s/e)G_n$, we see that the fluctuations in the long sample are relatively small:

$$\frac{\sqrt{\langle \Delta I_c^2 \rangle}}{\langle I_c \rangle} \sim \frac{\Delta}{\Delta_s} \lesssim 10^{-4}, \tag{7.52}$$

because the level spacing Δ is typically a fraction of a millikelvin, while the superconducting gap Δ_s is on the order of 10 K.[11] Altshuler and Spivak

[10] It has to be noted, however, that the detailed shape of $I_c(\phi)$ is different in this case from the one for the Josephson junction (Kulik and Omelyanchuk 1975, Beenakker 1992b). As an example, for a short sample, $L \ll \sqrt{\xi_0 l}$, and when $T \to 0$,

$$I_c(\phi) = \frac{e\Delta_0}{2\hbar} \sin\phi \sum_{n=1}^{N} \frac{T_n}{\sqrt{1 - T_n^2 \sin^2(\phi/2)}},$$

and for the ballistic case:

$$I_c(\phi) = \frac{e\Delta_0}{2\hbar} g_n \sin\frac{\phi}{2}.$$

[11] For the short junction, the relative fluctuation is of order $1/g_n$.

suggested various ways to depress the average critical current in order to make the fluctuations more observable. For example, a relatively strong magnetic field will depress $\langle I_c \rangle$ exponentially, while $\langle \Delta I_c^2 \rangle$ will decrease by a factor of 2 only. The latter is of a similar nature to the relative insensitivity of the conductance fluctuations and the h/e A–B conductance oscillation to a large magnetic field (chapter 5). Another interesting aspect of the sample-specific $\langle \Delta I_c^2 \rangle$ is its decay over the length scale L_ϕ (rather than ξ_N for $\langle I_c \rangle$). Since $L_\phi \gg \xi_N$ is possible (and usually holds), these fluctuations are relatively robust.

There is an interesting difference between the mesoscopic behavior of tunnel junctions and metallic weak links. This stems from the difference in the behavior of the transmission eigenvalues T_n. For the same average conductance, many small T_n's are relevant in the former case, while in the latter, most T_n's are close to zero or to unity (Dorokhov 1982, 1984, Imry 1986, Pendry et al. 1992) in the diffusive situation. (In the ballistic case, when the conductance g is quantized, this is even more extreme: Most of the T_n's vanish and g of them are equal to unity). This aspect, discussed by Beenakker et al. 1994, leads to a number of interesting physical consequences.

The period of 2π as function of $\chi_1 - \chi_2$ corresponds via eq. 7.44 to the period h/e as a function of the flux in the corresponding ring. A period of π in average quantities would correspond to $h/2e$. Earlier, Spivak and Khmelnitskii (1982) obtained such a dependence (period π) of the *average* normal conductivity of the *SNS* junction on $\chi_1 - \chi_2$. This π-period was also found in the dependence on $\chi_1 - \chi_2$ of a contribution to the average supercurrent amplitude $\langle I_c \rangle$ due to the electron–electron interactions by Altshuler et al. (1983). This is the direct analogue of the $h/2e$-periodic average persistent currents due to the interactions (Ambegaokar and Eckern 1991). Since this effect has a sign related to that of the effective interaction, the possibility of a "negative Josephson current amplitude" arises.[12] This may cause an instability (as can be seen easily from the Bloch argument described earlier) of the $\theta = 0$ self-consistent current solution, leading to two stable ground states with opposite nonzero trapped fluxes θ and $-\theta$, for $\theta_{ext} = 0$. When a voltage V is applied between the two superconductors, $\chi_1 - \chi_2$ will change with time according to eq. 7.37. Thus the π-periodic persistent current will oscillate with time with a basic period of $4eV/\hbar$, as was found by Spivak and Khmelnitskii (1982). The factor of 4 is obtained from two factors of 2: one from the Josephson relation and one from the ensemble averaging. De Vegvar et al. (1994) addressed some of these issues experimentally.

The question of h/e vs. $h/2e$ periodicity of the persistent currents in a ring having both N and S segments was considered by Büttiker and Klapwijk (1986). They found that in general the period is $h/2e$, except when the S region is short enough for normal electrons to tunnel through it. In this case, the sample-specific period of the excitation spectrum and the persistent current

[12] Such a negative amplitude may also occur due to a fluctuation, if $\langle I_c \rangle$ is depressed, as mentioned before.

becomes h/e. Nazarov (1994) constructed a very useful circuit-type theory to deal with more general N–S combinations.

We conclude this section by mentioning some interesting effects in narrow, *short* ($L \ll \sqrt{\xi_0 l}$), ballistic point contacts (which should be achievable with semiconducting components) and some novel mesoscopic phenomena in superconductors. Since the normal conductance in the former case is quantized in units of $e^2/\pi\hbar$, the supercurrent (eq. 7.39) becomes quantized in units of (Beenakker and van Houten 1991a,b, Beenakker 1991)

$$\Delta I_c = \frac{e\Delta_s}{\hbar}. \tag{7.53}$$

When due to a transmission resonance the conductance of a barrier becomes $e^2/\pi\hbar$ (e.g., problem 4 at the end of chapter 5), the Josephson current amplitude also develops a resonance (Glazman and Matveev 1989) with magnitude $e\Delta_s/\hbar$ (Beenakker and van Houten 1991c). Thus, the ratio of I_c to Δ_s at low temperatures is an integer multiple of a universal quantity for such situations.

Four further notable experimental findings for mesoscopic superconductors are:

1. The difference between an even and an odd number of electrons on a small superconducting island (Tuominen et al. 1992, Eiles et al. 1993a,b, Lafarge et al. 1993), occurring since only an even number of electrons participate in the superconducting state.
2. The demonstration (Elion et al. 1994, Joyez et al. 1994, Matters et al. 1995) of the phase–number uncertainty relation, achieved by appropriately coupling to a superconducting island.
3. Anomalies in the "Little–Parks" effect (which is the A–B oscillation of T_c in a cylindrical sample), observed by Vlohbergs et al. (1992).
4. Anomalies in the magnetic response of macroscopic cylinders having N and S walls in good contact (Visani et al. 1990).

The two latter effects still need a theoretical interpretation.

The field of mesoscopics with superconductivity is very promising for further interesting new physical effects. For recent reviews, see Bruder (1995) and articles in Hekking et al. (1994).

8

Noise in Mesoscopic Systems

1. INTRODUCTION

We shall be concerned here with three main types of noise phenomena:

1. Equilibrium or Nyquist–Johnson noise across a resistor (see eqs. A.9 and A.13–17).
2. Various nonequilibrium or shot noise phenomena around a steady state with a current flow.
3. Low-frequency, typically "$1/f$," noise due to slow changes of the resistance with time.

In the first two cases, the noise power is typically "white" (frequency-independent) over a sizable frequency range from zero to $1/\tau^*$, the cutoff frequency, which is the smaller of $k_B T/\hbar = 1/\beta\hbar$ (in this chapter we shall mostly reserve the notation T for the transmission coefficient) and $1/\tau$. τ is a characteristic time for the transport, for example, the transport mean free time for a classical resistor in equilibrium. In this case, and for $\beta\hbar \gg \tau$, the noise power is *linear in* ω for a constant conductance for $0 < (\beta\hbar)^{-1} \ll \omega \ll 1/\tau$. Concentrating on the current noise (which is measured in equilibrium by connecting a zero impedance "a.c." amperometer across the resistor), one considers (see Wax 1954 and Reif 1965 for general references) the Fourier transform of the current–current correlation function (which depends only on t, not on t'), the nature of the averaging denoted by the angular brackets will de discussed later

$$S_I(\omega) = \frac{1}{2\pi}\int_{-\infty}^{\infty} \langle \Delta I(t')\Delta I(t'+t)\rangle e^{-i\omega t}\, dt. \tag{8.1}$$

Here $\Delta I(t) = I(t) - \bar{I}$; where the d.c. average current, $\bar{I}$, vanishes for the equilibrium case. $S(\omega)$ is proportional to $|I(\omega)|^2$, $I(\omega)$ being the Fourier transform of $\Delta I(t)$; $S(\omega)$ is the noise power spectrum per unit angular frequency (see problem 1 at the end of this chapter for a precise treatment of these relationships). Commonly, $\langle \Delta I^2(t)\rangle \equiv \langle \Delta I(0)\Delta I(t)\rangle = \langle \Delta I^2(0)\rangle e^{-|t|/\tau^*}$, in simple cases. These include equilibrium noise for $\beta\hbar \ll \tau$ and classical shot noise, and the power spectrum for low frequencies ($\omega < 1/\tau^*$), $S_I(\omega)$, is given by

$$S_I(\omega) = \frac{1}{\pi}\langle \Delta I^2(0)\rangle \tau^*. \tag{8.2}$$

The fluctuation-dissipation theorem (eqs. A.9 and A.13–17 of appendix A) ensures that, in equilibrium, when $\bar{I} = 0$, R being the resistance, $\omega \ll 1/\tau^*$ and $\omega\beta\hbar \ll 1$,

$$S_I(\omega) = \frac{1}{\pi\beta R}. \tag{8.3}$$

The "engineering" definition of the noise spectrum employs the symmetrized version (cosine transform in eq. 8.1, not important for $\beta\hbar\omega \ll 1$) and a factor of 2 due to using only $\omega > 0$. The noise power at a temperature T, per unit frequency, f, is thus $4k_B T/R$. For a theoretical confirmation of the equilibrium noise in a mesoscopic sample see, for example, Entin-Wohlman and Gefen (1991).

The phenomenon of shot noise is due to the current in a steady-state nonequilibrium situation being carried by discrete charges e. An average $\bar{I}$ means that on the average $\bar{I}/e$ electrons are passing through the charge counter per unit time. Assuming that these single-charge events are uncorrelated, one obtains for the fluctuations of charge flowing *per unit time*, $\overline{\Delta I^2} = e\bar{I}$, and at low frequencies[1] it can be shown from (8.1), along the lines of problem 1, that

$$S_I(\omega) = \frac{\overline{\Delta I^2}}{2\pi} = \frac{e}{2\pi}\bar{I}. \tag{8.4}$$

This is the result for classical shot noise (see, e.g., van der Ziel 1986) with uncorrelated events, examples being the fluctuations of the rainfall in a small area due to the water coming in raindrops and electron current in vacuum tubes, neglecting space charges.[2] The quantum nature of the particles, simply

[1] Note that the shot-noise vanishes, in fact, for a continuous charge, $e \to 0$, for a given $\bar{I}$.

[2] Coulomb effects and inelastic scattering greatly reduce the shot noise in classical conductors. See Landauer 1993, 1994.

their being fermions or bosons, induces correlations which may modify eq. 8.4. A well-known phenomenon of this type is photon "bunching" as seen in the Hanbury-Brown and Twiss experiment (1956, 1957). Currents in real electrical wires may not have the full shot noise given by eq. 8.4, due to smoothing of the charge fluctuations by scattering and by Coulomb interactions. We shall consider here only the modifications due to the quantum nature of the particles.

Of course, the equilibrium noise is an *exact* result of equilibrium statistical mechanical fluctuation theory and it is valid for mesoscopics as well. It turns out, on the other hand, that quantum effects (for coherent conductors), both on the single particle level and due to Fermi statistics, make important corrections to the shot noise result. However, when $\bar{I}$ tends to zero, one can obtain the equilibrium noise as the limit of the quantum shot noise (Landauer 1989a, 1993).

The simplest case of shot noise, for a reservoir radiating, via an "obstacle" in a waveguide, into free space, will be discussed in section 2. The nontrivial modifications due to connecting another reservoir and the equilibrium limit will be considered in section 3.

In section 4 we shall discuss the low-fequency (typically $1/f$, f being the frequency) noise. This was found (Bernamot 1937, Dutta and Horn 1981) to be caused in many cases by infrequent changes of the resistance in time due to slow motion of atoms, or changes in their ionization state. Feng et al. (1986) suggested that a mesoscopic effect—change of the resistance of a phase-coherent piece of a sample due to a change of the impurity configuration—might be enough to account for the $1/f$ noise in macroscopic samples.

2. SHOT NOISE FOR "RADIATION" FROM A RESERVOIR

Consider a particle reservoir in equilibrium (chemical potential μ, inverse temperature β) radiating particles into free space via a waveguide—that is, a pipe or an ideal conductor connected to it through an opening. The waveguide has a total transmission coefficient T from the reservoir to the vacuum. T can be determined by obstacles placed in the guide and/or by imperfect impedance matching with free space. The opening connecting the reservoir to the pipe is small enough so that the emitted radiation does not disturb the equilibrium in the large reservoir for all times of interest. We also assume that the opening is engineered to have perfect transmission into the guide. This is depicted in Fig. 8.1. For simplicity, we consider a single-channel waveguide. The generalization to many channels is not difficult, and it is similar to that done in chapter 5. This generalization, as well as the one to many reservoirs, is given in the cited literature. We denote the density of states per unit length and per unit energy of the guide (going *away* from the reservoir) by $n_1(\epsilon)$. By the same assumptions as in the Landauer formulation, the outgoing states in the pipe are fed to an equilibrium population $\bar{f}(\epsilon)$ ($\bar{f}$ is the Fermi or Bose function, as the

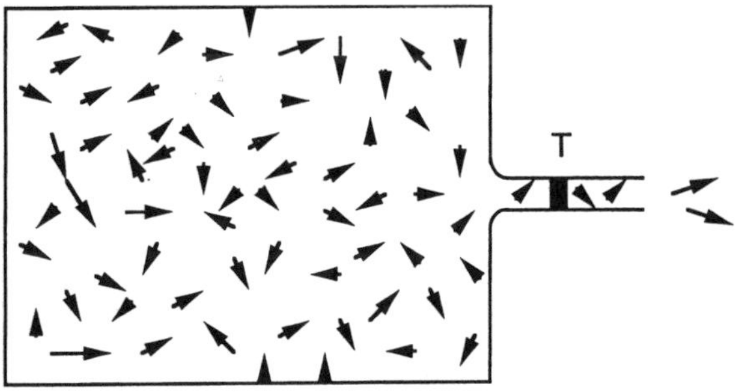

Figure 8.1 An equilibrium reservoir radiating into free space via a perfectly matched waveguide having a transmission T to the vacuum.

case may be). These assumptions are substantiated (Landau and Lifschitz 1959, p. 178) for a "black body" using the Liouville theorem as mentioned in chapter 5. For a small energy interval $\Delta\epsilon$, the average number of particles emanating per unit time from the reservoir into the pipe (in the general expressions below, through eq. 8.9, we shall consider *particle* current—the number of particles transmitted per unit time) is $I_0 = \nu f$, whose average is

$$\bar{I}_0 = \nu \bar{f}, \qquad \nu = v n_1(\epsilon)\Delta\epsilon, \tag{8.5}$$

v being the velocity along the pipe, f the instantaneous population and $\bar{f}$ its average. For massive particles the rate or "frequency of emission" ν (having nothing to do with the measured one, $\omega/2\pi$) is given[3] by $\nu = \Delta\epsilon/\pi\hbar$. ν happens to be the same also for unpolarized photons in a 1D pipe. We are now (Schwimmer and Imry 1994, unpublished) going to present a simple calculation of the fluctuations due to those of I_0, and of the transmission for given I_0. Out of I_0 such particles, the probability of exactly I passing the obstacle in unit time (note again that here I_0 is a time-fluctuating quantity due to fluctuations in the populations f; the average is given by eq. 8.5)

$$P_{I_0}(I) = \binom{I_0}{I} T^I (1-T)^{I_0 - I}. \tag{8.6}$$

This is the probability for I successes out of I_0 *independent* trials with a success probability T. The average and the standard deviation squared are given by I_0 times those per trial. The latter are given, respectively, by T and $T(1-T)$.

The average current going into the vacuum, denoting the probability to have an I_0 by $P(I_0)$, is

$$\bar{I} = \sum_{I_0=0}^{\infty} P(I_0) \sum_{I=1}^{I_0} I P_{I_0}(I) = \sum_{I_0} P(I_0) T I_0 = T\bar{I}_0 = \nu \bar{f} T. \tag{8.7}$$

[3] See eq. 5.14; we include a factor of 2 for a "spin" degeneracy.

Let us now calculate the fluctuations $\overline{\Delta I^2} = \overline{I^2} - \bar{I}^2$:

$$\overline{I^2} = \sum_{I_0} P(I_0) \sum_{I} I^2 P_{I_0}(I) = \sum_{I_0} P(I_0)[(TI_0)^2 + I_0 T(1-T)], \tag{8.8}$$

where we have used the fact (mentioned following eq. 8.6) that the variance of the binomial distribution (8.6) is $I_0 T(1-T)$. We need $\overline{I_0^2} = \sum_{I_0} P(I_0) \cdot I_0^2 = \bar{I}_0^2 + \sum_{I_0} P(I_0)\overline{\Delta I_0^2} = \bar{I}_0^2 + \nu\bar{f}(1 \mp \bar{f})$; to get the last equality we use the well-known result that for fermions (upper sign) or bosons (lower sign) the variance of f is given by $\bar{f}(1 \mp \bar{f})$ (Landau and Lifschitz 1959) and that the ν occupations are independent. Putting this into eq. 8.8 we obtain for the fluctuations of the current in unit time:

$$\overline{\Delta I^2} = T^2 \nu\bar{f}(1 \mp \bar{f}) + \nu\bar{f}T(1-T) = \nu\bar{f}T(1 \mp \bar{f}T). \tag{8.9}$$

This has the simple physical meaning that the *fluctuations in the final outgoing current per state are given by the usual Fermi/Bose result.* The latter is $\bar{f}T(1 \mp \bar{f}T)$, per outgoing state.

In the simplest version of the Hanbury-Brown and Twiss (1956) (see, e.g., Baym 1969, p. 431 for a pedagogical presentation) photon correlation experiments, fluctuations in the intensity seen by a photon counter looking at a source with $T = 1$ are larger than the classical result for independent events $\nu\bar{f}$ and in fact are given by $\nu\bar{f}(1 + \bar{f})$. The latter result appears in section 112.7 of Landau and Lifschitz (1959), and is attributed there as due to Einstein in 1909. We see that, operationally, having a transmission coefficient T between the source and the counter would alter the result in a very well-defined way (e.q. 8.9), governed by the average observed population per state $\bar{f}T$.

An interesting suggestion due to Martin and Landauer (1992) and Murphy (1992 unpublished) is to split an electron beam with $f = 1$ and no fluctuations into two beams, each having its effective $\bar{f} = 1/2$ and, thus, maximal $T\bar{f}(1 - T\bar{f})$ fluctuations. These fluctuations must be fully anticorrelated to agree with the vanishing fluctuations in the initial beam. This is a Fermion analogue to a two-counter Hanbury-Brown and Twiss experiment.

For electrons at zero temperature, the shot noise fluctuations of the emitted charge current per unit time due to a voltage V are given by adding the $eV/\Delta\epsilon$ independent contributions as above

$$\overline{\Delta I^2} = e^2(eV/\Delta\epsilon)\nu T(1-T) = eVG(1-T) = e\bar{I}(1-T). \tag{8.10}$$

This is equivalent to the classical result (eq. 8.4) multiplied by $(1 - T)$. Thus, this simplest type of shot noise *vanishes* for both $T = 0$ and $T = 1$. In these two limits and at zero temperature the transmission process has no probabilistic character. The result (8.10) for arbitrary T, which is rather easily generalizable to many channels, was first derived by Khlus (1987) and Lesovik (1989) (see also Yurke and Kochanski 1989, Büttiker 1990). When the conductance has quantized steps, as in a ballistic quantum point contact the shot noise will be maximal at the transitions between steps. Only then does a channel with $0 < T < 1$ open up. The result as presented in eqs. 8.8–8.9 clarifies that the total fluctuations are due to *two independent* reasons: the fluctuations in the populations of the states in the source, and simple statistical fluctuations around the average transmission T. Clearly, each incoming particle is either transmitted or not! This understanding suggests immediately a simple restatement of eq. 8.9: In each of the ν *independent* events, the "success" probability is $\bar{f}T$ and the variance is $\bar{f}T(1 - \bar{f}T)$. Multiplying by ν yields eq. 8.9. Our above short derivation of the latter (eqs. 8.7–8.9) just reiterates known results in probability theory.

This treatment also clarifies where the Fermi vs. Bose nature of the particles appears. Thus, we see traightforwardly the relationship with the Hanbury-Brown and Twiss fluctuations. For further references, see Beenakker and van Houten (1991b), Büttiker (1992a, b), Chen and Ting (1992), Hershfield (1992), Davies et al. (1992), Shimizu et al. (1992), Landauer and Martin (1992), Martin and Landauer (1992), de Jong and Beenakker (1992, 1994), Levitov and Lesovik (1993), Landauer (1993, 1995), Ueda and Shimizu (1993), Gurevich and Rudin (1996). Landauer, in the last-mentioned reference, discussed how the shot noise is reduced in a sample much larger than L_ϕ, and goes to zero in the macroscopic limit. This can be seen by mentally dividing the sample, as in chapter 5, into many coherent volumes. Notably, Beenakker and Büttiker (1992) and Nagaev (1992) found a universal modification of the shot noise for a quasi-1D diffusive coherent conductor by a factor of 1/3. This follows from a calculation relying on the fact that most of the transmission eigenvalues are 0 or 1 (Dorokhov 1984, Imry 1986, Pendry et al. 1992). For related experiments, see Li et al. (1990a, b), Kil et al. (1990), Washburn et al. (1992), and Liefrink et al. (1994a, b). The reduced shot noise in the quantum point contact was definitively observed by Reznikov et al. (1995). A new type of reduction at small transmissions, which may possibly be due to Coulomb correlations, was discovered by these authors and by Birk et al. (1995). The lower frequency behavior was obtained by Kumar et al. (1995).

3. THE EFFECT OF FLUCTUATIONS IN THE SINK, THE EQUILIBRIUM LIMIT

Obviously, when the particles from the source are radiated into a second reservoir (which we shall call "the particle sink"), instead of a vacuum, two additional effects must be considered. First, due to the quantum (Bose or Fermi) nature of particles, the fluctuating currents depend on the *populations of the final states*. Fluctuations in the latter present an additional reason for noise.[4] Second, particles will also be radiated backward from the sink into the source. The effect of this can easily be obtained by reversing the roles of the source and the sink. The source and the sink are incoherent (chapter 5) and these two contributions have to be added appropriately.

The effect of the population of the final state is given by multiplying the transmission coefficient T by $(1 \mp f')$ where f' is the population of the sink state at the given energy. A physical discussion of the effect of Pauli correlations for fermions along the section of the waveguide between the scatterer and the sink, employing a wave-packet picture, was given by Martin and Landauer (1992). The treatment below is equivalent to theirs; see also Muzykantskii and Khmelnitskii (1994).

For simplicity, we present the calculation of the current and its fluctuations for fermions. The current I_1 carried by each state is obtained as follows: Provided there is an electron on the left and a hole on the right $(f_l \equiv 1, f^1 \equiv f_r = 0)$, the probability to go from left to right is T and the current is given by $ev_F T/L$, L being the length of each section of the "waveguide," since v_F/L is the rate at which electrons from the left hit the barrier. For $f_l = 0, f_r = 1$, $I_1 = -ev_F T/L$. For $f_l = f_r = 0$ or $f_l = f_r = 1$, $I_1 = 0$. Thus, more generally, I_1 is given by

$$T\frac{v_F}{L}[f_l(1-f_r) - f_r(1-f_l) = T\frac{v_F}{L}(f_l - f_r).$$

For an energy interval $d\epsilon$, the total rate at which electrons, in all states in the interval $d\epsilon$ would hit the barrier is $v_F n_1(\epsilon)\, d\epsilon = \nu$ (see eq. 8.5). For each such trial the probability to go to the right is $x_+ \equiv f_l(l - f_r)T$ and to the left it is $x_- = f_r(1 - f_l)T$. The probability to do nothing is $1 - x_- - x_+$.

The probabilities x_+ and x_- to go to the right and to the left are per "event," and there are ν such independent events per unit time. For each of the events the average net current going to the right is $x_+ - x_-$ and the corresponding average square of the current is $x_+ + x_-$. Thus, the average current in the interval $d\epsilon$ in unit time is

$$\bar{I} = \nu T(f_l - f_r). \tag{8.11}$$

Its variance is $\nu(x_+ + x_-) - \nu(x_+ - x_-)^2$. Thus

$$\overline{\Delta I^2} = \nu T[f_l(1 - f_r) + f_r(1 - f_l)] - \nu T^2(f_l - f_r)^2, \tag{8.12}$$

which is the principal result for fermions.[5] To clarify this further, we shall now rederive it and, rather easily, find the approximate distribution function for the

[4] It is interesting that these populations cancel and do *not* play a role in ordinary (averaged current) transport, see chapter 5. However (as comparison of eqs. 8.11 and 8.12 below shows), they do influence the *fluctuations* of the current! This point has been emphasized and discussed by Landauer (1993, 1994).

[5] An equivalent expression is $\overline{\Delta I^2} = \nu T[f_l(1 - f_l) + f_r(1 - f_r) + (1 - T)(f_l - f_r)^2]$; this was obtained by Büttiker (1992b).

right- and left-moving currents. The probability per unit time to go n_+ times to the right and n_- times to the left is given by the trinomial distribution

$$P_\nu(n_+, n_-) = \frac{\nu!}{n_+!n_-!(\nu - n_+ - n_-)!} x_+^{n_+} x_-^{n_-} (1 - x_+ - x_-)^{1-n_+-n_-}, \qquad (8.13)$$

which is a simple generalization of the binomial one (see eq. 8.6). We display here its gaussian approximation, valid for large ν. Expanding ln P_ν around its maximum at $\bar{n}_+, \bar{n}_-$ (i.e., writing $n_\pm = \bar{n}_\pm + \delta_\pm$), for large ν, one finds, after some algebra, that P_ν is well approximated by a two-variable gaussian distribution,

$$P_\nu(n_+, n_-) = \tilde{N} \exp\left\{-\frac{1}{2\nu}\left[\frac{\delta_+^2}{x_+} + \frac{\delta_-^2}{x_-} + \frac{(\delta_+ + \delta_-)^2}{1 - x_+ - x_-}\right]\right\}, \qquad (8.14)$$

where $\tilde{N}$ is a normalizing factor and $\bar{n}_\pm = \nu x_\pm$, as expected. The various fluctuations, $\overline{\delta_\pm^2}$, $\overline{\delta_+\delta_-}$, can be most simply obtained (e.g., Landau and Lifschitz 1959) from the matrix inverse to that in the exponent of eq. 8.14. These results, and the correct higher-order moments obtained when one goes beyond the gaussian approximation, can also easily be obtained directly from eq. 8.13 using the generating function method (Levitov and Lesovik 1993). It follows from such elementary evaluations that the distribution of I is gaussian for $\nu \to \infty$, and in agreement with eq. 8.12,

$$\begin{aligned}\overline{\Delta I^2} &= \nu[x_+ + x_- - (x_- - x_+)^2] \\ &\to \nu\{T[\bar{f}_l(1 \mp \bar{f}_r) + \bar{f}_r(1 \mp \bar{f}_l)] \mp T^2(\bar{f}_l - \bar{f}_r)^2\},\end{aligned}$$

where in the last expression, which is the general *central result*, we have inserted (as the lower signs) the results for bosons, which can be obtained from somewhat more complicated calculations on a specific model (Schwimmer and Imry 1994 unpublished). The $\mp$ sign in the last term of eq. 8.15 may seem surprising: it is a generalization of the, very natural, $\mp$ sign in the last term of eq. 8.9 (see also Büttiker 1992b).

As a check we take the equilibrium case, $\bar{f}_l = \bar{f}_r = f$, finding:

$$\overline{\Delta I_{eq}^2} = 2\nu\bar{f}(1 \mp \bar{f})T = \frac{2\Delta\epsilon e^2}{\pi\hbar}\bar{f}(1 \mp \bar{f})T = 2G\Delta\epsilon\bar{f}(1 \mp \bar{f}). \qquad (8.16)$$

Noting that for fermions $d\bar{f}/d\epsilon = -\beta\bar{f}(1 - \bar{f})$ and integrating over energies we get

$$\overline{\Delta I_{eq}^2} = \frac{2G}{\beta}, \qquad (8.17)$$

which is the Nyquist–Johnson result. The idea to get the equilibrium noise from the shot noise was suggested and discussed in this connection by Landauer (1989a, 1993). Note that we have treated here only the static ($\omega \ll 1/\tau^*$) limit.

For the case where the "sink" on the right is effectively a vacuum (i.e., $\mu_r \to -\infty, f_r \to 0$ or $\beta(\epsilon - \mu_r) \gg 1$ for fermions) we find from eq. 8.13

$$\overline{\Delta I^2} = \nu T\bar{f}_l(1 \mp T\bar{f}_l), \tag{8.18}$$

in agreement with eq. 8.9, the simplest shot noise result. Equation 8.15 is the most general result for the low-frequency fluctuations, including both the equilibrium and the shot noise. We emphasize that it is entirely obtainable from an elementary treatment of the combined occupation probabilities in the source and in the sink, and the fluctuations due to the transmission of discrete particles through the "barrier." The former is determined by the Fermi or Bose nature of the particles (tending to the classical limit for $f \ll 1$ or $T \ll 1$) and the latter is due to the probabilistic nature of going through the barrier or passing through a disordered conductor.

While we have considered both the thermal equilibrium noise and the nonequilibrium shot noise using the Landauer picture, it is very likely that the results apply to any situation where the discrete nature of the charge carriers is relevant. For example, one may consider the tunnel-junction model, leading to the conductance expression of eq. 2.19. Two massive conducting pieces are weakly coupled (a sufficient condition for weak coupling is that the conductance between them is much smaller than e^2/h) via the "transfer Hamiltonian"

$$\mathcal{H}_T = \sum_{l,r} tc_r^\dagger c_l + t^* c_l^\dagger c_r. \tag{8.19}$$

$c_{l,r}^\dagger$ and $c_{l,r}$ are the creation and annihilation operators for electrons in the left- and right-hand conductor and t is the matrix element as in chapter 5. Consider the states (l, r) at a given energy, ϵ. By forming the commutator between the number operator $\hat{n}_l = c_l^\dagger c_l$ on the left-hand side and the Hamiltonian, and noting that $\dot{n}_l = \hat{I}_1$, the operator for particle current *into* l, one finds

$$\hat{I}_1 = (i/\hbar)(tc_r^\dagger c_l - t^* c_l^\dagger c_r). \tag{8.20}$$

We see immediately that

$$\overline{\hat{I}_1^2} = \frac{|t|^2}{\hbar}[\bar{f}_r(1 - \bar{f}_l) + \bar{f}_l(1 - \bar{f}_r)], \tag{8.21}$$

In the nonequilibrium case, the average current, given by eq. 2.19, and being of order, $|t|^2$, as well, must be taken into account too, in order to obtain the variance of I_1.

4. LOW-FREQUENCY (1/f) NOISE

In almost all situations where the voltage noise is measured across a current-carrying sample, it is found that at low frequencies the dominant noise increases as an inverse power of the frequency. The exponent is often very close to unity and therefore this noise, which is very different from the white (frequency-independent) noise, is called "$1/f$," or simply low-frequency noise. Its ubiquity makes the observation of the "white" (at low frequencies) shot noise much more difficult. It was soon recognized that the intensity of this $1/f$ noise is *proportional to the square* of the impressed d.c. current. A rather general presentation (Hooge 1969) is

$$\overline{\Delta V^2(\omega)} = \gamma \frac{V^2}{N_e f}, \tag{8.22}$$

where N_e is the total number of charge carriers and γ is typically $\sim 10^{-3}$. The $1/f$ noise may be interpreted as due to slow variations of the resistance, which lead at a constant current to slow voltage fluctuations. Thus, $1/f$ noise can show up as fluctuations in the Johnson–Nyquist noise (Voss and Clarke 1976; these authors developed a temperature-fluctuations model for $1/f$ noise, which has had some success but does not seem to have a general applicability). The evidence for the interpretation of $1/f$ noise as due in many cases to slow resistance fluctuations is presented in Dutta and Horn (1981). For a more recent review emphasizing the great variability and possible nonuniversality of $1/f$ noise, see Weissman (1988). We note that the shot noise is proportional to the current to the *first* power, and is thus distinguishable, in principle, from the $1/f$ noise.

In many systems (Ralls et al. 1984; Farmer et al. 1987; Ralls and Buhrman 1988) the conductance jumps between two or more locally stable values ("telegraph noise"), which may be interpreted as due to either the motion of a scatterer between two locally stable positions or to the ionization and deionization of an impurity (e.g., a donor or acceptor in a semiconducting system). It is shown below that $1/f$ noise follows when many such activated centers exist, with a smooth distribution of the barriers governing the transitions between the two states. The possible relevance of surface states in this connection was pointed out by McWhorter (1957). We remark parenthetically that it is immediately suggested that such processes should occur with a greatly enhanced intensity near a metal–insulator transition due to doping (Finkelstein and Imry 1992 unpublished). In fact, a huge, apparently universal, increase of the intensity of $1/f$ noise near metal–insulator transitions was found experimentally by Cohen et al. (1992).

It is easy to appreciate the main idea that resistance changes due to activated processes with an activation energy which is distributed more or less uniformly in a rather modest range should lead to $1/f$ noise (Bernamot 1937). Consider such an activated process characterized by an activation energy W, whose rate is given by

$$\frac{1}{\tau} \sim \frac{1}{\tau_0} e^{-\beta W} \tag{8.23}$$

and occurring around "impurity" sites whose concentration is n_i (say $n_i \sim 10^{17}/\mathrm{cm}^3$). Assume that this process leads to a resistance change of the order of ΔR. Suppose that the activation energy for such motions is approximately uniformly distributed between W_{min} and W_{max} (typically, these energies may be expected to be in the range of $\sim (10^{-1}$–$5)$ eV). For changes at frequency ω, the relevant barriers yielding $1/\tau \sim \omega$ are obviously of the order of (for a more quantitative consideration, see below)

$$W_{\mathrm{eff}} \sim \frac{1}{\beta} |\ln \omega\tau_0|. \tag{8.24}$$

To take a typical example, for $\tau_0 \sim 10^{-14}$ s and $\omega \sim 1 - 10^8$ Hz (eight decades!), at room temperature the relevant, and rather limited, range of W_{eff} is roughly $\frac{1}{2}$–1 eV, and it depends exceedingly weakly on the frequency and on τ_0. We approximate the distribution of W in the above range as a uniform one, $P(W) \sim 1/W_0$ (more generally, W_0 may vary slightly within the relevant frequency range). We find from eqs. 8.23 and 8.24 that the distribution of the characteristic rate or frequency is indeed of the inverse frequency type:

$$P(\omega) \sim P(W_{eff}) \frac{dW_{eff}}{d\omega} \sim \frac{1}{W_0 \beta \omega}. \tag{8.25}$$

The coefficient $1/W_0\beta$ can have values of the order of 10^{-3}–10^{-1}.

To obtain eq. 8.25 in a more systematic fashion (Bernamot 1937, see Dutta and Horn 1981), we consider first a process with a simple characteristic rate of $1/\tau$. It will cause the autocorrelation function of the resistance to decay like $\exp(-t/\tau)$. Thus, the resistance, or measured voltage, will have a noise spectrum of

$$S_W(\omega) = \frac{1/(\tau\pi)}{\omega^2 + (1/\tau)^2}. \tag{8.26}$$

The index W signifies that τ is governed by an activation barrier W. We next take a broad spectrum of τ values. We assume that the τ's have a distribution due mostly to that of W, $P(W)$. The total noise power will be

$$S_{tot}(\omega) = \int dW\ S_W(\omega)P(W) \tag{8.27}$$

(where we understand that S_W is taken with $1/\tau$ having W as its activation energy). Again, for a large but finite range of ω, the integral over W will be mainly in the regime defined by eq. 8.24 and $P(W)$ is approximated there by $1/W_0$ (W_0 is a constant having dimensions of energy). Since

$$S_W(\omega) = \frac{\tau_0 e^{\beta W}/\pi}{1 + \omega^2 \tau_0^2 e^{2\beta W}},$$

the integral is elementary and one obtains eq. 8.25. It is interesting to note (Pytte and Imry 1987) that the fluctuation-dissipation theorem (eq. A.16, or eq. 8.1) and eq. 8.25 imply that the imaginary part of the susceptibility, or $\sigma(\omega)$ is proportional to $P(W)$ at a value of W which depends logarithmically on frequency. Thus, the correction to $P(W)$ beyond the constant, which causes γ in eq. 8.22 to depend weakly on frequency, makes $\sigma(\omega)$ have a logarithmic frequency dependence. Further general properties of the response functions follow from the Kramers–Kronig relationships.

We remark that the $1/f$ dependence is not exact, since $P(W)$ at the appropriate range (eq. 8.24) will vary (albeit very weakly) with ω. Obviously, the $1/f$ dependence cannot apply for $f \to 0$ or $f \to \infty$, since the total integrated noise power must be finite. This means, of course, that arbitrarily large and small values of W are effectively cut off. This is, however, hardly a real issue, because of the logarithmic dependence (eq. 8.24).

It is far from straightforward even to estimate how a given relaxation process in the sample affects its resistance (an exception is that of "two-level systems" in metallic glasses (Ludviksson et al. 1984); we shall briefly come back to these later). The occurrence of the inverse volume or particle number (eq. 8.22) goes in the right direction, but one still needs a specific mechanism. Perhaps surprisingly, it appears that mesoscopic physics may come to the rescue in this respect, even for macroscopic systems (Feng et al. 1986).

Let us start with a mesoscopic system which is fully coherent (all its dimensions, L_i, satisfy $L_i \ll L_\phi, L_T$). We know that two distinct members of the impurity ensemble have their conductances differing by $\sim(e^2/\hbar)$, and the resistances differing by $\sim\bar{R}^2(e^2/\hbar)$. The conductance fluctuation is thus a "fingerprint" of the impurity configuration (this concept has been more frequently used concerning the magnetic field dependence, which we do not consider here). The question is immediately suggested: How much of a change in the impurity configuration is needed to bring about a full change of the fingerprint or the "ensemble member" and a concurrent change of the conductance by $\sim e^2/\hbar$?

The full diagrammatic calculation of this conductance change was done by Altshuler and Spivak (1985) and Feng et al. (1986). In the metallic limit

$l \ll L \ll \xi$, it was found that moving a *single* scattering center by the order of a Fermi wavelength is enough to accomplish the full change of the conductance by $e^2/\hbar$ in *one dimension* and by $e^2/(\hbar k_F l)$ *in a strictly two-dimensional system*. In 3D, this change of $e^2/\hbar$ is reduced by a factor $(k_F^2 lL)^{-1/2}$, where L is replaced by the thickness for thin wires and films. These results can be understood physically using the proportionality of the conductance to the transmission probability across the sample. The latter is given in terms of the absolute value squared of the sum of classical Feynman paths crossing the system (an unpublished quantitative evaluation of this representation was done by Argaman in 1993). It is very easy to see that the number of such diffusive paths is of the order $(L/l)^2$. In 2D, this implies that each path passes through a finite fraction of the sites and a change of 2π in the phase of paths passing through a single scatterer is enough to change the interference significantly. This is even stronger in 1D, where the paths pass many times through any site. In 3D, the probability of paths passing through each site is smaller by another l/L ratio, which reduces the effect, compared to 2D, by a further $\sqrt{l/L}$ factor.

It was suggested by Feng et al. (1986) (see also Pendry et al. 1986) that the above may have an implication for $1/f$ noise. In coherent systems, the above change of R provides the scales for the resistance changes to supply the prefactor in the $1/f$ noise. The reason the above mechanism produces reasonable changes of the resistance even for macroscopic samples ($L \gg L_\phi$) is the relatively weak, power-law, dependence of the conductance fluctuations on L/L_ϕ (Altshuler and Khmelnitskii 1985, Lee et al. 1987). A clear way to understand this physically is, as in chapters 2 and 4 (Imry 1986), to divide the large sample mentally into $(L/L_\phi)^d$ boxes of size L_ϕ. Each box is almost coherent and has roughly the "coherent" value (i.e., the value obtained for $L_\phi \gg L$) of the sensitivity of the conductance to impurity motions. The contributions of the boxes have to be added classically since the motion between two boxes is incoherent. Thus, the fluctuation squared is multiplied by $(L_\phi)^d/\mathrm{Vol}$ (notice the agreement with the volume dependence, eq. 8.22). For a further semiquantitative evaluation, one has to remember that the conduction occurs in a thermal strip of width $k_B T$ and $k_B T \gtrsim \hbar/\tau_\phi$ in all cases where a comparison was made.[6] The energy averaging in this thermal energy strip leads to a further power-law reduction of ΔG^2 by a $(\hbar)/k_B T\tau_\phi$ (Lee et al. 1987, Atshuler and Khmelnitskii 1985). Thus the final relative change of the conductance due to impurity motion is

$$\frac{(\delta G)^2}{G^2} = \frac{\delta G_{coh}^2}{G_{coh}^2} \frac{L_\phi^d}{\mathrm{Vol}} f(\beta\hbar/\tau_\phi). \tag{8.28}$$

Here the subscript *coh* signifies a box with size $\lesssim L_\phi$ and the function f is of $O(1)$/proportional to its argument, when the latter is larger/smaller than unity.

[6] In fact this inequality should be valid as long as the Fermi-liquid picture holds.

Such estimates can give the right order of magnitude to explain ordinary $1/f$ noise, under appropriate assumptions. To convert the above picture with that of eqs. 8.26–28 to the power spectrum of the conductance fluctuations, we write for a given τ, $\langle \Delta G(0) \Delta G(t) \rangle = \overline{\Delta G^2} e^{-t/\tau}$. Integrating over the distribution of the τ's (or W's) is done as before. For a 3D sample of size L, and taking $\overline{\Delta G^2_{L_\phi}} = (Ae^2/\hbar)^2$, we find, up to numerical coefficients, that the Hooge parameter (eq. 8.22) is given (taking $f = 1$) by

$$\gamma \sim \frac{A^2}{k_F l} \frac{L_\phi}{l} \frac{K_B T}{W_0}. \tag{8.29}$$

For $A \sim 1$, which necessitates $k_F l \sim 1$ in 3D, this is of a similar order of magnitude to experiment. Larger values of $k_F l$ will not change the result too much, because L_ϕ increases with $k_F l$.[7] It is perhaps fair to say that the quantitative general validity of this idea has not been confirmed yet, but that it appears to be relevant at least in many situations.

Further applications of these ideas may occur in metallic glasses, where two-level tunneling systems seem to be relevant to the low-temperature properties (including the propagation of acoustic waves). A classical theory of their contribution of $1/f$ noise was given by Ludviksson et al. (1984), but the above quantum effect (even at relatively high temperatures) may well be relevant (Feng et al. 1986). Somewhat related ideas on spin glasses, which we shall not discuss further here, were given very early by Altshuler and Spivak (1985).

Problems

1. This problem is aimed to make the notion of "power spectrum" very clear in relation to the average noise power in a frequency window of unity, which is the quantity measured experimentally.

   (a) **Definitions:** Suppose the signal $v(t)$ is measured over a long interval T (periodic boundary conditions are taken for simplicity, but that is immaterial). The Fourier representation of $v(t)$ is

$$v(t) = \frac{2\pi}{T} \sum_n v_n e^{2\pi int/T}. \tag{8.30}$$

   v_n is also denoted as $v_\omega, \omega = 2\pi n/T, v_\omega = v^*_{-\omega}$; where $v_n = (1/2\pi) \times \int_0^T v(t) e^{2\pi int/T}\, dt$. The correlation function of v, $K_v(t)$, may be defined for $t \ll T$ as the average of $v(t')v(t'+t)$ (either over an ensemble of signals, or over the initial time, for large enough T:

$$K_v(t) = \overline{v(t_0)v(t_0+t)} \equiv \frac{1}{T} \int_0^T v(t')v(t'+t)\, dt' \tag{8.31}$$

[7] However, as discussed above, the $1/f$ noise should increase near the metal–insulator transition, due to electronic rearrangement effects.

and the power spectrum is defined as:

$$v_\omega^2 \equiv \frac{1}{2\pi}\int K_v(t)e^{-i\omega t}\,dt. \tag{8.32}$$

(b) Prove that $|f_\omega|^2 = 2\pi T v_\omega^2$ (this is called "Wiener–Khintchin theorem").

(c) An experimentalist filters the signal over a frequency window $\Delta\omega$ around ω ($\Delta\omega \gg 2\pi/T$, but $\Delta\omega$ is so small that the variation of v_ω^2 over the interval $\Delta\omega$ is negligible). The filtered signal is denoted in obvious notation ($n \in \Delta\omega$ means: $(\omega - \Delta\omega/2) < 2\pi n/T < (\omega + \Delta\omega/2)$):

$$v_{\Delta\omega}(t) = \frac{2\pi}{T}\sum_{n\in\Delta\omega} v_n e^{2\pi i n t/T}. \tag{8.33}$$

Its averaged noise power is clearly

$$P_{\Delta\omega} = \frac{1}{T}\int_0^T [v_{\Delta\omega}(t)]^2\,dt. \tag{8.34}$$

Prove that it satisfies

$$P_{\Delta\omega} = \left(\frac{2\pi}{T}\right)^2 \sum_{n\in\Delta\omega} |v_n|^2. \tag{8.35}$$

Since the number of ω's in the interval $\Delta\omega$ is $T\Delta\omega/2\pi$, and using (b)

$$P_{\Delta\omega} = (2\pi)^2 v_\omega^2\,\Delta\omega. \tag{8.36}$$

Thus, $(2\pi)^2 v_\omega^2\,\Delta\omega$ is the averaged noise power in the appropriate frequency window $\Delta\omega$. Note that this is independent of T once $K_v(t)$ reaches its limit at large T.

2. Calculate the power spectrum of the voltage noise across a capacitor C, connected in parallel with a resistor R, in two ways; first, by applying the Nyquist theorem; second by looking at thermodynamic fluctuations of the capacitor's voltage and regarding R as providing a frequency scale. Discuss physically the two limits: $\omega RC \gg 1$ and $\omega RC \ll 1$.
3. Do the steps in deriving eq. (8.29).

9

Concluding Remarks

Many of the interesting and novel phenomena that occur in mesoscopic systems have been reviewed and discussed in this book. After a brief discussion in chapter 1 of the available experimental systems and "micro-" and "nano-" fabrication possibilities, the modification of the electronic properties, especially the transport, due to Anderson localization were considered in chapter 2. This is the first example where quantum interference has important consequences even macroscopically. Chapter 3 dealt with the general question of how inelastic scattering produces an uncertainty in the relative phase of partial waves, thereby eliminating their interference. An expression was produced which helps to evaluate the inelastic rate in terms of the dissipative response function of the system, which is usually known. As an application, the very nontrivial dephasing due to electron–electron interaction in the diffusive case was considered.

In chapter 4, equilibrium properties, mainly the response of a mesoscopic system to a magnetic field, were discussed, an instructive situation being the persistent current in a ring due to the Aharonov–Bohm (A–B) flux. The physics of these currents was explained, emphasizing their absolute stability, and estimates were made of their magnitudes for noninteracting electrons. This is a clear example of the omnipresent, important difference between a mesoscopic effect—the current in a given, specific ring—and the often much smaller result obtained when averaging over the impurity-ensemble is performed. In the latter case, the period is halved and the result depends on fine points, such as whether the electron number or the chemical potential is kept fixed when the flux is varied. Since the experimentally observed persistent currents are much larger than the above estimates, electron–electron interactions must be introduced. Some ideas on those interactions were presented, emphasizing the local charge neutrality concept, which is of general validity and relevance (including an application in chapter 5). As for explaining the persistent-current experiments, these ideas are still tentative and a more complete treatment is necessary.

Chapter 5 treated the most active part of mesoscopic physics at present—transport phenomena. The Landauer formulation was introduced as the basic paradigm, with both the two-terminal and the four-terminal versions. Many applications follow: the quantized conductance of ballistic "point contacts"; series addition of resistors leading to 1D and quasi-1D localization; parallel addition, including the h/e-periodic A-B oscillation in a ring; and many others. The example of the A-B ring is an excellent one for demonstrating a sample-specific effect. The ensemble averaging leads, as is well known by now, to the $h/2e$ oscillation. From the experimental observations and the understanding of the relevance of sample-to-sample specificity, the idea of reproducible conductance fluctuations emerged, with its universal magnitude for the two-terminal case. The latter was qualitatively and semiquantitatively discussed in relation to random-matrix universalities. Finally, we treated the multiterminal generalization of the Landauer formulation, using the ubiquitous four-terminal case as an example. Treating all the terminals on the same footing enables the proper Onsager symmetries, due to time-reversal invariance, to be obtained in agreement with the general picture of Casimir and with experiments.

The last three chapters treated what the author regards as the most interesting physical applications at present, emphasizing for the first two their convenient presentation within the general theoretical framework of the A-B Byers–Yang theorem. Chapter 6 contained a cursory description of the QHE, with a discussion of the 2D large-field electron dynamics in the presence of disorder, which is necessary to understand the integral QHE. The Laughlin picture for the 1/3 state yielding the 1/3 fractional Hall effect was briefly described. Exciting new ideas related to "composite Fermions" and the special role of a 1/2 filling were then mentioned.

In Chapter 7 some remarks were made on mesoscopic effects in superconductors and normal–super (*N–S*) combinations. The interesting situation where, in spite of the lack of long-range order, due to fluctuations, superconducting properties are present, to an extent, was considered. Then, some results on the extremely interesting mode of communication of phase information between two superconductors via a normal section were reviewed. The main mechanism is the Andreev reflection of electrons and holes on the *N–S* boundaries. Supercurrent can thus flow in the normal section, provided the latter is not much longer than L_ϕ. A correspondence with the normal persistent currents was indicated and some interesting properties of *S–N* and *S–N–S* junctions were pointed out. The physics of vortices was very briefly reviewed.

In the last chapter (8), noise phenomena were reviewed, starting with a summary of equilibrium, Johnson–Nyquist, noise. The shot noise in a current-carrying state is understood as due to a combination of fluctuations in the occupation numbers of the "emitting" and "collecting" reservoirs, and those due to the random nature of transmittance through the "resistor." Each discrete electron is either transmitted or not, with a probability given by the transmission coefficient, which is related to the conductance. The low-frequency "$1/f$" noise was reviewed, emphasizing its interpretation via discrete changes in the conductance due to the motion or change of the charge state of defects. A mesoscopic mechanism, relying on the change of the "fingerprint" of the sample due to the change of the impurity configuration, may supply the necessary mechanism for such resistance changes. These may, surprisingly, be relevant even for macroscopic systems. In the last three chapters, attempts were made to explain briefly some of the underlying physics for readers who are not conversant with these topics.

We have *not* treated here all the work on mesoscopics. There are many interesting and important subjects which were not reviewed either because their physics is more straightforward or because excellent reviews of them exist. Among these are optical effects (see Schmitt-Rink et al. 1989); the ballistic regime, described very well by Beenakker and van Houten (1991d); various resonant tunneling situations (e.g., problem 4 of chapter 5); and the Coulomb blockade (e.g., problem 5 of chapter 5; Grabert and Devoret 1992, Glattli and Sanquer 1994). The problem of Coulomb effects in resonant tunneling beyond simple approximations has some real subtleties which were not discussed in this book (see, e.g., Imry and Sivan 1994; another Fermi-level effect is discussed in appendix F). The whole fascinating subject of quantum effects in Josephson systems (Likharev 1986) due to the capacitive "charging" term (Anderson 1963) and the related detailed dynamics of charges and vortices (see some of the articles in Hekking et al. 1994), including interesting analogies with the QHE (see, e.g., Ao and Thouless 1994, Stern 1994 for recent references) is outside our scope. We have also refrained from treating the large body of work concerning analogies between electrons and classical waves (see, e.g., Anderson 1985, Genack et al. 1990, van Haeringen and Lenstra 1991, Pendry and MacKinnon 1992, Sheng 1995).

It is hoped that the interesting physics that can be encountered in the regime between the microscopic and macroscopic has been amply demonstrated here. At the expense of being superficial, one may say that the excursion into the interface between quantum and statistical physics has highlighted the following principal points:

1. Elastic and inelastic scattering are very different. The former gives the electron a well-defined, possibly complicated, phase. The latter induces a phase uncertainty which washes away quantum interference effects.
2. These interference effects thus exist up to the scale of L_ϕ and induce several quantum phenomena such as Anderson localization, various A-B oscillations and conductance fluctuations.
3. The sample-specific nature of mesoscopic systems leads to significant sample-to-sample-fluctuations in, for example, conductance and

orbital magnetic response. Some of these have magnitudes which are "universal" and this is related to universalities in the spectral characteristics of operators such as the Hamiltonian and the transmission matrix. It is expected that the spectrum of relaxation times should also show some universalities.

4. The phase coherence of the normal electrons on the scale of L_ϕ enables them to carry superconducting information. It is hoped that this will contribute to a new understanding of induced superconductivity, even in semiconductors.

Generally speaking, the mesoscopic and quantum interference effects on the single-electron level are by now rather well understood. The many novel effects due to interactions and to combinations of charges and magnetic fluxes should be the arena for principal new developments in this field. It must be recognized that, in spite of many insights, the fundamental problem of the interplay between the three basic electronic states of matter—insulators, conductors, and superconductors—is still far from being understood. This is true even without the further complication of various types of magnetic ordering. It is clear that mesoscopic physics has much to contribute in this respect.

Clearly, the impressive technological developments which have led and will undoubtedly continue to lead to smaller and smaller scales of nanostructures are fueled by the tendency to further miniaturize real electronic devices. This will go on independently of fundamental advances in mesoscopic physics. It would appear obvious that these advances must eventually give a positive crucial feedback to device technology. Without implying that many of the current mesoscopic device ideas will really work, it is clear that a further reduction of available sizes by, say, one-half to a full order of magnitude will bring the relevant temperature range to that of liquid nitrogen.[1] The changed rules of the game of electrical conduction will habe to be reckoned with and relevant device ideas based on quantum and/or Coulomb energy phenomena are very likely to be generated. Thus, while cautionary remarks on current ideas and extrapolations given by Landauer (1989b, 1990a) and Moore (1993) are very pertinent, we believe that it should not take long before mesoscopic physics, perhaps with superconductivity, will be studied by electronics engineers. After all, quantum mechanics and semiconductors had not been a standard part of the electronic engineering curriculum up to 30–40 years ago! At the same time, STM-related techniques are progressing quickly and may pave the way to addressing small structures, down to molecular scales. It is likely that many interesting and applicable phenomena are hidden in this low end of the mesoscopic scale.

[1] Smaller sizes, approaching the molecular limit, will bring the relevant range to room temperature. This should be doable in a controlled way by AFM-STM techniques, and does sometimes happen spontaneously under certain conditions (for a beautiful example, see Costa-Krämer et al. 1995).

Appendices

A. THE KUBO, LINEAR RESPONSE, FORMULATION

We develop here for completeness the linear response formulation for a system started at early times in its ground state $|g\rangle$ and described by a hamiltonian $\mathcal{H}$ which can be taken as time-independent. A good general reference is the book by Nozières (1963). The perturbation is monochromatic, without loss of generality, with frequency ω. Taking η as a positive infinitesimal, we take the perturbation to be

$$\mathcal{H}^{ex} = \lim_{\eta \to 0} \lambda e^{-i\omega t} \hat{A} e^{\eta t}. \tag{A.1}$$

We work in the Heisenberg representation with respect to $\mathcal{H}$ (i.e., in the interaction representation). Simple time-dependent perturbation theory tells us (e.g., Fetter and Walecka 1971, p. 173) that the change in $\langle B\rangle$, the expectation value of some operator B, due to the perturbation is

$$\begin{aligned}\delta\langle B(t)\rangle &= \frac{i}{\hbar}\int_{-\infty}^{t} dt'\, \langle g|[\mathcal{H}_H^{ex}(t'), B_H(t)]|g\rangle \\ &= \frac{i}{\hbar}\int_{-\infty}^{t} dt' \sum_n \{\langle g|\mathcal{H}_H^{ex}|n\rangle\langle n|B_H|g\rangle - \langle g|B_H|n\rangle\langle n|\mathcal{H}_H^{ex}|g\rangle\}\end{aligned} \tag{A.2}$$

where the n's are all the states of $\mathcal{H}$ (a complete set) and the subscript H signifies the Heisenberg representation with $\mathcal{H}$. Going to the Schrödinger representation (no subscript on operators), denoting $\hbar\omega_{ij} = E_i - E_j$ and using $\langle g|A_H|n\rangle = e^{i\omega_{gn}t}\langle g|A|n\rangle$, and so on, we find a monochromatic behavior of $\delta B(t) = \delta B_\omega e^{i\omega t}$, with

$$\delta B_\omega = \lim_{\eta\to 0}\frac{\lambda}{\hbar}\sum_n\left\{-\frac{\langle g|A|n\rangle\langle n|B|g\rangle}{\omega+\omega_{ng}+i\eta}+\frac{\langle g|B|n\rangle\langle n|A|g\rangle}{\omega-\omega_{ng}+i\eta}\right\}. \tag{A.3}$$

Since the response function is defined by $\delta B_\omega = \chi_{BA}(\omega)\lambda$, we find the response of B to A:

$$\chi_{BA}(\omega) = \lim_{\eta\to 0}\frac{1}{\hbar}\sum_n -\frac{A_{gn}B_{ng}}{\omega+\omega_{ng}+i\eta}+\frac{B_{gn}A_{ng}}{\omega-\omega_{ng}+i\eta}. \tag{A.4}$$

We shall apply this first to the polarizability of the system $\chi_{\beta\alpha}$, given by the dipole moment ex^β induced per unit electric field in the α direction, having a perturbation Hamiltonian eEx^α; α and β are here cartesian components, and we have in mind charged particles with a charge e.

$$\chi_{\beta\alpha}(\omega) = \lim_{\eta\to 0}\frac{e^2}{\hbar}\sum_n\left\{-\frac{x^\alpha_{gn}x^\beta_{ng}}{\omega+\omega_{ng}+i\eta}+\frac{x^\beta_{gn}x^\alpha_{ng}}{\omega-\omega_{ng}+i\eta}\right\}. \tag{A.5}$$

The complex dielectric constant is given by $1+4\pi\chi(\omega)$ and its imaginary part is $(4\pi\sigma_r(\omega)/\omega)$ where $\sigma_r(\omega)$ is the real conductivity. The complex conductivity is thus given by (alternatively, an equivalent formula can be obtained by representing E via a vector potential and calculating the resulting current)

$$-\sigma_{\beta\alpha}(\omega) = \lim_{\eta\to 0}\frac{e^2 i\omega}{\hbar}\sum_n\left\{-\frac{x^\alpha_{gn}x^\beta_{ng}}{\omega+\omega_{ng}+i\eta}+\frac{x^\beta_{gn}x^\alpha_{ng}}{\omega-\omega_{ng}+i\eta}\right\}. \tag{A.6}$$

In particular,

$$\mathrm{Re}\,\sigma_{\beta\alpha}(\omega) = \frac{we^2\pi}{\hbar}\sum_n\{\delta(\omega+\omega_{ng})x^\alpha_{gn}x^\beta_{ng} - \delta(\omega-\omega_{ng})x^\beta_{gn}x^\alpha_{ng}\}. \tag{A.7}$$

By (van Hove 1954) representing the δ functions as Fourier transforms of exponentials and doing inverse manipulations to those used in going from eq. A.2 to eq. A.3, one can express the conductivity as the Fourier transform of a commutator-type difference of correlation functions:

$$\mathrm{Re}\,\sigma_{\beta\alpha}(\omega) = \frac{we^2}{2\hbar}\int_{-\infty}^{\infty} dt\langle g|[x^\alpha(0), x^\beta(t)]|g\rangle e^{-i\omega t}$$

$$\xrightarrow[\text{finite } T]{} \frac{we^2}{2\hbar}\int_{-\infty}^{\infty} dt\langle[x^\alpha(0), x^\beta(t)]\rangle_T e^{-i\omega t}, \tag{A.8}$$

where the last expression in eq. A.8 generalizes the middle one by replacing the ground-state average by a thermal average, $\langle\ \rangle_T$, at finite temperatures. In equilibrium it is easy to obtain the detailed balance relationship, a ratio of $e^{-\beta\hbar\omega}(\beta = 1/k_BT)$ between the two terms due to $x^\beta x^\alpha$ and $x^\alpha x^\beta$ in eq. A.8, hence the *fluctuation-dissipation theorem*:

$$\mathrm{Re}\,\sigma_{\beta\alpha}(\omega) = \frac{e^2\omega\pi}{\hbar} S_{\beta\alpha}(\omega)[1 - e^{-\beta\hbar\omega}], \tag{A.9}$$

where (P_g being the thermal weight of the state g)

$$S_{\beta\alpha}(\omega) = \int_{-\infty}^{\infty} dt \langle x^\alpha(0) x^\beta(t)\rangle e^{i\omega t} = \sum_{g,n} x^\alpha_{gn} x^\beta_{ng} \delta(\omega - \omega_{ng}) P_g. \tag{A.10}$$

This is a general, powerful relationship between equilibrium fluctuation correlations and the dissipative response, including the dynamic case.

A case which will be of particular relevance to us is when the operators A and B are the Fourier transforms of the particle density, n_{-q} and n_q, where $n_q = \sum_j e^{ij\cdot r_j}$. The corresponding S is then called the dynamic structure factor, $S(q, \omega)$:

$$S(q,\ \omega) = \sum_{g,n} |\langle g|n_q|n\rangle|^2 \delta(\omega - \omega_{ng}) P_g = \frac{1}{2\pi}\int_{-\infty}^{\infty} dt \langle n_{-q}(0) n_q(t)\rangle e^{-i\omega t}. \tag{A.11}$$

$S(q, \omega)$ is proportional (Van Hove 1954) to the Born aproximation cross-section of inelastic scattering with momentum transfer $\hbar q$ and energy transfer $\hbar\omega$ of a test particle from the system. The detailed balance condition at finite T is

$$S(q,\ -\omega) = e^{-\beta\hbar\omega} S(q, \omega). \tag{A.12}$$

The fluctuation-dissipation theorem (see also Landau and Lifschitz 1959, Lifschitz and Pitaevskii 1980) relates $S(q,\ \omega)$ to the imaginary part of the exact inverse dielectric function of the system, at the same q and ω:

$$\mathrm{Im}\frac{1}{\epsilon(q,\ \omega)} = \frac{4\pi^2 e^2}{\hbar q^2} S(q, \omega)[1 - e^{-\beta\hbar\omega}] \tag{A.13}$$

(remembering that the Fourier transform of the Coulomb interaction is $(4\pi e^2/q^2)$).

To recast the fluctuation-dissipation theorem, eq. A.9 into the ordinary form we use the usual relationship between the matrix elements of $v = \dot{x}$ and those of x. Since the δ-function in eq. A.10 makes $\omega_{ng} = \omega$, we find for $k_BT \gg \omega$:

$$\operatorname{Re}\sigma_{xx}(\omega) = \frac{e^2}{2k_BT}\int_{-\infty}^{\infty} dt\langle v^x(0)v^x(t)\rangle e^{i\omega t} \tag{A.14}$$

By writing $\langle v^x(0)v^x(t)\rangle$ as the Fourier transform of $S_v(\omega)$—the power spectrum (or spectral density) of v (see, e.g., Reif 1965, section 15.15–16) we find:

$$\frac{1}{\text{Vol}}e^2 S_v(\omega) = \frac{k_BT}{\pi}\sigma(\omega), \tag{A.15}$$

where we have inserted an inverse volume factor (the volume was taken as unity up to eq. A.15.

This is equivalent to the well-known Nyquist–Johnson relationship between the current noise spectrum and the conductance, keeping in mind that an electron with a velocity v contributes a current ev/L in a system of length L. For a general ratio between $\hbar\omega$ and k_BT one gets

$$e^2 S_v(\omega) = \frac{\text{Vol}\,\hbar\omega}{\pi}\left(\tfrac{1}{2}+\frac{1}{e^{\beta\hbar\omega}-1}\right)\sigma(\omega) \tag{A.16}$$

and

$$S_I(\omega) = \frac{\hbar\omega}{2\pi}G(\omega)\coth\left(\frac{\beta\hbar\omega}{2}\right) \tag{A.17}$$

Only systems with time-reversal invariance and $\sigma(\omega) = \sigma(-\omega)$ were considered here.

B. THE KUBO–GREENWOOD CONDUCTIVITY AND THE EDWARDS–THOULESS RELATIONSHIPS

Writing (but being aware of the subtleties) the matrix elements of x in terms of those of v, noting, for example, that for positive ω, $\omega = \omega_{ng}$, one expresses the low-frequency real conductivity σ_{xx}, from eq. A.7 as

$$\operatorname{Re}\,\sigma(\omega) = \frac{\pi e^2}{\hbar\omega}\sum_n |v_{gn}|^2\delta(\omega-\omega_{ng}), \tag{B.1}$$

where the cartesian index x has been dropped. For noninteracting quasi-particles the excited states are particle–hole excitations where the hole can be created anywhere between E_F and $E_F - \hbar\omega$. Replacing the $|v_{gn}|^2$ in this small range by an average value $\overline{v^2}$, one obtains

$$\operatorname{Re}\,\sigma(\omega\to 0) = \frac{\pi e^2\hbar}{\text{Vol}}\overline{v^2}[N(0)]^2, \tag{B.2}$$

where $N(0)$ is the density of states per unit energy and Vol is the system's volume. $N(0) = n(0) \cdot \text{Vol}$. This is called the Kubo–Greenwood formula. Note that this is valid for a large enough system having effectively a continuous spectrum. How to handle the discrete spectrum in mesoscopies is discussed in chapter 5.

The conductivity (eq. B.1) is quantitatively related to a properly defined "sensitivity to boundary conditions" via an ingenious argument of Edwards and Thouless (1972). As shown in appendix C, a boundary condition of a phase change of ψ by ϕ on moving an electron across the system of length L is exactly equivalent to closing the system upon itself into a ring along L and applying through the hole of the ring a flux, Φ, given by eq. C.2. This amounts to a vector potential defined by $\oint A_x \, dx = \Phi$, where the relevant component is along the azimuthal direction, denoted here as x. The perturbation due to this, choosing a constant azimuthal vector potential A_x, is

$$\mathcal{H}'_\phi = \frac{e}{c}\hat{v}_x A_x = \frac{\hbar \hat{v}_x \phi}{L}. \tag{B.3}$$

The second-order shift of an eigenstate due to A or to the equivalent phase shift ϕ leads to the quantity

$$\frac{\partial^2 E_i}{\partial \phi^2} = \frac{\hbar^2 N}{mL^2} + 2\frac{\hbar}{L^2}\sum_{j \neq i} \frac{|\langle j|v_x|i\rangle|^2}{E_i - E_j}. \tag{B.4}$$

One may again replace the matrix element by its characteristic value $\overline{v^2}$. The $\sum_j$ tends to cancel the first ("diamagnetic") term and the order of magnitude of (B.4) is determined by the term with the smallest denominator, which is the level spacing

$$\Delta \equiv 1/N(0). \tag{B.5}$$

We define the Thouless energy E_c as $E_c \equiv \hbar D/L^2$,

$$E_c \equiv \pi^2 \left| \overline{\frac{\partial^2 E_i}{\partial \phi^2}} \right| \sim \left(\frac{\hbar}{L}\right)^2 \frac{\overline{v^2}}{\Delta}, \tag{B.6}$$

where π^2 was inserted for consistency with the above definition. $\overline{v^2}$ is parametrized in terms of σ, via eq. B.2. For a wire of cross-section A in the yz plane,

$$\frac{E_c}{\Delta} = \frac{\hbar}{e^2}\frac{\sigma A}{L}. \tag{B.7}$$

Remembering that $\sigma A/L$ is the conductance G, one finds

$$G = \frac{e^2}{\hbar}\frac{E_c}{\Delta}. \tag{B.8}$$

We identify E_c with V_L/π, V_L is the parameter V_L of section 3 of chapter 2 (cf. eq. 2.20).

This is the celebrated result that the dimensionless conductance is given by the ratio of the two energy parameters giving the sensitivity to boundary conditions and the level spacing.

C. THE AHARONOV–BOHM EFFECT AND THE BYERS–YANG AND BLOCH THEOREM

Consider a general doubly-connected system with an Aharonov–Bohm flux Φ through its opening (see the figure below). An important and very general theorem due to Byers and Yang (1961) and Bloch (1970) states that all physical properties of this "ring" are periodic in Φ with a period Φ_0. The proof proceeds by eliminating Φ with the gauge transformation

$$\psi' = e^{(ie/\hbar c)\sum_j \chi(r_j)}\psi \tag{C.1}$$

where r_j are the coordinates of the electrons and χ is defined by $\tilde{A}_\phi = \nabla\chi$, where $\tilde{A}_\phi$ is the vector potential whose curl is the Aharonov–Bohm (A-B) magnetic field (i.e., curl $\tilde{A} = 0$ in the material and $\oint \tilde{A}_\phi \cdot d\tilde{l}$ on a path circulating the ring's opening is equal to Φ). The gauge-transformed many-electron Schrödinger equation has $\tilde{A}_\phi = 0$. The price for this is, of course, that the transformed wavefunction, ψ', does not in general satisfy periodic boundary conditions around the ring. In fact, the phase of ψ' changes by

$$\phi = 2\pi\Phi/\Phi_0 \tag{C.2}$$

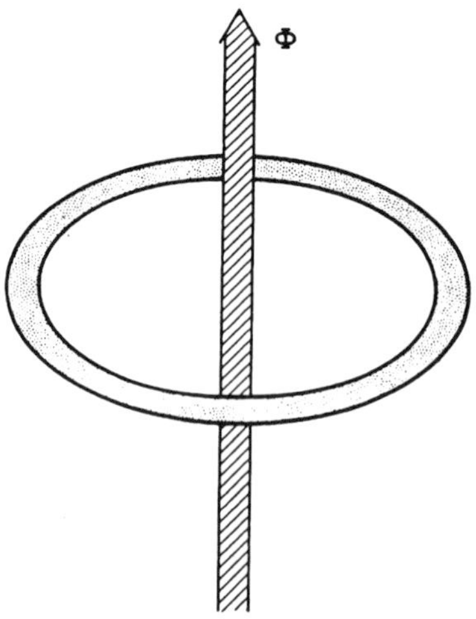

when one electronic coordinate is rotated once around the ring. This phase shift, due to a vector potential with a vanishing magnetic field on the electrons, obviously appears between every two paths encircling a flux. The fluxes Φ and $\Phi + n\Phi_0$ are *indistinguishable* since ϕ is meaningful only modulo 2π.

D. DERIVATION OF MATRIX ELEMENTS IN THE DIFFUSION REGIME

For a diffusing electron, one finds for the following matrix elements of $e^{i\mathbf{q}\cdot\mathbf{r}}$ between the exact eigenstates:

$$|\langle m|e^{i\mathbf{q}\cdot\mathbf{r}}|n\rangle|^2 = \frac{1}{\pi\hbar N(0)} \frac{Dq^2}{(Dq^2)^2 + \omega^2}; \qquad \omega \equiv \frac{E_n - E_m}{\hbar}, \tag{D.1}$$

where $N(0)$ is the density of states (DOS) of the system.

The simplest way to derive eq. D.1, for which we will have many uses later, is to note (Azbel, private communication, 1981) that it follows from the quasiclassical approximation for a diffusing particle. In this approximation the transition probabilities—the l.h.s. of eq. D.1 times the DOS—in the quantum case are equal to the Fourier transforms of the correlation function of the appropriate classical quantity with frequencies $(E_n - E_m)/\hbar$. For $e^{i\mathbf{q}\cdot\mathbf{r}(t)}$ (taking $\mathbf{r}(0) = 0$) the classical average is given by

$$\langle e^{i\mathbf{q}\cdot\mathbf{r}(t)}\rangle = e^{-(1/2)q^2\langle r^2(t)\rangle = e^{-q^2Dt}}, \tag{D.2}$$

the Fourier transform of which yields the well-known Lorentzian on the r.h.s. of eq. D.1. Another derivation of eq. D.1 (Abrahams et al. 1981, Kaveh and Mott 1981, McMillan 1981, Imry et al. 1982) uses the appropriate dynamic structure factor, $S(q, \omega)$, for diffusion, which is proportional to the r.h.s. of eq. D.1 and is given in terms of the desired matrix elements squared. This "diffusion pole" appears naturally in the perturbative theories.

E. CAREFUL TREATMENT OF DEPHASING IN 2D CONDUCTORS AT LOW TEMPERATURES

We do the 2D $\int d^2k$ in eq. 3.32, neglecting numerical factors of order unity. After angular integration we obtain

$$\int_0^{k_m} \frac{dk}{k}(1 - J_0(kx_{12})) \simeq \ln(k_m x_{12}), \tag{E.1}$$

where J_0 is a zero order Bessel function, $x_{12}(t) = |x_1(t) - x_2(t)|$ and $k_m \sim (K_B T/\hbar D)^{1/2}$ is the upper cutoff of k, discussed preceding eq. 3.36.

Thus, τ_ϕ is given (after remembering that for a very thin film of thickness d, $\int dk_z$ is replaced by $2\pi/d$ times the $k_z = 0$ contribution) by

$$\frac{1}{\tau\phi} \sim \frac{e^2 k_B T}{\hbar^2 \sigma d} \ln\left(\sqrt{\frac{kT}{\hbar D}} x_{12}\right) \sim \frac{e^2 k_B T}{\hbar^2 \sigma d} \ln\left(\frac{k_B T \tau_\phi}{\hbar}\right), \tag{E.2}$$

where we put $\sqrt{D\tau_\phi}$ for the characteristic x_{12} and neglect the constant within the log. Solving this equation by iterations, we obtain

$$\frac{1}{\tau\phi} \sim e^2 \frac{k_B T}{\hbar^2 \sigma d} \ln\left(\frac{\sigma d}{e^2 \hbar}\right) \sim \frac{e^2 k_B T}{\hbar^2 \sigma d} \ln(k_F^2 l d) \frac{k_B T}{g_\square} \ln g_\square, \tag{E.3}$$

where $g_\square$ is the dimensionless "conductance per square" of the film, $g_\square = (\sigma d \hbar)/e^2$, appearing both in the coefficient and inside the log. This agrees with thin-film result of Altshuler, Aronov and Khmelnitskii (see, for example, Altshuler and Aronov 1985, eq. 4.47a).

F. ANOMALIES IN THE DENSITY OF STATES (DOS)

It is well known (see, e.g., Kittel 1963) that for the usual electron gas with Coulomb interactions in the Hartree–Fock approximation, the exchange terms yield anomalies in the self-energy and in the resulting DOS (a logarithmic vanishing of the latter at E_F). These logarithmic singularities are spurious; for example, they are eliminated if the screened Coulomb interaction is used. It is found, however, that in the electron gas with even a weak disorder, a true weak singularity of the single-particle $n(E)$ at E_F should exist (Altshuler and Aronov 1979, see also McMillan 1981) and it is, in fact, found, experimentally (Abeles et al. 1975, McMillan and Mochel 1981, Dynes and Garno 1981, Imry and Ovadyahu 1982a, Hertel et al. 1983, White et al. 1985). This effect becomes more conspicuous at low dimensions, $d \leq 2$.

We remind ourselves of the Hartree–Fock theory (see, e.g., Kittel 1963); the self-consistent approximate Schrödinger equation reads

$$\frac{p^2}{2m}\phi_m + \int d^d y\, V(x-y) \sum_l f_l \phi_l^*(y)\phi_l(y)\phi_m(x)$$
$$- \int d^d y\, V(x-y) {\sum_l}' f_l \phi_l^*(y)\phi_l(x)\phi_m(y) = E_m \phi_m(y) \tag{F.1}$$

where E_m and ϕ_m are the one-electron energy levels and orbitals, respectively; f_l are the occupations; $\sum_l'$ signifies summation over spins parallel to that of m; V is the electron–electron interaction, the second term is the (direct) Hartree contribution and the third is the (exchange) Fock one. The energies are

$$E_m = \epsilon_m + \Sigma_m, \tag{F.2}$$

ϵ_m being the unperturbed noninteracting electron energies and Σ_m is the self-energy given in the Hartree–Fock approximation by

$$\Sigma_m = \sum_l f_l \langle ml|V|ml\rangle - \sum_l{}' f_l \langle ml|V|lm\rangle, \tag{F.3}$$

where $\langle ml|V|ml\rangle = \int\int d^dx\, d^dy\, \phi_m^*(x)\phi_m(x)V(x-y)\phi_l^*(y)\phi_l(y)$ and $\langle ml|V|lm\rangle$ is the same with x and y interchanged in the arguments of ϕ_l and ϕ_m (but not in ϕ_l^* and ϕ_m^*).

In the usual electron gas without disorder, the Hartree term just cancels exactly the contribution of the uniform positive background. This cancelation will not occur in our disordered system. We shall nevertheless concentrate on the exchange terms $\sum_m^{ex}$. More complete calculations including the direct term and using dynamic screening multiply the result we shall get by important numerical factors, but do not change the nature of the singularity.

We use the Fourier representation of V:

$$V(r) = \frac{1}{(2\pi)^3}\int d\boldsymbol{q}\, V_q e^{i\boldsymbol{q}\cdot\boldsymbol{r}}, \tag{F.4}$$

thus

$$\langle ml|V|lm\rangle = \frac{1}{(2\pi)^3}\int d\boldsymbol{q} |\langle m|e^{i\boldsymbol{q}\cdot\boldsymbol{r}}|l\rangle|^2 V_q. \tag{F.5}$$

In the simplest and least sophisticated approximation, we use the statically screened Coulomb interaction for V_q and eq. D.1 for the matrix elements squared. The interesting behavior follows from the small q limit where $V_q = (4\pi e^2/q^2)\cdot \Lambda^2 q^2 = 1/n(0)$ and Λ is the screening length (note that the electron charge e cancels out). Thus as a function of the unperturbed energy, ϵ,

$$\Sigma^{ex}(\epsilon) = -\frac{1}{(2\pi)^3}\int_{\epsilon' < \epsilon_F} d\epsilon' \int d^dq\, \frac{1}{\pi\hbar N(0)} \frac{Dq^2}{(Dq^2)^2 + (\epsilon - \epsilon')^2/\hbar^2}. \tag{F.6}$$

We note that the change in the DOS is given by

$$\frac{dn}{dE} = \frac{dn}{d\epsilon}\frac{d\epsilon}{dE} = \frac{dn}{d\epsilon}\bigg/\left(1 + \frac{d\Sigma}{d\epsilon}\right), \tag{F.7}$$

where $dn/d\epsilon = n(\epsilon)$ is the unperturbed DOS. Taking the derivative of eq. F.6, $\Sigma^{ex\prime}$ is seen to have a mild singularity at $\epsilon = E_F$ in 3D. It is convenient to accentuate the singularity by taking another derivative. We first note that changing the ϵ' integration in eq. F.6 to the variable $(\epsilon' - E_F)$, the upper limit on $\epsilon - \epsilon'$ is $(\epsilon - E_F)$. Thus, the first derivative is just the integrand:

$$\Sigma^{ex}(\epsilon) = -\frac{1}{8\pi^4 n(0)\hbar}\int dq \frac{Dq^2}{(Dq^2)^2 + \omega^2}, \tag{F.8}$$

where here $\hbar\omega \equiv \epsilon - E_F$. We thus find

$$\Sigma^{ex\prime\prime}(\epsilon) = \frac{1}{4\pi^4 n(0)\hbar}\int d^d q \frac{Dq^2\omega}{[(Dq^2)^2 + \omega^2]^2} = -\frac{\text{const}}{\hbar D^{3/2}\omega^{1/2} n(0)}, \tag{F.9}$$

where the last equality holds in 3D. Thus, Σ' in 3D will have a singular contribution which is a numerical constant, C times $\omega^{1/2}/\hbar n(0) D^{3/2}$:

$$n(E) = n_0(E)[1 + C_1\omega^{1/2}/\hbar n(0)D^{3/2}] \tag{F.10}$$

Thus, the DOS has a square-root singularity at E_F, whose total relative amplitude (for $\hbar\omega \sim E_F$, say) is of the order of $1/(k_F l)^{3/2}$. The calculation we have presented here (McMillan 1981) is a simplification of the low-order systematic one by Altshuler and Aronov (1979). That calculation is only valid when the DOS correction is small (or $k_F l \gg 1$). In this case, there is a good agreement with the experiments.

An interesting aspect of the DOS anomalies is their dimensionality dependence. For a system which is thin enough to be effectively 2D, the $\sqrt{\omega}$ singularity is replaced by a stronger, logarithmic, one. To be effectively 2D, it turns out that the thickness d, of the film should be smaller than the characteristic length (see discussion following eq. 3.38) $L_\omega \sim \sqrt{D/\omega}$ (we take $\hbar\omega \gg kT, \tau_\phi^{-1}$). Both this crossover and the logarithmic behavior have been observed experimentally (Imry and Ovadyahu 1982a). The singularity is stronger in 1D, but we emphasize that these results are valid only as long as the relative correction to the DOS is small.

G. QUASICLASSICAL THEORY OF SPECTRAL CORRELATIONS

We consider, following Argaman et al. (1993), based on Berry (1985), a metallic particle of volume $V = L^d$ and are interested in its level statistics near the Fermi energy E_F. The metallic limit means that eq. 4.38 is satisfied, but the consideration below has a more general range of validity. By approximating the Feynman sum for the propagator (Fourier transform of Green's function G) as sum over periodic classical paths, j, Gutzwiller (1971) obtained the following "trace formula" for a "chaotic" system, whose classical paths are all unstable,

$$n(E) = \sum_j A_j e^{iS_j/\hbar}, \tag{G.1}$$

(we employed this method in getting eq. 4.16).

Here S_j is the classical action along the periodic path j (apart from Maslow-index corrections) and the coefficient A_j is due to the gaussian integral around the (unstable) orbit:

$$A_j = \frac{1}{2\pi\hbar} T_j |\det(M_j - I)|^{-\frac{1}{2}}. \tag{G.2}$$

T_j is the period of the orbit j and M_j the "monodromy matrix" (giving, in linear approximation, the transformation of the $(2d - 2)$-dimensional vector measuring the deviation from the orbit in phase space after one period). The factor T_j reflects the fact that the path can start and end at all points on the orbit and all such contributions are coherent for a given orbit.

One is interested in the level-density correlation function (eq. 4.33) and in its Fourier transform $\tilde{K}(E, t)$, which can be called "the spectral structure factor" (the averaging in the definition of K can be over a suitable range of E or on the "impurity ensemble" in the case of a disordered system).

There are serious mathematical questions having to do with the Gutzwiller sum not being absolutely convergent. Moreover, it is clear that the quasiclassical approximation fails at long enough times, due to the exponential proliferation of the orbits as t increases. There are strong indications that the limiting time is longer (Tomsovic and Heller 1993) than what naive estimates give; possibly it is of order $\hbar/\Delta$. We shall thus employ a long time cut off τ_ϕ (and a small energy cutoff, $\gamma \sim \tau_\phi^{-1}$, due physically to, for example, dephasing processes). Taking $\Delta \ll \gamma \ll E_c$ appears to be necessary (this is also based on a comparison with Altshuler and Shklovskii 1986). The quasiclassical method is for sure not valid for energy scales below Δ. Since the interesting scales here are $|\epsilon - \epsilon'| \sim \Delta$, E_c, which are of order $\hbar^d$, $\hbar$, respectively, and much smaller than E_F, a quasiclassical approximation relying on $\hbar$ being small is suggested. Berry (1985) used the classical expression $T_j = \partial S_j / \partial E$ in the semiclassical approximation, keeping only the "diagonal" terms[1] in the double sum over paths. He obtained for the spectral structure factor $\tilde{K}(t)$,

$$\tilde{K}(E, t) = \sum_j |A_j|^2 \delta(t - T_j). \tag{G.3}$$

Berry used a classical sum-rule due to Hannay and d'Almeida (1984) that the classical probability to return to the origin in time t is the sum over all such orbits

$$P_{cl}(E, t) = \sum_j P_j(E, t) = \sum_j \delta(t - T_j) T_j |\det(M_j - I)|^{-1}, \tag{G.4}$$

[1] For systems with time-reversal symmetry ($H = 0$, no magnetic impurities) there is an additional factor of 2 here. This is due to the existence, as in the weak localization case, of pairs of time-reversal orbits whose A_j's in eq. (G.1) have to be added coherently.

(here the reason for the appearance of the same matrix $(M_j - I)$ as before, is the transformation of the volume element in phase space). He then proved that the quasiclassical spectrum of a classically ergodic system satisfies random matrix correlations ($\tilde{K}(\epsilon) \propto \epsilon^{-2}$) over scales smaller than an energy of order $\hbar$ (see below). It has to be noted that in order to apply the above theory to a classically ergodic system and for $t \lesssim \hbar/\Delta$, the (unstable) orbits and gaussian regimes around them should be regarded as effectively separated.

Argaman et al. (1993) (see also Doron et al. 1992) rewrote the Berry calculation in the following way. From eqs. G.2–G.4 one obtains

$$\tilde{K}(t) = \frac{t}{h^2} \frac{d\Omega}{dE} P_{cl}(t), \tag{G.5}$$

where $\Omega(E)$ is the (purely classical) phase space volume for given energy E and the factor $d\Omega/dE$ is for correct normalization (see Argaman et al. 1993). This is an extremely useful expression and, remarkably, uses the same classical probability as, for example, the "weak localization" quantum correction to transport (chapter 2). For the simple example of a single particle which performs diffusive motion in the classical limit in a volume $V = L^d$, and which is not localized in the quantum case (so the dimensionless conductance $g \cong E_c/\Delta \gg 1$) one obtains (noting that the particle diffuses isotropically on an equal energy surface and thus the return probability density in phase space is $1/4\pi$ of that in x-space),

$$P_{cl}(t) \propto \begin{cases} 1, & t \gtrsim \hbar/E_c \sim \dfrac{L^2}{D} \\ \dfrac{V}{(4\pi D t)^{d/2}}, & t \lesssim \hbar/E_c. \end{cases} \tag{G.6}$$

Fourier transforming eqs. G.5 and G.6, using only times $\ll \hbar/\Delta$ we obtain very simply the Altshuler–Shklovskii results for the spectral correlations:

$$K(\epsilon) \propto \begin{cases} -\epsilon^{-2} & \gamma \lesssim \epsilon \lesssim E_c \\ \dfrac{V}{(\hbar D)^{d/2}} \epsilon^{d/2-2} & E_c \lesssim \epsilon \lesssim \hbar/\tau_{cl}. \end{cases} \tag{G.7}$$

The semiclassical interpretation of these results is obvious. The low-energy (0D) range is in agreement with Berry and with RMT. (For example, to obtain the spectral rigidity, one calculates the fluctuations in the number of levels in an interval W. The double integral of $K(\epsilon)$ for $\epsilon \lesssim E_c$ indeed yields the well-known $\log W/\gamma$ result, with the correct coefficient). The cut-off has to be properly introduced, for example, by multiplying $K(t)$ by $e^{-\gamma|t|}$. This introduces a positive portion of $K(\epsilon)$ for $\epsilon \lesssim \gamma$ which cancels the infrared behavior of

$-1/\epsilon^2$. The limiting energy for the RMT behavior, E_c, is of order $\hbar$, as expected and the new, Altshuler-Shklovskii, behavior is obtained for $\epsilon > E_c$. Different power-laws will be obtained for scale-dependent classical diffusion processes (chapter 2; Derrida and Pomeau 1982).

H. DETAILS OF THE FOUR-TERMINAL FORMULATION

We write eq. 5.26 in full:

$$\begin{pmatrix} -T_1 & T_{12} & T_{13} & T_{14} \\ T_{21} & -T_2 & T_{23} & T_{24} \\ T_{31} & T_{32} & -T_3 & T_{34} \\ T_{41} & T_{42} & T_{43} & -T_4 \end{pmatrix} \begin{pmatrix} \mu_1 \\ \mu_2 \\ \mu_3 \\ \mu_4 \end{pmatrix} = \begin{pmatrix} I_1 \\ I_2 \\ I_3 \\ I_4 \end{pmatrix}, \tag{H.1}$$

where $T_i = \sum_{j \neq i} T_{ij} = \sum_{j \neq i} T_{ji}$, for $i = 1, \ldots, 4$. Since a vector with equal μ's yields zero current, this is really a set of three independent equations. We also have $\sum I_i = 0$.

The four-terminal situation consists of taking $I_1 = -I_3 = J_1$, $I_2 = -I_4 = J_2$. These currents determine all the three independent voltages $\mu_i - \mu_j$. But, since we have only two independent variables, J_1 and J_2, we can express J_1 and J_2 in terms of, for example, $eV_1 = \mu_1 - \mu_3$ and $eV_2 = \mu_2 - \mu_4$ as in eqs. 5.27 and 5.31. In particular, in an ordinary resistance measurement with 1 and 3 being the current terminals and 2 and 4 the voltage terminals, one takes $J_2 = 0$ and, as in eq. 5.32, with $k = 1$, $l = 3$, $m = 2$, $n = 4$, $R_{13,24} = V_2/J_1$. The easiest way to determine the α's in eq. 5.27 is by applying some nonzero V_1 and $V_2 = 0$, which can be done by choosing the set of μ's in eq. 5.26 to be $(1, 0, -A, 0)$ where $V_1 = 1 + A$. We then enforce $I_1 = -I_3$, $I_2 = -I_4$ by choosing A correctly, and then obtain J_1 and J_2. This determines α_{11} and α_{21}. α_{12} and α_{22} may be determined by a similar procedure involving μ_2 and μ_4. We find immediately

$$A = \frac{T_{31} - T_1}{T_{13} - T_3} = \frac{T_{21} + T_{41}}{T_{23} + T_{43}} \tag{H.2}$$

and, for example,

$$J_2 = I_2 = T_{21} - T_{23}A = \frac{T_{21}T_{43} - T_{23}T_{41}}{T_{23} + T_{43}}, \tag{H.3}$$

$$V_1 = 1 + A = \frac{S}{T_{23} + T_{43}}, \tag{H.4}$$

where $S = T_{21} + T_{41} + T_{23} + T_{43}$, as in eq. 5.29. By definition $-\alpha_{21} = J_2/V_1$. The ratio of eq. H.3 (and a similar equation for J_1) to eq. H.4 indeed yields α_{21} and α_{11} in agreement with eq. 5.28:

$$\alpha_{11} = -T_1 + \frac{(T_{12} + T_{14})(T_{21} + T_{41})}{S}, \tag{H.5}$$

$$-\alpha_{21} = \frac{T_{21}T_{43} - T_{23}T_{41}}{S}. \tag{H.6}$$

It must be understood that although we chose to express the two J's in terms of the two V's as in eq. 5.27, they also depend, for example, on $\mu_2 - \mu_4$. This is why we were not free to apply V_1 by an arbitrary choice of μ_1 and μ_3 with the correct difference.

I. UNIVERSALITY OF THE CONDUCTANCE FLUCTUATIONS IN TERMS OF THE UNIVERSAL CORRELATION OF TRANSMISSION EIGENVALUES

This discussion is based on the two-terminal Landauer conductance expression (5.16). It is convenient (Imry 1986) to cast the r.h.s. in terms of the eigenvalues of the transfer matrix T, following Pichard and Sarma (1981a, b) and Pichard (1984).

$$\operatorname{tr} tt^\dagger = \operatorname{tr} \frac{2}{TT^\dagger + (TT^\dagger)^{-1} + 2I}. \tag{I.1}$$

I is the unit matrix, t is the $N_\perp \times N_\perp$ transmission matrix, giving the transmitted waves on the right of the scattering system whose conductance is being considered, in terms of the incident waves on the left. The $2N_\perp \times 2N_\perp$ transmission matrix T can be expressed in terms of t and the reflection matrix r, defined in (5.9). T gives both left- and right-moving wave amplitudes on the r.h.s. of the obstacle in terms of those on the left. Thus, T has a multiplicative property: T for series addition is given by the product of the T's for the components. (For a more complete treatment of these issues, see Pichard 1984, Pendry 1989.) It thus makes sense, and it is substantiated by a theorem due to Oseledec (1968), that for large system length L (measured here in atomic units) the eigenvalues (Dorokhov 1982, 1984) appearing in the denominator of eq. I.1 will be exponential in L, thus eq. I.1 can be written as

$$g = \sum_n \frac{1}{1 + \cosh \lambda_n} = \sum_n \frac{1}{1 + \cosh (L\mu_n)}, \tag{I.2}$$

where λ_n are the eigenvalues of the whole system and μ_n can be thought of as those "per unit length," or as the inverse localization lengths. The smallest of them gives the inverse of the physical localization length, ξ. Arranging λ_n by increasing magnitude, one may say that N_{eff} of them are smaller than or equal to unity and all the others are quickly making exponentially small contributions. Thus

$$N_{eff} = g, \tag{I.3}$$

which was used in the argument leading to eq. 5.60. A fundamental property of the eigenvalues is their "repulsion" (preventing degeneracies) which tends to reduce the fluctuations of N_{eff}. In fact, since eq. I.2 expresses g as what is called a "linear statistic" of the eigenvalues (i.e., a sum over n of a smooth function of the λ_n), one may appeal to results of Dyson (1962) and Mehta and Dyson (1963) (see also Mehta 1967) according to which the fluctuations of g are of order of unity provided one postulates that the eigenvalues obey RMT correlations. The latter assumption can be justified using the global "maximum entropy principle" for the distribution of the eigenvalues (Mello et al. 1988) and it does yield a constant universal value for $\langle \Delta g^2 \rangle$. It was appreciated by Beenakker (1993), however, that the result, although being very close (ratio of 16/15) was not equal to the correct one, as obtained diagrammatically for quasi-1D systems (this is the appropriate case since one is taking a finite $N_\perp$ and very long L). Very recently, Beenakker and Rejaei (1993) and, independently, Chalker and Maĉedo (1993) solved the problem by noting that the global maximum entropy rule (Mello and Pichard 1989) is only a mean-field type approximation to the correct distribution given by the solution of an appropriate Fokker–Planck-type equation. This approximation is an excellent one but it is not exact. Going beyond it provides the necessary correction to give the precise values of $\langle \Delta g^2 \rangle$ in quasi-1D systems for all known symmetries. At higher dimensions too, the RMT assumption gives the correct order of magnitude, but not the precise value, for $\langle \Delta g^2 \rangle$. However, even without quantitative validity, the repulsion of the transmission eigenvalues is the qualitative physical reason for the universality of $\langle \Delta g^2 \rangle$.

J. THE CONDUCTANCE OF BALLISTIC "POINT CONTACTS"

Consider two massive reservoirs connected by an ideal conducting wire having $N_\perp$ channels, that is, $T_{ij} = T'_{ij} = \delta_{ij}$, $R_{ij} = R'_{ij} = 0$. In this case, $G_c = N_\perp(e^2/\pi\hbar)$ and the resistance per contact per channel is $(\pi\hbar/2e^2)$, as in the single channel case. The situation here is identical to that of a small narrow orifice (i.e., a "point contact"; Jansen et al. 1983) between two large conductors. The resistance of such a "ballistic" contact, with a cross-section A such that $A \ll l^2$, where l is the mean free path in the conductors (the reservoirs in our case), has been calculated by Sharvin (1965). His result is

$$R_{orifice} = 4\rho l/3A, \tag{J.1}$$

where ρ is the resistivity of the conductor. We note that this result which is easily obtained in a ballistic quasiclassical kinetic theory is independent of l. Since the number of channels (not including spin degeneracy, which we included in the channel conductance, eq. 5.16) in an orifice of area A is $Ak_F^2/4\pi$, the resistance of the orifice with a conductor becomes (using $\rho = 3\pi^2\hbar/e^2k_F^2l$)

$$R_c = \pi\hbar/(e^2N_\perp), \tag{J.2}$$

which is the same as the resistance $1/G_c$ alluded to above. We emphasize that this resistance has nothing to do with the resistance of the "wire" itself and it exists also when the conductor connecting the two reservoirs is ideal (no scattering). It is just due to the geometry, the ideal orifice being a "bottleneck' between the two conductors, in each of which the electrons are in equilibrium.

For 2D systems the number of channels is given by Wk_F/π, where W is the width of the orifice. Both the ballistic 2D and 3D results can be summarized by

$$G_c = \frac{e^2}{\pi\hbar}N_\perp, \tag{J.3}$$

where $N_\perp$ is the number of channels not including spin (Imry 1986). Alternatively one may include the spin in $N_\perp$ and then the ideal channel conductance is $e^2/2\pi\hbar$. Wharam et al. (1988) and van Wees et al. (1988) have independently discovered this experimentally for point contacts in GaAs 2D quantum-well systems. It is rather nontrivial to identify the conditions to have the full conductance (eq. J.3). This is briefly discussed following eq. 5.16. More work is needed for one- (and few-) atom contacts.

References

Special Issue of IBM Journal of Research and Development **32**, no. 3 (May 1988).

Special Issue of IBM Journal of Research and Development **32**, no. 4 (July 1988).

Special Issue of Physics Today, June 1993 on Optics of Nanostructures.

Abeles, B., Sheng, P., Coutts, M. D. and Arie, Y. (1975) Adv. Phys. **23**, 407.

Abrahams, E. and Lee, P. A. (1986) Phys. Rev. **B33**, 683.

Abrahams, E., Anderson, P. W., Licciardello, D. C. and Ramakrishnan, T. V. (1979) Phys. Rev. Lett. **42**, 673.

Abrahams, E., Anderson, P. W., Lee, P. A. and Ramakrishnan, T. V. (1981) Phys. Rev. **B24**, 6783.

Abrikosov, A. A. (1988) Fundamentals of the Theory of Metals, North-Holland, Amsterdam.

Adkins, C. J. (1977) Phil. Mag. **36**, 1285.

Aharonov, Y. and Bohm, D. (1959) Phys. Rev. **115**, 485.

Akkermans, E., Montambaux, G., Pichard, J.-L and Zinn-Justin, J. (eds.) (1995) Mesoscopic Quantum Physics, Les Houches Session LXI. Elsevier, Amsterdam.

Altland, A. and Gefen, Y. (1995) Phys. Rev. **51**, 10 671.

Altland, A., Iida, S., Müller-Groeling, A. and Weidenmüller, H. A. (1992) Ann. Phys. **219**, 148.

Altshuler, B. L. (1985) Pis'ma Zh. Eksp. Teor. Fiz. **41**, 530 [JETP Lett. **41**, 648 (1985)].

Altshuler, B. L. and Aronov, A. G. (1979) J. Eksp. Teor. Fiz. **77**, 2028 [Sov. Phys. JETP **50**, 968 (1980)].

Altshuler, B. L. and Aronov, A. G. (1985), in Electron–Electron Interactions in Disordered Systems, Efros, A. L. and Pollak, M., eds. North-Holland, Amsterdam, p. 1.

Altshuler, B. L. and Khmelnitskii, D. E. (1985) JETP Lett. **42**, 359.

Altshuler, B. L. and Lee, P. A. (1988) Physics Today **41**, 36.

Altshuler, B. L. and Shklovskii, B. (1986) Sov. Phys. JETP **64**, 127.

Altshuler, B. L. and Spivak, B. Z. (1985) JETP Lett. **42**, 447.

Altshuler, B. L. and Spivak, B. Z. (1987) Sov. Phys. JETP **65**, 343.

Altshuler, B. L., Aronov, A. G. and Lee, P. A. (1980a) Phys. Rev. Lett. **44**, 1288.

Altshuler, B. L., Khmel'nitskii, D., Larkin, A. I. and Lee, P. A. (1980b) Phys. Rev. **B22**, 5142.

Altshuler, B. L., Aronov, A. G. and Spivak, B. Z. (1981a) JETP Lett. **33**, 94.

Altshuler, B. L., Aronov, A. G. and Khmelnitskii, D. E. (1981b) Solid State Commun. **39**, 619.

Altshuler, B. L., Aronov, A. G. and Khmelnitskii, D. E., (1982a) J. Phys. **C15**, 7367.

Altshuler, B. L., Aronov, A. G., Khmelnitskii, D. E. and Larkin, A. I. (1982b), in Quantum Theory of Solids, Lifschitz, I. M., ed. Mir Publishers, Moscow, p. 130.
Altshuler, B. L., Aronov, A. G., Spivak, B. Z., Sharvin D. Yu and Sharvin, Yu V. (1982c) JETP Lett. **35**, 588.
Altshuler, B. L., Kravtsov, V. E. and Lerner I. V. (1982d) Zh. Eksp. Teor. Fiz. **91**, 2276 [Sov. Phys. JETP **64**, 1352 (1982)].
Altshuler, B. L., Khmelnitskii, D. E. and Spivak, B. Z. (1983) Solid State Commun. **48**, 841.
Altshuler, B. L., Gefen, Y. and Imry, Y. (1991a) Phys. Rev. Lett. **66**, 88.
Altshuler, B. L., Lee, P. A. and Webb, R. A. (eds.) (1991b) Mesoscopic Phenomena in Solids, North-Holland, Amsterdam.
Altshuler, B. L., Gefen, Y., Imry, Y. and G. Montambaux (1993) Phys. Rev. **B47**, 10335.
Ambegaokar, V. and Baratoff, A. (1963) Phys. Rev. Lett. **10**, 486.
Ambegaokar, V. and Eckern, U. (1990) Phys. Rev. Lett. **65**, 381. (1991) **67**, 3192.
Ambegaokar, V., Halperin, B. I. and Langer, J. S. (1971) Phys. Rev. **B64**, 2612.
Anderson, P. W. (1958) Phys. Rev. **109**, 1492.
Anderson, P. W. (1963), in Lectures on the Many-Body Problem, Ravello, 1963, E. R. Caianello, ed. Academic Press, New York, 1964, p. 113.
Anderson, P. W. (1967), in The Josephson Effect and Quantum Coherence Measurements in Superconductors and Superfluids. Progress in Low Temperature Physics, Gorter, C. J., ed. North-Holland, Amsterdam.
Anderson, P. W. (1981) Phys. Rev. B **23**, 4828.
Anderson, P. W. (1985), Phil. Mag. **B52**, 505.
Anderson, P. W., Thouless, D. J., Abrahams, E. and Fisher, D. S. (1980) Phys. Rev. B **22**, 3519.
Ando, T. (1983) J. Phys. Soc. Japan **52**, 1740.
Ando, T. (1984) J. Phys. Soc. Japan **53**, 3101, 3126.
Ando, T., Fowler, A. B. and Stern, F. (1982) Rev. Mod. Phys. **59**, 437.
Andreev, A. F. (1964) Zh. Eksp. Teor. Fiz. **46**, 1823 [Sov. Phys. JETP **19**, 1228; ibid. **22**, 455].
Andreev, A. F. (1966) Zh. Eksp. Teor. Fiz. **51**, 1510 [Sov. Phys. JETP **24**, 1019].
Ao. P. and Thouless, D. J. (1993) Phys. Rev. Lett. **70**, 2158.
Ao, P. and Thouless, D. J. (1994) Phys. Rev. Lett. **72**, 128.
Aoki, H. and Ando, T, (1981) Solid State Commun. **38**, 1079.
Aoki, H., Tsukada, M., Schluter, M. and Levy, F. (eds.) (1992) New Horizons in Low-Dimensional Electronic Systems. Kluwer, Dordrecht.
Argaman, N. and Imry, Y. (1993) Physica Scripta **T49**, 333.
Argaman, N., Smilansky, U. and Imry, Y. (1993) Phys. Rev. **B47**, 4440.
Aronov, A. G. and Sharvin, Yu V. (1987) Rev. Mod. Phys. **59**, 755.
Arovas, D. P., Scrieffer, J. R., Wilczek, F. and Zee, A. (1985) Nucl. Phys. **B251**, 117.
Ashcroft, N. and Mermin, N. D. (1976) Solid State Physics. Holt, Rinehart and Winston, New York.
Aslamazov, L. G. and Larkin, A. I. (1974) Zh. Eksp. Teor. Fiz. **67**, 647 [Sov. Phys. JETP **40**, 321 (1975)].
Aslamazov, L. G., Larkin, A. I. and Ovchinnikov, Yu. N. (1969) Sov. Phys. JETP **28**, 171.
Averin, D. V. and Likharev, K. K. in Altshuler et al. (1991b), p. 173.
Avouris, P. and Lyo, I.-W. (1994) Science **264**, 942.

Azbel, M. Ya. (1973) Phys. Rev. Lett. **31**, 589.
Azbel, M. Ya. (1981) J. Phys. C. **14**, L225.
Azbel, M. Ya. (1983) Solid State Commun. **45**, 527.
Azbel, M. Ya. (1993) Phys. Rev. **B48**, 4592.
Azbel, M. Ya and Soven, P. (1983) Phys. Rev. B. **27**, 831.
Balian, R., Maynard, R. and Toulouse, G. (eds.) (1979), Ill-Condensed Matter. North-Holland, Amsterdam.
Bardeen, J. (1961) Phys. Rev. Lett. **6**, 57.
Bastard, G., Brum, J. A. and Ferreira, R. (1991), in Solid State Physics **44**, 229.
Baym, G. (1969) Quantum Mechanics. Benjamin, New York.
Beenakker, C. W. J. (1991) Phys. Rev. Lett. **67**, 3836; (1992) **68**, 1442E.
Beenakker, C. W. J. (1992a) Phys. Rev. **B46**, 12 841.
Beenakker, C. W. J. (1992b) in Fukuyama and Ando (1992), p. 235.
Beenakker, C. W. J. (1993) Phys. Rev. Lett. **70**, 1155.
Beenakker, C. W. J. (1994) Phys. Rev. **B49**, 2205.
Beenakker, C. W. J. (1995) in Akkermans et al. (1995), p. 259.
Beenakker, C. W. J and Büttiker, M. (1992) Phys. Rev. **B46**, 1889.
Beenakker, C. W. J. and Rejaei, B. (1993) Phys. Rev. Lett. **71**, 3693.
Beenakker, C. W. J. and van Houten, H. (1991a) Phys. Rev. Lett. **66**, 3056.
Beenakker, C. W. J. and van Houten, H. (1991b) Phys. Rev. **B43**, 12 066.
Beenakker, C. W. J. and van Houten, H. (1991c) in SQUID '91, Proc. 4th Int. Conf. on Superconducting and Quantum Effect Devices and Their Applications, H. Koch and H. Lübbig, eds. Springer, Berlin.
Beenakker, C. and van Houten, H. (1991d), in Solid State Physics, **44**, 1.
Beenakker, C. W. J., Rejaei, B. and Melsen, J. A. (1994) Phys. Rev. Lett. **72**, 2470.
Ben-Jacob, E. and Gefen, Y. (1985) Phys. Lett. **108A**, 289.
Benoit, A. D., Washburn, S., Umbach, C. P., Laibovitz, R. B. and Webb, R. A. (1987a) Phys. Rev. Lett. **57**, 1765.
Benoit, A. D., Umbach, C. P., Laibovitz, R. B. and Webb, R. A. (1987b), Phys. Rev. Lett. **58**, 2343.
Berezinskii, V. L. (1971) Sov. Phys. JETP **32**, 493.
Bergman, D. J. (1983) Private communication; the author is indebted to D. J. Bergman for this suggestion.
Bergman, D. J., Ben-Jacob, E., Imry, Y. and Maki, K. (1983) Phys. Rev. **A27**, 3345.
Bergmann, G. (1982) Z. Phys. **B48**, 5.
Bergmann, G. (1984) Phys. Rep. **107**, 1.
Bergmann, G., Bruynseraede, Y. and Kramer, B. (eds.) (1984) Localization, Interaction and Transport Phenomena in Impure Metals. Springer-Verlag, Heidelberg.
Berkovits, R. and Avishai, Y. (1995a) Europhys. Lett. **29**, 475.
Berkovits, R. and Avishai, Y. (1995b) Phys. Rev. Lett. **76**, 261.
Bernamot, J. (1937) Ann. Phys. (Leipzig) **7**, 71.
Berry, M. V. (1985) Proc. Roy. Soc. London, Ser. A **400**, 229.
Binder, K. (ed.) (1984) (Application of Monte-Carlo Methods in Statistical Physics. Springer-Verlag, Berlin.
Binning, G., Rohrer, H., Gerber, Ch. and Weibel, E. (1982) Phys. Rev. Lett. **49**, 57.
Birk, M., de Jong, M. J. M. and Schönenberger, C. (1995) Phys. Rev. Lett. **75**, 1610.
Bishop, D. J., Tsui, D. C. and Dynes, R. C. (1980) Phys. Rev. Lett. **44**, 1153.
Bishop, D. J., Licini, J. C. and Dolan, G. J., (1985) Appl. Phys. Lett. **46**, 1000.

Blatter, G., Feigelman, M. V., Geshkenbein, V. B., Larkin, A. I. and Vinokur, V. M. (1994) Rev. Mod. Phys. **66**, 1125.
Bloch, F., (1930) Z. Phys. **61**, 206.
Bloch, F. (1968) Phys. Rev. Lett. **27**, 1241; Phys. Rev. **165**, 415.
Bloch, F. (1970) Phys. Rev. B. **2**, 109.
Blonder, M. (1984) Bull. Am. Phys. Soc. **29**, 535.
Blonder, M., Tinkham, M. and Klapwijk, T. M. (1982) Phys. Rev. **B25**, 4515.
Bohigas, O., Giannonni, M. J. and Schmidt, C. (1984) Phys. Rev. Lett. **52**, 1.
Borland, R. E. (1963) Proc. Roy. Soc. London **A274**, 529; (1968) Proc. Phys. Soc. London **28**, 926.
Böttger H. and Bryksin, V. V. (1985) Hopping Conduction in Solids. Academic Verlag, Berlin; VCH Verlag, Mannheim.
Bouchiat, H. and Montambaux, J. (1989) J. Phys. (Paris) **50**, 2695.
Brandt, N. B., Bogachek, E. N., Gitsu, D. V., Gogadze, G. A., Kulik, I. O., Nikolaeva A. A. and Ponomarev, Ya. G. (1976) JETP Lett. **24**, 273.
Brandt, N. B., Bogachek, E. N., Gitsu, D. V., Gogadze, G. A., Kulik, I. O., Nikolaeva, A. A. and Ponomarev, Ya. G. (1982) Sov. J. Low Temp. Phys. **8**, 358.
Bratus, E. N., Shumeiko, V. S. and Wendin G. (1995) Phys. Rev. Lett. **74**, 2110.
Brody, T. A., Flores, J., Frech, J. B., Mello, P. A., Pandey, A. and Wong, S. S. M. (1981) Rev. Mod. Phys. **53**, 385.
Broers, A. N. (1989) in Reed and Kirk (1989), p. 421.
Browne, D. A. and Nagel, S. R. (1985) Phys. Rev. **B32**, 8424.
Browne, D. A., Carini, J. P., Muttalib, K. A. and Nagel, S. R. (1984) Phys. Rev. **B30**, 6798.
Bruder, C. (1995) Mesoscopic Superconductivity. Habilitation thesis, University of Karlsruhe; to appear in Superconductivity Review **1**(4) (1996).
Bruder, C., Fazio, R. and Schön (1994) Physica **B203**, 240; Phys. Rev. **B50**, 12 766.
Buot, F. (1993) Phys. Rep. **254**, 74.
Büttiker, M. (1985a) in Hahlbohm and Lübbig (1985), p. 429.
Büttiker, M. (1985b) Phys. Rev. B **32**, 1846.
Büttiker, M. (1986a) Phys. Rev. **B33**, 3020.
Büttiker, M. (1986b) Phys. Rev. Lett. **57**, 1761.
Büttiker, M. (1988) IBM J. Res. Dev. **32**, 317.
Büttiker, M. (1990) Phys. Rev. Lett. **65**, 2901.
Büttiker, M. (1992a) Phys. Rev. Lett. **68**, 843.
Büttiker, M. (1992b) Phys. Rev. **B46**, 12485.
Büttiker, M. (1993) J. Phys. Cond. Matter **5**, 9361.
Büttiker, M. and Imry, Y. (1985) J. Phys. C. L467.
Büttiker, M. and Klapwijk, T. M. (1986) Phys. Rev. **36**, 5114.
Büttiker, M., Imry, Y. and Landauer, R. (1983a) Phys. Lett. **96A**, 365.
Büttiker, M., Harris, E. P. and Landauer, R. (1983b) Phys. Rev. B. **28**, 1268.
Büttiker, M., Imry, Y. and Azbel, M. Ya (1984) Phys Rev. A **30**, 1982.
Büttiker, M., Imry, Y., Landauer, R. and Pinhas, S. (1985) Phys. Rev. B **31**, 6207.
Byers, N. and Yang, C. N. (1961) Phys. Rev. Lett. **7**, 46.
Caldeira, A. O. and Leggett, A. J. (1983) Ann. Phys. **149**, 374.
Capasso, F. and Datta, S. (1990) Physics Today **43**, 74.
Carini, J. P., Muttalib, K. A. and Nagel, S. R. (1984) Phys. Rev. Lett. **53**, 102.
Casimir, H. B. J. (1945) Rev. Mod. Phys. **17**, 343.
Castaing, B. and Nozières, P. (1980) J. Phys. (Paris) **41**, 701.

Castaing, B. and Nozières, P. (1985) Private communication. The author is indebted to Drs. Castaing and Nozières for a discussion on this point.
Castellani, C., di Castro, C. and Pelliti, L. (1981) Disordered Systems and Localization. Springer-Verlag, Berlin.
Cerdeira, H. A., Kramer, B. and Schön, G. (eds.) (1995) Quantum Dynamics of Submicron Structures, NATO ASI Ser. E, vol. 291. Kluwer, Dordrecht.
Chakraborty, T. and Pietiläinen, P. (1988) The Fractional Quantum Hall Effect. Springer-Verlag, Berlin.
Chakravarty, S. and Schmid, A. (1986) Phys. Rep. **140**, 193.
Chalker, J. T. and Coddington, P. D. (1988) J. Phys. **C21**, 2665.
Chalker, J. T. and Maĉedo, A. M. S. (1993) Phys. Rev. Lett. **71**, 3693.
Chambers, R. G. (1960) Phys. Rev. Lett. **5**, 3.
Chandrasekhar, V., Rooks, M. J., Wind, S. and Prober, D. E. (1985) Phys. Rev. Lett. **15**, 1610.
Chandrasekhar, V., Webb, R. A., Brady, M. J., Ketchen, M. B., Gallagher. W. J. and Kleinsasser, A. (1991) Phys. Rev. Lett. **67**, 3578.
Chen, L. Y. and Ting, C. S. (1992) Phys. Rev. **B46**, 4714.
Cheung, H. F., Gefen, Y. and Riedel, E. K. (1988) IBM J. Res. Dev. **32**, 359.
Cheung, H. F., Gefen, Y. and Riedel, E. K. (1989) Phys. Rev. Lett. **52**, 587.
Cohen, O., Ovadyahu, Z. and Rokni, M. (1992) Phys. Rev. Lett. **69**, 3555.
Costa-Krämer, J. L., Garcia, N., Garcia-Mochales, P. and Serena, P. A. (1995) Surf. Sci. **342**, L1144.
Courtois, H., Gandit, P. and Pannetier, B. (1994) in Glattli and Sanquer (1994), p. 85.
Courtois, H., Gandit, P., Mailly, D. and Pannetier, B. (1995) Phys. Rev. Lett. **76**, 130.
Crommie, M. F., Lutz, C. P. and Eigler, D. M. (1993) Science **262**, 218; Phys. Rev. **B48**, 2851; Nature **363**, 524.
Czycholl, G. and Kramer, B. (1979) Solid State Commun. **32**, 945.
Datta, S., Melloch, M., Bandyopadhyay, S., Noren, R., Vaziri, M., Miller, M. and Reifenberger, R. (1986) Phys. Rev. Lett. **55**, 2344.
Davies, J. H., Hyldegaard, P., Hershfield, S. and Wilkins, J. W. (1992) Phys. Rev. **B46**, 9620.
Davies, R. A., Pepper, M. and Kaveh, M. (1983) J. Phys. **C16**, L285.
De Gennes, P. G. (1965) Superconductivity in Metals and Alloys. Benjamin, New York.
de Gennes, P. G. (1966) in 1965 Tokyo Lectures in Theoretical Physics, Vol. 1, R. Kubo, ed. W. A. Benjamin, Inc., New York, p. 117.
de Jong, M. J. M. and Beenakker, C. W. J. (1992) Phys. Rev. **B46**, 13 400.
de Jong, M. J. M. and Beenakker, C. W. J. (1994) in Glattli and Sanquer (1994), p. 427.
de Jong, M. J. M. and Beenakker, C. W. J. (1995) Phys. Rev. **B51**, 16 867.
de Jong, M. J. M. and Molenkamp, W. (1995) Phys. Rev. **B51**, 13 389.
de Vegvar, P. G. N., Fulton, T. A., Mallison, W. H. and Miller, R. E. (1994) Phys. Rev. Lett. **73**, 1416.
Delsing, P., Chen, C. D., Haviland, D. B. and Claeson, T. (1994) Phys. Rev. Lett. **194**, 993.
den Hartog, S. G., Kapteyn, C. M. A., van Wees, B. J. and Klapwijk, T. M. (1995) Groningen, preprint.
Derrida, B. and Pomeau, Y. (1982) Phys. Rev. Lett. **48**, 627; see also: Sinai, Ya. G., In Proc. 6th Int. Conf. on Mathematical Physics, Berlin 1981. Springer, Berlin.
des Cloizeaux, J. (1965) J. Phys. Chem. Solids **26**, 259.

Deutscher, G. and de Gennes, P. G. (1969), in Superconductivity, vol. 2, Parks, R. O., ed. Marcel Dekker, New York, p. 1005.
de Vegvar, P. G. N., Fulton, T. A., Mallison, W. H. and Miller, R. E. (1994) Phys. Rev Lett. **73**, 1416.
Devoret, M. (1995) in Akkermans et al. (1995).
Dingle, R. B. (1952) Proc. Phys. Soc. A. **212**, 47.
Dolan, G. J. and Osheroff, D. D. (1979) Phys. Rev. Lett. **43**, 721.
Dolan, G. J., Licini, J. C. and Bishop, D. J. (1986) Phys. Rev. Lett. **56**, 1493.
Domany, E. and Sarker, S. (1979) Phys. Rev. **B20**, 4726.
Donnely, R. J., Quantized Vortices in He^{II}. Cambridge UP, Cambridge.
Dorokhov, O. N. (1982) Pis'ma J. Eksp. Teor. Fiz. **36**, 259 [JETP Lett. **36**, 318].
Dorokhov, O. N. (1984) Solid State Commun. **51**, 381.
Doron, E., Smilansky, U. and Dittrich, T. (1992) Physica **B179**, 1.
Du, R., Stormer, H., Tsui, D., Pfeiffer, L., West, K. (1993) Phys. Rev. Lett. **70**, 2944.
Dupuis, N. and Montambaux, G. (1991) Phys. Rev. **B43**, 14 390.
Dutta, P. and Horn, P. M. (1981) Rev. Mod. Phys. **53**, 497.
Dynes, R. C. and Garno, J. (1981) Phys. Rev. Lett. **46**, 137.
Dyson, F. J. (1962) J. Math. Phys. **3**, 140, 157, 166.
Ebisawa, H. (1992) in Fukuyama and Ando, p. 273.
Echternach, P. M., Gershenson, M. E., Bozler, H. M., Bogdanov, A. M. and Nilsson, B. (1993) Phys. Rev. **B46**, 11 516.
Eckern, Z. (1991) Z. Phys. **B82**, 393.
Eckern, Z. and Schmid, A. (1989) Phys. Rev. **B39**, 6441.
Economou, E. N. (1990) Green's Functions in Quantum Physics, 2nd ed. Springer-Verlag, Berlin.
Economou, E. N. and Soukoulis, C. M. (1981a) Phys. Rev. Lett. **46**, 618.
Economou, E. N. and Soukoulis, C. M. (1981b) Phys. Rev. Lett. **47**, 972.
Edwards, J. T. and Thouless, D. J. (1972) J. Phys. **C5**, 807.
Efetov, K. B. (1982) Zh. Eksp. Teor. Fiz. **82**, 872 [Sov. Phys. JETP **55**, 514].
Efetov, K. B. (1983) Adv. Phys. **32**, 53.
Efros, A. L. and Pollak, M. (eds.) (1985) Electron–electron Interaction in Disordered Systems. North-Holland, Amsterdam.
Eiler, W. (1985) Solid State Commun. **56**, 11.
Eiles, T. M., Martinis, J. M. and Devoret, M. H. (1993a), in Geerligs et al., p. 210.
Eiles, T. M., Martinis, J. M. and Devoret, M. H. (1993b), Phys. Rev. Lett. **70**, 1862.
Elion, W. J., Geerligs, L. J. and Mooij, J. E. (1992) Phys. Rev. Lett. **69**, 2971.
Elion, W. J., Wachters, J. J., Sohn, L. L. and Mooij, J. E. (1993) Phys. Rev. Lett. **71**, 2311.
Elion, W. J., Matters, M. Geigenmüller, U. and Mooij, J. E. (1994) Nature **371**, 594.
Engquist, H. L. and Anderson, P. W. (1981) Phys. Rev. B **24**, 1151.
Entin-Wohlman, O. and Gefen, Y. (1989) Europhys. Lett. **8**, 477.
Entin-Wohlman, O. and Gefen, Y. (1991) Ann. Phys. (NY) **206**, 68.
Entin-Wohlman, O., Hartsztein, K. and Imry, Y. (1986) Phys. Rev. **B34**, 921.
Entin-Wohlman, O., Imry, Y. and Sivan, U. (1989) Phys. Rev. **B40**, 8342.
Entin-Wohlman, O., Imry, Y., Aronov, A. G. and Levinson, Y. (1995a) Phys. Rev. **B51**, 11 584.
Entin-Wohlman, O., Aronov, A. G., Levinson, Y. and Imry, Y. (1995b) Phys. Rev. Lett. **75**, 4094.

Esaki, L. (1984) Proc. 17th Int. Conf. on the Physics of Semiconductors, San Francisco. Springer-Verlag, New York, p. 473.
Esaki, L. (1986) IEEE J. Quantum Electron. **QE-22**, 1611.
Farmer, K. R., Rogers, C. T. and Buhrman, R. A. (1987) Phys. Rev. Lett. **58**, 2255.
Fazio, R. and Schön, G. (1991) Phys. Rev. **B43**, 5307.
Fazio, R., Geigenmueller, U. and Schön, G. (1991a) in Quantum Fluctuations in Mesoscopic and Macroscopic Systems, H. A. Cerdeira, ed. World Scientific. Singapore.
Fazio, R., Bruder, C. and Schön, G. (1991b) in Glattli and Sanquer (1994), p. 49.
Feng, S. and Lee, P. A. (1991) Science **251**, 633.
Feng, S., Lee, P. A. and Stone, A. D. (1986) Phys. Rev. Lett. **56**, 1960; (2772-E).
Fetter, A. L. and Walecka, J. O. (1971) Quantum Theory of Many-Particle Systems. McGraw-Hill, New York.
Feynman, R. P. and Vernon, F. L. (1963) Ann. Phys. NY **24**, 118.
Feynman, R. P., Leighton, R. B. and Sands, M. (1965) The Feynman Lectures on Physics. Addison-Wesley, Reading, MA. vol. iii, p. 21.14.
Fisher, D. S. and Lee, P. A. (1981) Phys. Rev. **B23**, 6851.
Fisher, M. E., (1967) Rep. Progr. Phys. **30**, 1391.
Fisher, M. E., (1971) in Proceedings of the Varenna International Enrico Fermi School, course 51, M. S. Green, ed. Academic Press, New York.
Fisher, M. E. and Langer, J. S. (1967) Phys. Rev. Lett. **20**, 615.
Fowler, A. B., Fang, F. F., Howard, W. E. and Stiles, P. J. (1966) Phys. Rev. Lett. **16**, 901.
Fowler, A. B., Hartstein, A. and Webb, R. A. (1982) Phys. Rev. Lett. **48**, 196.
Friedman, L. R. and Tunstall, D. P. (eds.) (1978), The Metal–Nonmetal Transition in Disordered Systems. SUSSP, Edinburgh.
Frydman, A. and Ovadyaha, Z. (1996) Europhys. Lett. **33**, 217.
Fukuyama, H. (1980) J. Phys. Soc. Japan **48**, 2169; (1981a) **50**, 3407.
Fukuyama, H. (1981b) in Nagaoka and Fukuyama (1982), p. 89.
Fukuyama, H., (1983) in Goldman and Wolf (1983), p. 161.
Fukuyama, H. (1985) in Efros and Pollak (1985), p. 155.
Fukuyama, H. and Ando, T. (1992) Transport Phenomena in Mesoscopic Systems, Proc. 14th Taniguchi Symposium, Shima, Japan, 1991. Springer-Verlag, Berlin.
Fukuyama, H. and Yoshioka, H. (1992) in Aoki et al. (1992), p. 369.
Fulton, T. A. and Dolan, G. J. (1987) Phys. Rev. Lett. **59**, 109.
Furusaki, A., (1992) in Fukuyama and Ando (1992), p. 255.
Furusaki, A. and Tsukada, M. (1991) Solid State Commun. **78**, 290.
Furusaki, A., Takayanagi, H. and Tsukuda, M. (1991) Phys. Rev. Lett. **67**, 132.
Gantmakher, V. F. and Levinson, Y. (1987) Carrier Scattering in Metals and Semiconductors. North-Holland, Amsterdam.
Geerligs, L. J., Harmans, G. J. P. M. and Kouwendhoven, L. P. (1993) (eds.) The Physics of Few-Electron Nanostructures. North-Holland, Amsterdam.
Gefen, Y. and Thouless, D. E., (1993) Phys. Rev. **B49**,
Gefen, Y., Imry, Y. and Azbel, M. Ya (1984a) Phys. Rev. Lett. **52**, 129.
Gefen, Y., Imry, Y. and Azbel, M. Ya (1984b) Surf. Sci. **142**, 203.
Genack, A. Z., Garcia, N., Li, J., Polkosnik, W. and Drake, J. M. (1990) Physics **A168**, 387.
Gijs, M., van Haesendonk, C. and Bruynseraede, Y. (1984) Phys. Rev. Lett. **52**, 5069; (1985); Phys. Rev. B. **30**, 2964.

Ginzburg, V. L. and Landau, L. D. (1950) J. Exp. Teor. Fiz. **20**, 1064. Translated in: Tev Haar, D. (ed.) (1965) Men of Science, Pergamon, Oxford.

Giordano, N. Gilson, W. and Prober, D. E. (1979) Phys. Rev. Lett. **43**, 725.

Giulianni, G. F. and Quinn, J. J. (1982) Phys. Rev. **B26**, 4421.

Glattli, D. C. and Sanquer, M., eds. (1994) Coulomb and Interference Effects in Small Electronic Structures. Editions Frontieres Paris.

Glazman, L. I. and Matveev, K. A. (1989) JETP Lett. **49**, 659.

Glazman, L. I., Lesouik, G. B., Khmelnitskii, D. E. and Shekhter, R. I. (1988) JETP Lett. **48**, 238.

Goldman, A. M. and Wolf, S. A. (eds.) (1983) Percolation, Localization and Superconductivity, Nato Advanced Science Institutes, Series B: Physics, vol. 109. Plenum Press, New York.

Goldman, V., Su, B. and Jain, J. K. (1994) Phys. Rev. Lett. **72**, 2065.

Gordon, J. M. (1984) Phys. Rev. B **30**, 6770.

Gorkov, L. P. and Eliashberg, G. M. (1965) Sov. Phys. JETP **21**, 940.

Gorkov, L. P., Larkin, A. I. and Khmelnitskii, D. E. (1979) JETP Lett. **30**, 288.

Gorter, C. J. (1936) Physica **3**, 503.

Gorter, C. J. and Kronig, R. (1936) Physica **3**, 1009.

Gossard, A. C. (1986) IEEE J. Quantum Electron. **QE-22**, 1649.

Grabert, H. and Devoret, M. H. (eds.) (1992) Single Charge Tunneling, Coulomb Blockade Phenomena in Nanostructures, Nato ASI, Series B: Physics, vol. 294. Plenum Press, New York.

Greenwood, J. (1958) Proc. Phys. Soc. London **71**, 585.

Gunther, L. (1989) J. Low Temp. Phys. **77**, 15; (1990) ibid. **79**, 225 (erratum).

Gunther, L. and Gruenberg, L. (1972) Phys. Lett. **38A**, 463.

Gunther, L. and Imry, Y. (1969) Solid State Commun. **7**, 1391.

Gurevich, V. L. and Rudin, A. M. (1996) Phys. Rev. **B53**, 10078.

Gutzwiller, M. C. (1971) J. Math. Phys. **12**, 343.

Hahlbuhm, H. D. and Lübbig, H. (eds.) (1985) Proceedings of the Third International Conference on Superconducting Quantum Devices, Berlin. de Gruyter, Berlin [contains references on SQUIDS].

Haldane, F. D. M. (1983) Phys. Rev. Lett. **51**, 605.

Haldane, F. D. M. and Rezayi, E. H. (1988) Phys. Rev. Lett. **60**, 956.

Halperin, B. I. (1982) Phys. Rev. **B25**, 2182.

Halperin, B. I. (1984) Phys. Rev. Lett. **52**, 1583; **52**, 2390(E).

Halperin, B. I. and McCumber, D. E. (1970) Phys. Rev. **B1**, 1054.

Halperin, B. I. and Nelson, D. R. (1979) Phys. Rev. **B19**, 2457.

Halperin, B. I., Lee, P. A. and Read, N. (1993) Phys. **B47**, 7312.

Hanbury-Brown, R. and Twiss, R. Q. (1956) Nature **177**, 27.

Hanbury-Brown, R. and Twiss, R. Q. (1957) Proc. Roy. Soc. London **A242**, 300; **A243**, 291.

Hannay, J. H. and Ozorio de Almeida, A. M. (1984) J. Phys. **A17** 3429. In this work the classical sum rule was derived. The physical interpretation in terms of a probability was given by U. Smilansky, S. Tomsovic and O. Bohigas, (1991) J. Phys. **A25**, 3261 and by Argaman et al. 1993.

Harrison, W. A. (1970) Solid State Theory. McGraw-Hill, New York.

Haviland, D. B., Liu, Y. and Goldman, A. M. (1989) Phys. Rev. Lett. **62**, 2180.

Hebard, A. F. and Paalanen (1985) Phys. Rev. Lett. **54**, 2155.

Heiblum, M., Nathan, M. I., Thomas, D. E. and Knoedler, C. M. (1985) Phys. Rev. Lett. **55**, 2200.
Hekking, F. W. J. and Nazarov, Yu. V. (1994) Phys. Rev. **B49**, 6847.
Hekking, F. W. J., Schön, G. and Averin, D. V. (eds) (1994) Mesoscopic Superconductivity, Proceedings of the NATO ARW in Karlsruhe. Physica **B203**, 201.
Herman, M. A. and Sitter, H. (1989) Molecular Beam Epitaxy, Fundamentals and Current Status. Springer Series in Material Sciences **7**. Springer, Berlin.
Hershfield, S. (1992) Phys. Rev. **B46**, 7061.
Hertel, G. H., Bishop, D. J., Spencer, E. G., Rowell, J. M. and Dynes, R. C. (1983) Phys. Rev. Lett. **50**, 743.
Hikami, S., Larkin, A. I. and Nagaoka, Y. (1981) Progr. Theor. Phys. **63**, 707.
Hofmann, S. and Kummel, R. (1993) Phys. Rev. Lett. **70**, 1319.
Hofstetter, E. and Schreiber, M. (1993) Europhys. Lett. **21**, 933.
Hohenberg, P. C. (1967) Phys. Rev. **158**, 383.
Holstein, T. (1959) Ann. Phys. (NY) **8**, 325, 343.
Holstein, T. (1961) Phys. Rev. **124**, 1329.
Hooge, F. N. (1969) Phys. Lett. **A29**, 139.
Howard, R. E. and Prober, D. E. (1982), in VLSI Electronics: Microstructure Science (Academic, New York) vol. 5, chap. 9. [deals with nanofabrication].
Hubbard, J. (1964) Proc. Roy. Soc. London **A277**, 237.
Hughes, R. J. F., Nicholls, J. T., Frost, J. E. F., Einfield, E. F., Pepper, M., Ford, C. J. B., Ritchie, D. A., Jones, G. A. C., Kogan, E. and Kaveh, M. (1994) J. Phys. Cond. Matter **6**, 4769.
Hui, V. C. and Lambert, C. J. (1993) Europhys. Lett. **23**, 203.
Hund, F. (1938) Ann. der Physik **32**, 102.
Imry, Y. (1969a) Ann. Phys. (NY) **51**, 1.
Imry, Y. (1969b) in Proceedings of the 1969 Stanford Conference on Superconductivity, F. Chilton, ed. North-Holland, Amsterdam, p. 344.
Imry, Y. (1969c) Phys. Lett. **29A**, 82.
Imry, Y. (1977) Phys. Rev. B **15**, 4478.
Imry, Y. (1980a) Phys. Rev. Lett. **44**, 467.
Imry, Y. (1980b) Phys. Rev. B **21**, 2042.
Imry, Y. (1981a) Phys. Rev. B **24**, 1107.
Imry, Y. (1981b) J. Appl. Phys. **52**, 1817.
Imry, Y. (1983a) J. Phys. C **16**, 3501.
Imry, Y. (1983b) in Goldman and Wolf (1983), p. 189.
Imry, Y. (1985) Invited lecture at the Freudenstadt German Physical Society.
Imry, Y. (1986a) Europhys. Lett. **1**, 249.
Imry, Y. (1986b) in Directions in Condensed Matter Physics, Memorial Volume to S.-k Ma, Grinstein, G. and Mazenko, G., eds. World Scientific, Singapore, p. 102.
Imry, Y. (1988) Physica **B152**, 295.
Imry, Y. (1991), in Kramer (1991), p. 221.
Imry, Y. (1995a) Europhys. Lett. **30**, 405.
Imry, Y. (1995b) in Akkermans et al. (1995), p. 181.
Imry, Y. and Bergman, D. J. (1971) Phys. Rev. A **3**, 1416.
Imry, Y. and Gunther, L. (1971) Phys. Rev. **B3**, 3939.
Imry, Y. and Ovadyahu, Z. (1982a) Phys. Rev. Lett. **49**, 841.
Imry, Y. and Ovadyahu, Z. (1982b) J. Phys. **C15**, L327.
Imry, Y. and Shiren, N. (1986) Phys. Rev. **B33**, 7992.

Imry, Y. and Sivan, U. (1994) Solid State Commun. **92**, 83.
Imry, Y. and Strongin, M. (1981) Phys. Rev. **B24**, 6353, contains many prior references.
Imry, Y., Bergman, D. J., Deutscher, G. and Alexander, S. (1973) Phys. Rev. A **7**, 744.
Imry, Y., Gefen, Y. and Bergmann, D. J. (1982) Phys. Rev. Lett. **B26**, 3436.
Iordanskii, S. V. (1982) Solid State Commun. **43**, 1.
Ismail, K., Meyerson, B. S. and Wang, P. J. (1991) Appl. Phys. Lett. **58**, 2117; **59**, 973.
Jackiw, R. (1977) Rev. Mod. Phys. **49**, 681.
Jain, J. K. (1989) Phys. Rev. Lett. **63**, 199.
Jain, J. K. (1990) Phys. Rev. **B40**, 8079.
Jain, J. K. and Kivelson, S. A. (1988) Phys. Rev. Lett. **60**, 1542; Phys. Rev. **B37**, 4111, 4726.
Jalabert, R. A., Pichard, J.-L. and Beenakker, C. W. J. (1993) Europhys. Lett. **24**, 1.
Jansen, N. J. M., van Gelder, A. P., Duif, A. M. Wyder, P. and d'Ambrumenil, N. (1983) Helv. Phys. Acta **56**, 209.
Jiang, H. W., Johnson, C. E., Wong, K. L. and Hannahs, S. T. (1993) Phys. Rev. Lett. **71**, 1439.
John, S., Sompolinsky, H. and Stephen, M. J. (1982) Phys. Rev. B **27**, 5592.
Johnson, M. and Girvin, S. (1979) Phys. Rev. Lett. **43**, 1447.
Josephson, B. D. (1962) Phys. Lett. **1**, 251.
Josephson, B. D. (1965) Adv. Phys. **14**, 419.
Joyez, P., Lafarge, P., Filipe, A., Esteve, D. and Devoret, M. (1994) Phys. Rev. Lett. **72**, 2458.
Joynt, R. and Prange, R. E. (1984) Phys. Rev. **B29**, 3303; J. Phys. **C27**, 4807.
Kagan, Yu. and Leggett, A. J. (eds.) (1992) Quantum Tunnelling in Condensed Media. North-Holland, Amsterdam.
Kamenev, A. and Gefen, Y. (1993) Phys. Rev. Lett. **70**, 1976.
Kamenev, A. and Gefen, Y. (1994) Phys. Rev. **B49**, 14474.
Kamenev, A. and Gefen, Y. (1995) Int. J. Mod. Phys. **B9**, 751.
Kamenev, A. and Gefen, Y. (1996) WIS preprint.
Kamenev, A., Reulet, B., Bouchiat, H. and Gefen, Y. (1994) Europhys. Lett. **28**, 391.
Kanda, A. and Kobayashi, S-I. (1995) J. Phys. Soc. Japan (Letters) **64**, 19.
Kanda, A., Katsumoto, S. and Kobayashi, S-I. (1994) J. Phys. Soc. Japan **63**, 4306.
Kang, W., Störmer, H. L., Pfeiffer, L. N., Baldwin, K. W. and West, K. W. (1993) Phys. Rev. Lett. **71**, 3850.
Kapon, E., Huang, D. M. and Bhat, R. (1989) Phys. Rev. Lett. **63**, 430.
Kastalsky, A., Green, L. H., Barner, J. B., Bhat, R. (1990) Phys. Rev. Lett. **64**, 958.
Kastalsky, A., Kleinsasser, A. W., Greene, L. H., Bhat, R., Milliken, F. P. and Harbison, J. P. (1991) Phys. Rev. Lett. **67**, 3026.
Katsumoto, S. (1995) J. Low Temp. Phys. **98**, 287.
Kaveh, M. and Mott, N. F. (1981) J. Phys. C. **14**, L67.
Kaveh, M., Uren, M. J., Davies, R. A. and Pepper, M. (1981) J. Phys. **C14**, L413.
Kawabata, A. (1980) J. Phys. Soc. Japan **49**, 628; (1981) J. Phys. Soc. Japan **50**, 2461.
Kawaguchi, Y. and Kawaji, S. (1982) Surface Sci. **113**, 5051 and references therein.
Kazarinov, R. F. and Luryi, S. (1982) Phys. Rev. **B25**, 7626.
Keldysh, L. V. and Kopaev, Yu. V. (1965) Sov. Phys. Solid-State **6**, 2219.
Khlus, V. A. (1987) Sov. Phys. JETP **66**, 1243.
Khmelnitskii, D. E. (1983) Sov. Phys. JETP Lett. **38**, 552.
Khmelnitskii, D. E. (1984a) Phys. Lett. **106A**, 182.
Khmelnitskii, D. E. (1984b) Physica **126B**, 235.

Kil, A. J., Zijlstra, R. J. J. Schuurmans, M. F. H. and Andre, J. P. (1990) Phys. Rev. **41**, 5169.
Kirk, W. P. and Read, M. A. (eds.) (1992) Nanostructures and Mesoscopic Systems, Proc. of the 1991 Santa Fe Int. Symposium. Academic Press, Boston.
Kiss, L. B., Kertész, J. and Hajdu, J. (1990) Z. Phys. **B81**, 299.
Kittel, C. (1963) Quantum Theory of Solids. Wiley, New York.
Kittel, C. (1986) Introduction to Solid State Physics. Wiley, New York.
Kivelson, S. A., Lee, D. H. and Zhang, S. C. (1992) Phys. Rev. **B46**, 2223.
Kleinsasser, A. W., Jackson, T. N., McInturff, D., Rammo, F., Pettit, G. D. and Woodall, J. M. (1989) Appl. Phys. Lett. **55**, 1909.
Knox, R. S. (1963) Solid State Phys. Suppl. **5**, 100.
Kohn, W. (1964) Phys. Rev. **133**, A171.
Kohn, W. (1965), in Physics of Solids at High Pressures, C. T. Tomizuka and R. M. Emrick, eds., Academic Press, New York, p. 561.
Kohn, W. and Sham, L. J. (1965) Phys. Rev. **140**, A1133.
Kohn, W. and Vashishta, P. (1985) in Theory of the Inhomogeneous Electron Gas, S. Lundquist and N. H. March, eds. Plenum, Press, New York.
Kosterlitz, J. M. and Thouless, D. J. (1973) J. Phys. **C6**, 1181.
Kramer, B., Bergmann, G. and Bruynseraede, Y. (eds.) (1985) Transport Phenomena, Springer Series in Solid State Sciences **61**. Springer, Berlin.
Kramer, B. and MacKinnon, A. (1993) Rep. Prog. Phys. **56**, 1469.
Kramer, B., ed. (1991) Quantum Coherence in Mesoscopic Systems, NATO ASI Series no 254. Plenum Press, New York.
Kramers, H. A. (1940) Physica **7**, 284.
Kubo, R. (1957) J. Phys. Soc. Japan **12**, 570.
Kubo, R. (1962) J. Phys. Soc. Japan **17**, 975.
Kulik, I. O. (1969) Zh. Eksp. Teor. Fiz. **57**, 1745. [Sov. Phys. JETP **30**, 944].
Kulik, I. O. (1970a) JETP Lett, **11**, 275.
Kulik, I. O. (1970b) Zh. Ebsp. Teor. Fiz. **58**, 2171 [Sov. Phys. JETP **31**, 1172].
Kulik, I. O. and Omelyanchuk, A. N. (1975) Sov. Phys. JETP Lett. **21**, 96; (1977) Sov. Phys. J. Low Temp. Phys. **3**, 459; (1978) ibid. **4**, 142; (1984) ibid. **10**, 158.
Kulik, I. O. and Yanson, K. (1972) The Josephson Effect in Superconductive Tunneling Structures. Israel Program of Scientific Translations, Jerusalem.
Kumar, A., Saminadayar, L., Glattli, D. C., Jin, Y. and Etienne, B. (1995), Phys. Rev. Lett. **76**, 2778.
Ladan, F. R. and Maurer, C. R. (1983) C. R. Acad. Sci. **297**, 227.
Lafarge, P., Joyez, P., Esteve, D., Urbina, C. and Devoret, M. H. (1993) Phys. Rev. Lett. **70**, 994.
Laibowitz, R. (1983) in Percolation, Localization and Superconductivity, A. M. Goldman and S. A. Wolf, eds. Nato Advanced Science Institutes, Series B: Phyysics, 109.
Lambert, C. J. (1991) J. Phys. Cond. Matter **3**, 6579.
Lambert, C. J. (1993) J. Phys. Cond. Matter **5**, 707.
Lambert, C. J. (1994) Physica **B203**, 201.
Lambert, C. J. and Robinson, S. J. (1993) Phys. Rev. **B48**, 10391.
Lambert, C. J., Hui, V. C. and Robinson, S. J. (1993) J. Phys. Cond. Matter **5**, 4187.
Landau, L. D. and Lifschitz, E. M. (1959) Statistical Physics. Pergamon Press, London.
Landau, L. D. and Lifschitz, E. M. (1960a) Quantum Mechanics. Pergamon Press, London.

Landau, L. D. and Lifschitz, E. M. (1960b) Electrodynamics of Continuous Media. Pergamon Press, London.
Landauer, R. (1957) IBM J. Res. Fev. **1**, 223.
Landauer, R. (1970) Phil. Mag. **21**, 863.
Landauer, R. (1975) Z. Physik **B24**, 247.
Landauer, R. (1978) Proc. 1977 Ohio State University Conf. Electrical Transport and Optical Properties of Inhomogeneous Media. AIP Conf. Proc. **40**, J. C. Garland and D. Tanner, eds. AIP, New York.
Landauer, R. (1985) in Localization, Interaction and Transport Phenomena, G. Bergmann and Y. Bruynseraede, eds. Springer, New York, p. 38.
Landauer, R. (1987) Z. Phys. **B68**, 212.
Landauer, R. (1988) IBM J. Res. Dev. **32**, 306.
Landauer, R. (1989a) Physica **D38**, 226.
Landauer, R. (1989b) in Reed and Kirk (1989), p. 17.
Landauer, R. (1989c) J. Phys. Cond. Matter **1**, 8099.
Landauer, R. (1990a) Physica **A168**, 75.
Landauer, R. (1990b) in Analogies in Optics and Microelectronics, W. van Haeringen and D. Lenstra, eds. Kluwer, Dordrecht, p. 243.
Landauer, R. (1993) Phys. Rev. **47**, 16 427.
Landauer, R. (1995) in Proceedings of the Conference on Fundamental Problems in Quantum Theory, D. Greenberger, ed. New York Acad. Sci., New York, p. 419.
Landauer, R. and Büttiker, M. (1985) Phys. Rev. Lett. **54**, 2049.
Landauer, R. and Helland, J. C. (1954) J. Chem. Phys. **22**, 1655.
Landauer R. and Martin, T. (1992) Physica **B182**, 288.
Landauer, R. and Swanson, J. A. (1961) Phys. Rev. **121**, 1668.
Lang, N. D. (1987) Phys. Rev. **B36**, 8173.
Langenberg, D. N. (1969) in Tunneling Phenomena in Solids, E. Burstein and S. Lundquist, eds. Plenum Press, New York, p. 519.
Langer, J. S. (1971) Ann. Phys. **65**, 53.
Langer, J. and Ambegaokar, V. (1967) Phys. Rev. **164**, 498.
Langer, J. S. and Neal, T. (1966) Phys. Rev. Lett. **16**, 984.
Langreth, D. C. and Abrahams, E. (1981) Phys. Rev. B. **24**, 2978.
Larkin, A. I. and Khmelnitskii, D. E. (1982) Usp. Fiz. Nauk **136**, 336. [Sov. Phys. Usp. **25**, 185].
Laughlin, R. B. (1981) Phys. Rev. **B23**, 5632.
Laughlin, R. B. (1983) Phys. Rev. Lett. **50**, 1395.
Laughlin, R. B. (1988) Phys. Rev. Lett. **60**, 2677.
Laughlin, R. B., (1993) Phys. Rev. Lett. **50**, 1395; Phys. Rev. **B27**, 3383.
Lee, P. A. (1980) J. Noncryst. Solids **35**, 21.
Lee, P. A. (1984) Phys. Rev. Lett. **53**, 2042.
Lee, P. A. and Fisher, D. S. (1981) Phys. Rev. Lett. **47**, 882.
Lee, P. A. and Ramakrishnan, T. V. (1985) Rev. Mod. Phys. **57**, 287 [on localization and transport in disordered systems.]
Lee, P. A. and Stone, A. D. (1985), Phys. Rev. Lett. **55**, 1622.
Lee, P. A., Stone, A. D. and Fukuyama, H. (1987) Phys. Rev. **B35**, 1039.
Lenssen, K. M. H., Jeekel, P. C. A., Harmans. C. J. P. M., Mooij, J. E., Leys, M. R., Woltar, J. H. and Holland, M. C. (1994) in Coulomb and Interference Effects in Small Electronic Structures, D. C. Glattli and M. Sanquer, eds. Editions Frontieres, Paris, p. 63.

Lerner, I. V. and Imry, Y. (1995) Europhys. Lett **29**, 49.
Lesovik, G. B. (1989) Pis'ma Zh. Eksp. Teor. **49**, 515 [JETP Lett. **49**, 594 (1989)].
Levi, A. F. J. et al. (1990) Physics Today **43**, 58 [on devices].
Levin, H., Libby, S. B. and Pruisken, A. M. M. (1983) Phys. Rev. Lett. **51**, 1915.
Levitov, L. S. and Lesovik, G. B. (1993) JETP Lett. **58**, 230 and unpublished.
Levy, L. P., Dolan, G., Dunsmuir, J. and Bouchiat, H. (1990) Phys. Rev. Lett. **64**, 2074.
Li, Y. P., Tsui, D. C. Heremans, J. J., Simmons, J. A. and Weiman, G. W. (1990a) Appl. Phys. Lett. **57**, 774.
Li, Y. P., Zaslavsky, A., Tsui, D. C., Santos, M. and Shayegan, M. (1990b) Phys. Rev. **B41**, 8388.
Licciardello, D. C. and Thouless, D. J. (1978) J. Phys. **C8**, 4159; (1978) **C11**, 925.
Licini, J. C., Dolan, G. J. and Bishop, D. J. (1985a) Phys. Rev. Lett. **54**, 1585.
Licini, J. C., Bishop, D. J., Kastner, M. A. and Melngailis, J. (1985b) Phys. Rev. Lett. **55**, 2987.
Liefrink, F., Dijkhuis, J. I., de Jong, M. J. M., Molenkamp, L. W. and van Houten, H. (1994a) Phys. Rev. **B49**, 14066.
Liefrink, F., Dijkhuis, J. I. and van Houten, H. (1994b) Semiconductor Sci. Technol. **9**, 2178.
Lifschitz, I. M. and Kirpichenkov, V. Ya. (1979) Sov. Phys. JETP **50**, 499.
Lifshitz, E. M. and Pitaevskii, L. P. (1980) Statistical Physics, Vol. 2. Pergamon Press, New York. (See, in particular, secs. 75–77.)
Likharev, K. K. (1979) Revs. Mod. Phys. **51**, 101.
Likharev, K. K. (1986) Dynamics of Josephson Junctions and Circuits. Gordon and Breach, New York.
Likharev, K. K. and Zorin, A. B. (1985) J. Low Temp. Phys. **59**, 347.
Little, W. A. (1967) Phys. Rev. **156**, 396.
Liu, Y. and Price, J. C. (1994), Physica **194**, 1351.
London, F. (1937) J. Phys. Radium **8**, 397.
Loss, D. and Martin, T. (1993) Phys. Rev. **B47**, 4916.
Ludviksson, A., Kree, R. and Schmid, A. (1984) Phys. Rev. Lett. **52**, 950.
MacDonald, A. H. (1995) in Akkermans et al. (1995), p. 659.
MacDonald, A. H. and Girvin, S. M. (1988) Phys. Rev. **B38**, 6295.
MacDonald, D. K. C. (1962) Noise and Fluctuations. Wiley, New York.
Macêdo, A. M. S. and Chalker, J. T. (1992) Phys. Rev. **B46**, 14985.
Macêdo, A. M. S. and Chalker, J. T. (1994) Phys. Rev. **B49**, 4695.
MacKinnon, A. and Kramer, B. (1981) Phys. Rev. Lett. **47**, 1546.
McMillan, W. L. (1981), Phys. Rev. **B24**, 2739.
McMillan, W. L. and Mochel, J. (1981) Phys. Rev. Lett. **46**, 556.
McWhorter, A. L. (1957) in Semiconductor Surface Physics, R. H. Kingston, ed. University of Pennsylvania Press, Philadelphia, p. 207.
Mahan, G. D. (1990) Many-particle Physics. Plenum Press, New York.
Mailly, D., Chapelier, C. and Benoit, A. (1993) Phys. Rev. Lett. **70**, 2020.
Martin, T. and Landauer, R. (1992) Phys. Rev. **B45**, 1742.
Marmorkos, L. K., Beenakker, C. W. J. and Jalabert, R. A. (1993) Phys. Rev. **B48**, 2811.
Matters, M., Elion, W. J. and Mooij, J. E. (1995) University of Delft preprint.
Mehta, M. L. (1967) Random Matrices. Academic Press, New York.
Mehta, M. L. and Dyson, F. J. (1963) J. Math. Phys. **4**, 713.
Mello, P. A. (1988) Phys. Rev. Lett. **60**, 1089.

Mello, P. A. (1988) Phys. Rev. Lett. **60**, 1089.
Mello, P. A. (1990) J. Phys. **A23**, 4061.
Mello, P. A. and Pichard, J. L. (1989) Phys. Rev. **B40**, 5276.
Mello, P. A., Pereyra, P. and Kumar, N. (1988) Ann. Phys. **181**, 290.
Mercereau. J. E. (1969). in Tunneling Phenomena in Solids, E. Burstein and S. Lundquist, eds. Plenum Press, New York, p. 461.
Merzbacher, E. (1961) Am. J. Phys. **30**, 237.
Meyerson, B. S., Himpsel, F. J. and Uram, K. J. (1990) Appl. Phys. Lett. **57**, 1034.
Miller, A. and Abrahams, E. (1960) Phys. Rev. **120**, 745.
Milnikov, G. V. and Sokolov, I. M. (1988) JETP Lett. **48**, 536.
Mohanty, P., Jariwala, E. M. Q., Ketchen, M. B. and Webb, R. A. (1995) in Quantum Coherence and Decoherence, K. Fujikawa and Y. A. Ono, eds., North Holland, Amsterdam, p. 191.
Montambaux, G., Bouchiat, H., Sigeti, D. and Friesner, R. (1990) Phys. Rev. **B42**, 7647.
Mooij, J. E. (1973) Phys. Stat. Sol. **A17**, 521.
Mooij, J. E. and Schön, G. (1992) in Grabert and Devoret (1992), p. 275.
Mooij, J. E., van Wees, B. J., Geerligs, L. J., Peters, M. Fazio, R. and Schön, G. (1990) Phys. Rev. Lett. **65**, 645.
Moore, G. (1993) Bull. Am. Phys. Soc. **38**, 298.
Mott, N. F. (1960) Phil. Mag. **19**, 835.
Mott, N. F. (1966) Phil. Mag. **13**, 989.
Mott, N. F. (1970) Phil. Mag. **22**, 7.
Mott, N. F., (1974) Metal-Insulator Transitions. Taylor & Francis, London.
Mott, N. F. and Davis, G. A., (1979) Electronic Properties of Noncrystalline Materials, 2d ed. Clarendon Press, Oxford.
Mott, N. F. and Twose, W. D. (1961) Adv. Phys. **10**, 107.
Mühlschlegel, B. (1983) in Percolation, Localization and Superconductivity, A. M. Goldman and S. A. Wolf, eds. Nato Advanced Science Institutes, Series B: Physics, 109.
Mühlschlegel, B., Scalapino, D. J. and Denton, R. (1972) Phys. Rev. **B6**, 1767.
Müller-Groeling, A., Weidenmuller, H. A. and Lewenkopf, C. H. (1993) Europhys. Lett. **22**, 193.
Müller-Groeling, A. and Weidenmuller, H. A. (1994) Phys. Rev. **B49**, 4752.
Murat, M., Gefen, Y. and Imry, Y. (1986) Phys. Rev. **B34**, 659.
Muttalib, K. A., Pichard, J.-L. and Stone, A. D. (1987) Phys. Rev. Lett. **59**, 2475.
Muzykantskii, B. A. and Khmelnitskii, D. E. (1994) Phys. Rev. **B50**, 3982.
Nagaev, K. E. (1992) Phys. Lett. **A109**, 103.
Nagaoka, Y. and Fukuyama, H. (eds.) (1982) Anderson Localization. Springer-Verlag, Berlin.
Nakano, H. and Takayanagi, H. (1991) Solid State Commun. **80**, 997.
Nazarov, Yu. V. (1994) Phys. Rev. Lett. **73**, 1420.
Newbower, R. S., Beasley, M. R. and Tinkham, M. (1972) Phys. Rev. **B5**, 864.
Nguyen, V. L., Spivak, B. Z. and Shklovskii, B. I. (1985a) JETP Lett. **41**, 42.
Nguyen, V. L., Spivak, B. Z. and Shklovskii, B. I. (1985b) Sov. Phys. JETP **62**, 1021.
Nozières, P. (1963) Interacting Fermi Systems. Benjamin, New York. [See also Pines and Nozières (1989).]
Oakeshott, R. B. S. and McKinnon, A. (1994) J. Phys. **C6**, 1513.
Oh, S., Zyuzin, A. Yu. and Serota, R. A. (1991) Phys. Rev. **44**, 8858.

Oseledec, V. I. (1968) Trans. Moscow Math. Soc. **19**, 197.
Ovadyahu, Z. and Imry, Y. (1983) J. Phys. C **16**, L471.
Ovadyahu, Z. and Imry, Y. (1985) J. Phys. C **18**, L19.
Pauling, L. (1936) J. Chem. Phys. **4**, 673.
Payne, M. C. (1989) J. Phys. Condens. Matter **1**, 4931.
Peierls, R. (1955) Quantum Theory of Solids. Oxford University Press, Oxford, p. 29 (comment attributed to W. Shockley).
Pendry, J. B. (1989) J. Phys. Cond. Matter **2**, 3273, 3287.
Pendry, J. B. and MacKinnon, A. (1992) Phys. Rev. Lett. **69**, 2772.
Pendry, J. B., MacKinnon, A. and Castaño, E. (1986) Phys. Rev. Lett. **57**, 2983.
Pendry, J. B., MacKinnon, A. and Roberts, P. J. (1992) Proc. Roy. Soc. London **A437**, 67.
Pepper, M. and Uren, M. J. (1982) J. Phys. C **15**, L617.
Peshkin, M. and Tonomura, A. (1989) The Aharonov–Bohm Effect. Springer-Verlag, Berlin.
Petrashov, V. T., Antonov, V. N., Maksimov, S. V. and Shaikhaidarov, R. Sh. (1993a) Sov. Phys. JETP Lett. **58**, 49.
Petrashov, V. T., Antonov, V. N., Delsing, P. and Claeson, T. (1993b), Phys. Rev. Lett. **70**, 347.
Petrashov, V. T., Antonov, V. N., Delsing, P. and Claeson, T. (1995), Phys. Rev. Lett. **74**, 5268.
Pfeiffer, L. N., Stormer, H. L., Baldwin, K. W., West, K. W., Goni, A. R., Pinczuk, A., Ashoori, R. C., Dignam, M. M. and Wegscheide, W. (1993) J. Crystal Growth **840**, 127.
Pichard, J. L. (1984) Thesis, University of Paris, Orsay, No. 2858.
Pichard, J. L. (1991) in Quantum Coherence in Mesoscopic Systems, B. Kramer, ed., Proceedings of the 1990 Nato ASI. Plenum Press, New York, p. 369.
Pichard, J. L. and Sarma, G. (1981a) J. Phys. C **14**, L127.
Pichard, J. L. and Sarma, G. (1981b) J. Phys. (Paris) **10**, 4.
Pines, D. and Nozières, P. (1989) The Theory of Quantum Liquids. Addison-Wesley, Reading, M.A., Advanced Book Classics.
Pollak, M. (1970) Discuss Faraday Soc. **50**, 13; (1971) Proc. Roy. Soc. London **A325**, 383; Phil. Mag. **23**, 519.
Pollak, M. (1972) J. Noncryst. Solids **11**, 1.
Pooke, L., Paquin, N., Pepper, M. and Gundlach, A. J. (1989) Phys. Cond. Matter **1**, 3289.
Prange, R. E. in Prange and Girvin (1990) pp. 1, 69.
Prange, R. E. and Girvin, S. M. (1990) The Quantum Hall Effect, 2d ed. Springer-Verlag, Berlin.
Prober, D. (1983), in Goldman and Wolf (1983), p. 231.
Pruisken, A. M. M. (1984) Nucl. Phys. **B235**, 277.
Pruisken, A. M. M. (1985) Phys. Rev. **B32**, 1311; in Kramer et al. (1985), p. 188; Phys. Rev. **B31**, 416; in Prange and Girvin (1990), p. 177.
Pytte, E. and Imry, Y. (1987) Phys. Rev. **B35**, 1465.
Rajaraman, R. (1982) Solitons and Instantons. Elsevier, Amsterdam.
Ralls, K. S. and Buhrman, R. A. (1988) Phys. Rev. Lett. **60**, 2434.
Ralls, K. S., Skocpol, W. J., Jackel, L. D., Howard, R. E., Fetter, L. A., Epworth, R. W. and Tennant, D. M. (1984) Phys. Rev. Lett. **52**, 118.
Raveh, A. and Shapiro, B. (1992) Europhys. Lett. **19**, 109.

Raveh, A. and Shapiro, B. (1992) Europhys. Lett. **19**, 109.
Razeghi, M. (1989) The MOCVD Challenge. Adam Hilger, Bristol UK and Philadelphia PA.
Reed, M. A. and Kirk, W. P. (eds.) (1989) Nanostructure Physics and Fabrication, Proc. Texas A & M Int. Symposium. Academic Press, Boston.
Reed, N. (1994) Semicond. Sci. Technol. **9**, 1859.
Reif, F. (1965) Fundamentals of Statistical and Thermal Physics. McGraw-Hill, New York.
Rezayi, E. and Read, N. (1994) Phys. Rev. Lett. **72**, 900.
Reznikov, M., Heiblum, M. Shtrikman, H. and Mahalu, D. (1995) Phys. Rev. Lett. **75**, 3340.
Rice, T. M. (1965) Phys. Rev. **A140**, 1889.
Salem, L. (1966) The Molecular Orbital Theory of Conjugated Systems. Benjamin, New York.
Sample, H. H., Bruno, W. J., Sample, S. B. and Sichel, E. K. (1987) J. Appl. Phys. **61**, 1079.
Scalapino, D. J. (1993) Phys. Rev. **B47**, 7995.
Scalapino, D. J. (1969), in Tunneling Phenomena in Solids, E. Burstein and S. Lundquist, eds. Plenum Press, New York, p. 477.
Scalapino, D. J., Fye, R. M., Martins, N. J., Wagner, J. and Hanke, W. (1991) Phys. Rev. **B44**, 6909.
Scalapino, D. J., Sears, M. and Ferrel, R. A. (1972) Phys. Rev. **B6**, 3409.
Schmid, A. (1969) Phys. Rev. **180**, 627.
Schmid, A. (1974) Z. Phys. **271**, 251.
Schmid, A. (1988) Ann. Phys. (NY) **170**, 333.
Schmid, A. (1991) Phys. Rev. Lett. **66**, 80; **66**, 1379(E).
Schmitt-Rink, S., Chemla, D. S. and Miller, D. A. B. (1989) Adv. Phys. **38**, 89.
Schön, G. and Zaikin, A. D. (1990) Phys. Rep. **190**, 237.
Shalgi, A. and Imry, Y. (1995) in Akkermans et al. (1995), p. 229.
Shapiro, B. (1982) Phys. Rev. **25**, 4266.
Shapiro, B. (1983a) Phys. Rev. Lett. **50**, 747.
Shapiro, B. (1983b) Ann. Isr. Phys. Soc. **V**, 367.
Shapiro, B. and Abrahams, E. (1981) Phys. Rev. **B24**, 4889.
Sharvin, Yu V. (1965) Zh. Exp. Teor. Fiz. **48**, 984; (1965) Sov. Phys. JETP **21**, 655.
Sharvin, D. Yu and Sharvin, Yu V. (1981) JETP Lett. **34**, 272.
Shelankov, A. L. (1984) Fiz. Tverd. Tela **26**, 1615; [Sov. Phys. Solid State **26**, 981 (1984)].
Sheng, P. (ed.) (1990) Scattering and Localization of Classical Waves in Random Media. World Scientific, Singapore.
Sheng, P. (1995) Introduction to Wave Scattering, Localization and Mesoscopic Phenomena, chapter 8, Academic Press, Boston.
Shimizu. A., Ueda, M. and Sakaki, H. (1992) Proc. 4th Int. Symp. on Foundations of Quantum Mechanics, Physical Society of Japan, Tokyo.
Shklovskii, B. I. and Efros, A. L. (1971) Zh. Exp. Teor, Fiz. **60**, 867 [Sov. Phys. JETP **33**, 469 (1971)].
Shklovskii, B. I. and Efros, A. L. (1984) Electronic Properties of Doped Semiconductors, Springer Series in Solid State Sciences, vol. 45. Springer, Berlin, pp. 210–216.
Shmidt, V. V. (1966) JETP Lett. **3**, 89.
Shtrikman, S. and Thomas, H. (1965) Solid State Commun. **3**, 147.

Simmons, J. A., Wei, H. P., Engel, L. W., Tsui, D. C. and Shayegan, M. (1989) Phys. Rev. Lett. **63**, 1731.
Simmons, J. A., Wei, H. P., Engel, L. W., Tsui, D. C. and Shayegan, M. (1991) Phys. Rev. **B99**, 12 933.
Sivan, U. and Imry Y. (1986) Phys. Rev. **B33**, 55.
Sivan, U. and Imry, Y. (1987) Phys. Rev. **B35**, 6079.
Sivan, U. and Imry Y. (1988) Phys. Rev. Lett. **61**, 1001.
Sivan, U., Imry, Y. and Hartsztein, C. (1989) Phys. Rev. **B39**, 1242.
Sivan, U., Milliken, F. P., Milkove, K., Rishton, S., Lee, Y., Hong, J. M., Hoegli, V., Kern, D. and de Franza, M. (1994a) Europhys. Lett. **25**, 605.
Sivan, U., Imry, Y. and Aronov, A. G. (1994b) Europhys. Lett. **28**, 115.
Skocpol, W. J., Jackel, L. D., Howard, R. E., Mankiewich, P. M. and Tennant, P. M. (1986) Phys. Rev. Lett. **65**, 2865.
Slevin, K., Pichard, J.-L. and Muttalib, K. A. (1993) J. de Physique **3**, 1387.
Smith, H. J. and Craighead, N. G. (1990), Physics Today **43**, 24 [on nanofabrication].
Spivak, B. Z. and Khmelnitskii, D. E. (1982) Pis'ma Zh. Eksp. Teor. Fiz. **35**, 334 [JETP Lett. **35**, 412, (1982)].
Spivak, B. Z. and Zyuzin, A. Yu (1991), in Altshuler et al. (1991).
Stein, J. and Krey, U. (1979) Z. Phys. **B34**, 287.
Stein, J. and Krey, U. (1980) Z. Phys. **B37**, 18.
Stern, A. (1994) Phys. Rev. **B50**, 10 092.
Stern, A., Aharonov, Y. and Imry, Y. (1990a) Phys. Rev. **A40**, 3436.
Stern, A., Aharonov, Y. and Imry, Y. (1990b) in Kramer, (1991) p. 99.
Stern, F. (ed.) (1982) Electronic properties of Two Dimensional Systems. North-Holland, Amsterdam.
Stone, A. D. (1985) Phys. Rev. Lett. **54**, 2692.
Stone, A. D. and Imry, Y. (1986) Phys. Rev. Lett. **56**, 189.
Stone, A. D., Mello, P. A., Muttalib, K. A. and Pichard, J.-L. (1991) in Mesoscopic Phenomena in Solids, Altshuler, B. L., Lee, P. A. and Webb, R. A. eds. North-Holland, Amsterdam, p. 369.
Störmer, H. L., Du, R. R., Kay, W., Tsui, D. C., Pfeiffer, L. N., Baldwin, K. W. and West, K. W. (1994) Semicond. Sci. Technol. **9**, 1853.
Su, W. P. and Schrieffer, J. R. (1981) Phys. Rev. Lett. **46**, 738.
Sze, S. M. (1986) Physics of Semiconducting Devices, 2d ed. Wiley, New York.
Szopa, M. and Zipper, E. (1995) Int. J. Mod. Phys. **B9**, 161.
Takagi, S. (1992) Solid State Commun. **81**, 579.
Takane, Y. and Ebisawa, H. (1991) J. Phys. Soc. Japan **60**, 3130.
Takane, Y. and Ebisawa, H. (1992) J. Phys. Soc. Japan **61**, 1685; **61**, 2858.
Takane, Y. and Otani, H. (1994) J. Phys. Soc. Japan, **63**, 3361.
Takane, Y. (1994) J. Phys. Soc. Japan **63**, 2849, 2668, 4310.
Thornton, T. J., Pepper, M., Ahmed, H., Andrews, D. and Davies, J. J. (1986) Phys. Rev. Lett. **56**, 1198.
Thouless, D. J. (1970) J. Phys. **C3**, 1559.
Thouless, D. J. (1977) Phys. Rev. Lett. **39**, 1167.
Thouless, D. J. (1990) Phys. Rev. **B40**, 12 034.
Thouless, D. J. and Gefen, Y. (1991) Phys. Rev. Lett. **66**, 806.
Thouless, D. J. and Kirkpatrick, S. (1981) J. Phys. C. 235.
Tighe, T., Johnson, A. T. and Tinkham, M. (1991) Phys. Rev. **B44**, 10 286.

Tighe, T., Tuominnen, M. T., Hergenrother, J. M. and Tinkham, M. (1993) Phys. Rev. **B47**, 1145.

Tinkham, M. (1975) Introduction to Superconductivity. R. E. Krieger, Malabar, FL; 2nd edition (1996) McGraw-Hill, New York.

Tomsovic, S. and Heller, E. (1993) Phys. Rev. E47, 282.

Tonomura, A., Matsuda, T., Suzuki, R., Fukuhara, A., Osakabe, N., Umezaki, H., Endo, J., Shinogawa, K., Sugita, Y. and Fujiwara, H. (1982) Phys. Rev. Lett. **48**, 1443.

Toyozawa, Y. (1961) Progr. Theor. Phys. **26**, 29.

Trivedi, N. and Browne, D. A. (1988) Phys. Rev. **B38**, 9581.

Trugman, S. A. (1983) Phys. Rev. **B27**, 7539.

Tsui, D. C., Stormer, H. L. and Gossard, A. C. (1982) Phys. Rev. Lett. **48**, 1559; Phys. Rev. **B25**, 1405.

Tuominen, M. T., Hergenrother, J. M., Tighe, T. S. and Tinkham, A. (1992) Phys. Rev. Lett. **69**, 1997.

Ueda, M. and Simizu, A. (1993) J. Phys. Soc. Japan **62**, 2994.

Ulloa, S., MacKinnon, A., Castaño, E. and Kirczenow, G. (1992) Handbook on Semiconductors; vol. 1, Basic properties of Semiconductors, Landsberg, P. I., ed. (North-Holland, Amsterdam. p. 864.

Umbach, C. P., Washburn, S., Laibowitz, R. B. and Webb, R. A. (1984) Phys. Rev. B. **30**, 4048.

Uwaha, M. and Noziéres, P. (1985) J. de Physique **46**, 109.

van der Merwe, J. H. (1963) J. Appl. Phys. **34**, 117.

van der Paauw, L. J. (1958) Philips Res. Rep. **13**, 1.

van der Zant, H. S. J., Fritschy, F. C., Orlando, T. P. and Mooij, J. E. (1991a) Phys. Rev. Lett. **66**, 2531.

van der Zant, H. S. J., Geerligs, L. G. and Mooij, J. E. (1991b) in Kramer (1991) p. 511.

van der Zant, H. S. J., Fritschy, F. C., Orlando, T. P. and Mooij, J. E. (1992a) Europhys. Lett. **18**, 343.

van der Zant, H. S. J., Geerligs, L. G. and Mooij, J. E. (1992b) Europhys. Lett. **19**, 541.

van der Ziel, A. (1986) Noise in Solid State Devices and Circuits. Wiley, New York.

van Haeringen, W. and Lenstra, D. (eds.) (1991) Analogies in Optics and Microelectronics, Proceedings of the Conference. Physica **187**.

van Houten, H. and Beenakker, C. W. J. (1991) Physica **B175**, 187.

van Hove (1954) Phys. Rev. **95**, 249.

van Kampen, N. G. (1981) Stochastic Processes in Physics and Chemistry. North-Holland, Amsterdam.

van Otterlo, A., Wagenblast, K. H., Fazio, R. and Schön. G. (1993) Phys. Rev. **B48**, 3316.

van Ruitenbeck, J. M. and van Leeuwen, D. A. (1991) Phys. Rev. Lett. **67**, 640.

van Vleck, J. H. and Weisskopf, V. F. (1945) Rev. Mod. Phys. **17**, 227.

van Wees, B. J. (1988) Phys. Rev. Lett. **60**, 848.

van Wees, B. J. (1990a) Phys. Rev. Lett. **65**, 255.

van Wees, B. J. (1990b) Phys. Rev. **B44**, 2264.

van Wees, B. J. (1993) Bull. Am. Phys. Soc. **38**, 492.

van Wees, B. J., Van Houten, H., Beenakker, C. W. J., Williamson, J. G., Kouendhoven, L. P., van der Marel, D. and Foxon, C. T. (1988) Phys. Rev. Lett. **60**, 848.

van Wees, B. J., de Vries, P., Magnee, P. and Klapwijk, T. M. (1992) Phys. Rev. Lett. **69**, 510.
van Wees, B. J., Dimoulas, A., Heida, J. P. Klapwijk, T. M. Graaf, W. v. d. and Borghs G. (1994) Physica **B203**, 285.
Visani, P., Mota, A. C. and Polini, A. (1990) Phys. Rev. Lett. **65**, 1514.
Vloberghs, H., Moschalkov, V. V., van Haesendonk, C., Jonckheere, R. and Bruynseraede, Y. (1992) Phys. Rev. Lett. **62**, 1268.
Volkov, A. F. (1994) Physica **B203**, 267.
Vollhardt, D. and Wölfle, P. (1980) Phys. Rev. **B22**, 4678.
Vollhardt, D. and Wölfle, P. (1982) Phys. Rev. Lett. **48**, 699.
Von Klitzing, K. (1982) Europhysics News **13**, 3.
Von Klitzing, K., Dorda, G. and Pepper, M. (1980) Phys. Rev. Lett. **45**, 494.
Voss, R. F. and Clarke, J. (1976) Phys. Rev. **B13**, 556.
Wang, T., Clark, K. P., Spender, G. F., Mack, A. M. and Kirk, W. P. (1994) Phys. Rev. Lett. **72**, 709.
Washburn, S., Umbach, C. P., Laibowitz, R. B. and Webb, R. A. (1985) Phys. Rev. **B32**, 4789, and unpublished results.
Washburn, S., Haug, R. J., Lee, K. V. and Hong, J. M. (1992) Phys. Rev. **B44**, 3875.
Wax, N., ed. (1954) Noise and Stochastic Processes. Dover, New York.
Webb, R. A. and Washburn, S. (1986) Adv. Phys. **35**, 375.
Webb, R. A. and Washburn, S. (1988) Physics Today **41**, 46.
Webb, R. A., Fowler, A. B., Hartstein, A. and Wainer, J. J. (1986) Surf. Sci. **170**, 14.
Webb, R. A., Washburn, S., Umbach, C. P. and Laibowitz, R. B. (1984) in Bergmann et al. (1984), p. 121.
Webb, R. A., Washburn, S., Umbach, C. P. and Laibowitz, R. B. (1985a) in Hahlbohm and Lübbig, 1985, p. 561.
Webb, R. A., Washburn, S., Umbach, C. P. and Laibowitz, R. B. (1985b) Phys. Rev. Lett. **54**, 2696.
Webb, W. W. and Warburton, R. J. (1968) Phys. Rev. Lett. **20**, 461.
Wegner, F. (1976) Z. Phys. **25**, 327; (1980) Phys. Rep. **67**, 151.
Wegner, F. (1979) Z. Phys. **B35**, 207.
Wei, H. P., Tsui, D. C. and Pruisken, A. M. M. (1986) Phys. Rev. **B33**, 1488.
Weissman, M. B. (1988) Rev. Mod. Phys. **60**, 537.
Wharam, D. A., Thornton, T. J., Newbury, R., Pepper, M., Ahmed, H., Frost, J. E. F., Husko, D. G., Peacock, D. C., Ritchie, D. A. and Jones, G. A. C. (1988) J. Phys. **C21**, L209.
White, E., Dynes, R. C. and Garno, J. P. (1985) Phys. Rev. **B31**, 1174.
Wiesmann, H., Gurvitch, M., Lutz, H., Gosh, A., Schwartz, B., Allen, P. B. and Strongin, M. (1977) Phys. Rev. Lett. **38**, 782.
Wigner, E. P. (1951) Ann. Math. **53**, 36; (1955) **62**, 548.
Willett, R., Ruel, M., West, K. and Pfeiffer, L. (1993a) Phys. Rev. Lett. **71**, 3846.
Willett, R., Ruel, R., Paalanen, M., West, K. and Pfeiffer, L. (1993b) Phys. Rev. **B47**, 7344.
Wind, S., Rooks, M. J., Chandrasekhar, V. and Prober, D. E. (1986) Phys. Rev. Lett. **57**, 633.
Yacoby, A. and Imry, Y. (1990) Phys. Rev. **B41**, 5341, and references therein for conductance quantization.
Yacoby, A., Heiblum, M., Mahalu, D. and Shtrikman, H. (1995) Phys. Rev. Lett. **74**, 4047.

Yacoby, A., Stormer, H. L., Baldwin, K. W., Pfeiffer, L. N. and West, K. W. (1996) Bell Labs preprint, to be published in Solid State Comm.
Yamada, R.-I. and Kobayashi, S.-I. (1995) J. Phys. Soc. Japan (Letters) **64**, 360.
Yang, C. N. (1962) Rev. Mod. Phys. **34**, 694.
Yang, C. N. (1989) in Proc. Int. Symp. on Foundations of Quantum Mechanics, Kobayashi, S.-I., Ezawa, H., Murayama, Y. and Nomura, S, eds. The Physical Society of Japan, p. 383.
Yennie, D. R. (1987) Rev. Mod. Phys. **59**, 781.
Yoffe, A. F. and Regel, A. R. (1960) Progr. Semicon. **4**, 237.
Yoshioka, D. and Fukuyama, H. (1992) in Fukuyama and Ando (1992), p. 221.
Yoshioka, H. and Fukuyama, H. (1990) J. Phys. Soc. Japan **59**, 3065.
Yoshioka, H. and Fukuyama, H. (1992) in Fukuyama and Ando (1992), p. 263.
Yurke, B. and Kochanski, G. P. (1989) Phys. Rev. **41**, 8184.
Zaikin, A. D. (1992) J. Low Temp. Phys. **88**, 373.
Zaikin, A. D. (1994) Physica **B203**, 255.
Zaĭtsev, A. V. (1980) Zh. Eksp. Teor. Fiz. **78**, 221 [Sov. Phys. JETP **51**, 111]; (1980) **79**, 2016 (E) [**52**, 1018(E)]; (1984) **86**, 1742 [**59**, 1015].
Zener, C. (1930) Proc. Roy. Soc. A. **137**, 636.

Index

AB conductance oscillations, 109–116, 117
a.c. Josephson effect, 69–70, 131, 160–165
Aharonov–Bohm (AB) effect, 39, 61, 65, 70, 109–116, 129–132, 137, 143, 148, 152–156, 160–162, 175, 191, 200
Altshuler–Aronov density of states anomalies, 193, 202–204
Anderson localization, 13–18, 21–34, 133–138
Andreev reflection, 75, 148, 167–175, 184
anomalous diffusion, 31-32, 37
aspect ratio, 119

Bardeen tunnel junction formula, 22
beta function, 27–31
Byers–Yang and Bloch theorem, 61, 65, 69, 70, 109–116, 129–132, 137, 148, 152–156, 160–162, 175, 200

canonical vs. grand-canonical system, 78–82
Casimir–Onsager four-probe relationships, 103–108, 207–208
channel number, 70, 96, 99, 117
 effective, 121–122
charge neutrality, 86–88, 100
charging energy, 123, 166, 193
coherence length, superconducting, 150, 169, 173
coherent volumes, 25, 32, 121, 188
composite Fermions, 144–146
conductance fluctuations, 116–118
 time dependence and low-frequency noise, 189–190
 universal, 5, 120–122, 208–209
conductance formula for *SN* junction, 171
contact resistance, 94
continuous vs. discrete spectrum, 90
Cooper channel renormalization, 85
Coulomb blockade, 123, 166

Debye relaxation absorption, 92
defects, 15, 178, 185–189
density of states, 16, 202–204
dephasing, 38–46, 77
 and inelastic scattering, 56–59
 by electron–electron interactions, 46–56
dephasing length, 24, 25, 32, 53, 77, 188, 193

dephasing time, 24, 43, 53
destruction of interference, 38–46
disorder, 12–13, 15
dynamic structure factor, 48, 196–197

edge states, 131–132
Edwards–Thouless relationships, 22–23, 198–199
Einstein relation, 17, 23–24
elastic vs. inelastic scattering, 3, 12–13, 67–68, 70, 94, 193
electrochemical potential, 94, 99–100, 101–103
electron–electron interactions, 5–6, 46–56, 86–88, 100, 139–146
energy averaging, 77, 114–117, 121
energy price of a domain wall, 61
ensemble average, 4, 64, 71–74, 76–82, 93, 107–109, 112–116, 192
equilibrium (Johnson–Nyquist) noise, 176–177, 196–198
extended state in mid-Landau level, 135–137

fine-structure constant, 13, 129, 151, 153
fingerprint, 4, 117–118, 187
fixed point, 30, 138
fluctuation–dissipation theorem, 46, 48, 196–198
fluctuations
  mesoscopic, 4, 72, 116–118, 120–122, 208–209
  superconducting, 152–159
  thermodynamic, 62–64, 152–159, 176–177
flux-dependence of levels, 66–72
flux quantization, 152, 154, 159
flux sensitivity, 65, 72, 76, 165
frequency-dependent conductivity, 20–21, 122

gates, 6, 10, 71, 148
gauge transformation, 129–133, 146, 200
Ginzburg–Landau theory, 149–156, 162
granularity, 32–33

h/*e* vs. *h*/2*e*, 112–116, 79, 82, 174
Hall effect, 124–125
  fractional QHE, 139–146
  integer QHE, 127–31
Hanbury-Brown and Twiss effect, 178, 180
harmonic generation, 73
hopping conductivity, 18–21

impurity ensemble, 4
induced flux, 153–155
insulator–conductor–superconductor characterization, 62, 154

Josephson current fluctuations, 173
Josephson current from Andreev processes, 169–170, 172
Josephson effect, 69–70, 131, 158–159, 160–165

Kohn relationship, 62, 75
Kramers–Kronig relationships, 187
Kubo formula, 23, 90–92, 195–198
Kubo–Greenwood formula, 23, 90–92, 198–200

Landau levels, 126–127
Landauer formulation, 93–107
  multichannel, 96–103
  multiprobe, 103–107, 207–208
  single channel, 93–95
Laughlin argument for IQHE, 130–131
Laughlin quasiparticle, 143
Laughlin wavefunction, 140–142
level repulsion, 68, 204–206
level spacing, 68–72, 75, 77, 82–83, 91, 132, 204–207
level width, 21–22, 69, 90–92, 204–206
linear response theory, 23, 90–92, 195–198
lithography, 8–10
  e-beam, 9–10
local charge neutrality, 84–88, 93
localization in a strong magnetic field, 133–139
  scaling picture, 138, 146
localization length, 24, 30–31, 136
longitudinal e.m. fluctuations, 50
long-range order, 60–61, 150, 155–159
low frequency (1/*f*) noise, 185–190

macroscopic wave function, 149, 154
magnetic length 35, 134
magnetoconductance, 20, 33, 35–37
Mathiessen rule, breakdown, 12
matrix elements, 15, 21–23, 57, 90, 91, 97, 102, 184, 198
MBE, 5, 7–8
mesoscopic, 4
mesoscopic model for 1/*f* noise, 186–190
metal–insulator transition, 14, 17, 30–32, 185
metastable state, lifetime, 158

minimum metallic conductivity, 13–14, 18, 31
mobility edges, 16

negative TCR, 12, 18–21, 25, 29–32, 34–35
noninvasive potential measurement, 100–103
"normal coherence length", ξ_N, 143, 173, 174

one (and "quasi" one)-dimensional conductor, 14, 21–26, 28–30, 51–53, 68, 107–109, 120
one-dimensional localization, 68, 107–109
Onsager symmetry, 103–108, 207–208
orbital magnetic response, 65, 70, 72, 84

parallel addition of quantum resistors, 4, 109–112
Pauli constraint, 58–59
penetration depth, 151
percolation, 20, 32, 136–137
periodicity in the flux, 61, 65, 69, 70, 109–116, 129–132, 137, 148, 152–156, 160–162, 175, 200
persistent current, 65–88, 152–154, 157–159, 170
phase fluctuations, 152–157
phase operator, exp($i\phi$), 42
phase-slip center 157–158, 162
phase uncertainty, 38–45
phonon scattering, 53
photon fluctuations and correlations, 178–180
photon, optical analogies, 5
photon, phonon field, 44–45
point contact, 94, 99, 171, 209–210
power-law conductance (as function of temperature), 25
power spectrum, 177, 189, 198
proximity effect, 147, 164, 169, 173

quantized conductance, 94, 99, 194, 209–210
quantum conductance unit, 13, 127–129
quantum dot, 55, 65
quantum of flux, 61, 65, 69, 70, 109–116, 126–127, 129–132, 136–137, 143, 144–145, 148, 152–156, 160–162, 165, 169, 175, 200
quantum transport, 12–38, 89–124
quantum wire, 8–9, 23–25

radiation from reservoir, 97, 178–181
random-matrix theory, 83–84, 122, 208
Rayleigh fluctuations, 121
reflectionless tunneling, 172
resonant tunneling, 122–123, 172
return probability, 36–37, 77, 83
rings, AB, 39–40, 65–89, 151–162, 165, 174, 200

sample-specific effects, 72, 116–117
scale-dependent diffusion, 31–32, 37, 207
scaling, finite-size, 63
scaling laws, 63
scaling theory, 26–34, 138
scattering-matrix, 97, 170, 208
scattering rate, 56
screening, 84–88, 93
  magnetic, 151
semiclassical approximation, 36–37, 75–78, 82–84, 87–88, 135, 201, 204–207
sensitivity to boundary conditions, 22, 23, 62, 75
series addition of quantum resistors, 4, 107–109, 122
shot noise, 177–184
sink, 125, 181–183
SN junction, 171–172
SNS junction, 75, 168–170, 172–175
source, 125, 178–180
spectral correlations, rigidity, 81, 83–84, 122, 208
STM–AFM, 5, 9, 11, 194
superconducting quantum interference device (SQUID), 71–74, 165
systems for experiments, 6–11

T vs. *T*/*R*, 93–94, 99
temperature fluctuations, 62–63
thermal activation, 18–21, 158
thermal length, 116, 121, 143, 173–174
thermodynamic limit, 1, 61–62, 68
Thouless energy, 21–23, 55, 72, 76–77, 82, 83, 85, 116, 199–200
Thouless parameter, 21–23, 199–200
Thouless picture, 21–23, 75, 199–200
time-reversal symmetry, 36–37, 103–107, 205
trace formula, 204–206
trace in the environment, 38–43
transfer matrix, 208
translational symmetry, 126
two-dimensional, 8, 14, 28–30, 51–52, 124, 127, 129, 132, 134, 144

two-terminal vs. four terminal, 93–103

unitarity, 42, 45, 97
universal conductance fluctuations (UCF), 120–122, 208–209
universality, 5, 99, 120–122, 208–209

variable-range hopping, 18–21
vortices, 165–167

weak links, 147–148, 160, 164, 168–169, 172–174
weak localization, 34–37, 77, 111–114
Wiener–Khintchin theorem, 190

Yofe–Regel criterion, 13

Zener transitions, 70